AF574910

# SOLAR CELLS FOR PHOTOVOLTAIC GENERATION OF ELECTRICITY

# SOLAR CELLS FOR PHOTOVOLTAIC GENERATION OF ELECTRICITY

## Materials, Devices and Applications

Marshall Sittig

NOYES DATA CORPORATION
Park Ridge, New Jersey, U.S.A.
1979

Library of Congress Catalog Card Number: 79-17070
ISBN: 0-8155-0765-8
Printed in the United States

Published in the United States of America by
Noyes Data Corporation
Noyes Building, Park Ridge, New Jersey 07656

Library of Congress Cataloging in Publication Data

Sittig, Marshall.
Solar cells for photovoltaic generation of electricity.

(Energy technology review ; no. 48)
Bibliography: p.
Includes indexes.
1. Solar batteries--Patents. 2. Patents--United States. I. Title. II. Series.
TK2960.S53 621.47'5'0272 79-17070
ISBN 0-8155-0765-8

# Foreword

The detailed, descriptive information in this book is based on U.S. patents, issued since January 1970, that deal with solar cells suitable for the photovoltaic generation of electricity. Also additional material is presented concerning the principles and economics of producing electricity by means of photovoltaic cells.

This book serves a double purpose in that it supplies detailed technical information and can be used as a guide to the U.S. patent literature in this field. By indicating all the information that is significant, and eliminating legal jargon and juristic phraseology, this book presents an advanced, technically oriented review of recent advances in the photovoltaic generation of electricity.

The U.S. patent literature is the largest and most comprehensive collection of technical information in the world. There is more practical, commercial, timely process information assembled here than is available from any other source. The technical information obtained from a patent is extremely reliable and comprehensive; sufficient information must be included to avoid rejection for "insufficient disclosure." These patents include practically all of those issued on the subject in the United States during the period under review; there has been no bias in the selection of patents for inclusion.

The patent literature covers a substantial amount of information not available in the journal literature. The patent literature is a prime source of basic commercially useful information. This information is overlooked by those who rely primarily on the periodical journal literature. It is realized that there is a lag between a patent application on a new process development and the granting of a patent, but it is felt that this may roughly parallel or even anticipate the lag in putting that development into commercial practice.

Many of these patents are being utilized commercially. Whether used or not, they offer opportunities for technological transfer. Also, a major purpose of this book is to describe the number of technical possibilities available, which may open up profitable areas of research and development. The information contained in this book will allow you to establish a sound background before launching into research in this field.

Advanced composition and production methods developed by Noyes Data are employed to bring these durably bound books to you in a minimum of time. Special techniques are used to close the gap between "manuscript" and "completed book." Industrial technology is progressing so rapidly that time-honored, conventional typesetting, binding and shipping methods are no longer suitable. We have by-passed the delays in the conventional book publishing cycle and provide the user with an effective and convenient means of reviewing up-to-date information in depth.

The table of contents is organized in such a way as to serve as a subject index. Other indexes by company, inventor and patent number help in providing easy access to the information contained in this book. There is also a short bibliography of the U.S. Government reports used in this book.

Some of the illustrations in this book may be less clear than could be desired; however, they are reproduced from the best material available to us.

## 15 Reasons Why the U.S. Patent Office Literature Is Important to You —

1. The U.S. patent literature is the largest and most comprehensive collection of technical information in the world. There is more practical commercial process information assembled here than is available from any other source.

2. The technical information obtained from the patent literature is extremely comprehensive; sufficient information must be included to avoid rejection for "insufficient disclosure."

3. The patent literature is a prime source of basic commercially utilizable information. This information is overlooked by those who rely primarily on the periodical journal literature.

4. An important feature of the patent literature is that it can serve to avoid duplication of research and development.

5. Patents, unlike periodical literature, are bound by definition to contain new information, data and ideas.

6. It can serve as a source of new ideas in a different but related field, and may be outside the patent protection offered the original invention.

7. Since claims are narrowly defined, much valuable information is included that may be outside the legal protection afforded by the claims.

8. Patents discuss the difficulties associated with previous research, development or production techniques, and offer a specific method of overcoming problems. This gives clues to current process information that has not been published in periodicals or books.

9. Can aid in process design by providing a selection of alternate techniques. A powerful research and engineering tool.

10. Obtain licenses — many U.S. chemical patents have not been developed commercially.

11. Patents provide an excellent starting point for the next investigator.

12. Frequently, innovations derived from research are first disclosed in the patent literature, prior to coverage in the periodical literature.

13. Patents offer a most valuable method of keeping abreast of latest technologies, serving an individual's own "current awareness" program.

14. Copies of U.S. patents are easily obtained from the U.S. Patent Office at 50¢ a copy.

15. It is a creative source of ideas for those with imagination.

# Contents and Subject Index

# Introduction

In an era of steadily increasing population and increasing world energy consumption we are faced with a series of problems limiting our future energy supplies. These include:

> Increased aggressiveness on the part of OPEC countries resulting in price increases and proposals to limit production to maintain their price levels.
>
> Increased militancy on the part of antinuclear activists protesting the construction and start-up of new nuclear facilities. This has been coupled with government disclosures of safety inadequacies of nuclear plants located in areas possibly prone to earthquakes, as well as the questions posed by the Three-Mile Island accident.
>
> Problems of conversion of oil- and gas-burning industrial facilities to coal. These include change-over economics, availability of coal including labor union problems and pollution problems involved with the combustion of high-sulfur coal.

All of the foregoing problems combine to place increased emphasis on alternative energy sources such as solar energy which is available internationally, safe, and relatively free from environmental impact.

In this book, solar energy will be restricted to the proposition of producing useful energy directly from sunlight and will not include indirect use of solar energy such as in wind machines or biomass conversion (10).

Further, this volume will concentrate solely on photovoltaic systems and will not include solar collectors in which thermal energy is recovered by solar heating of circulating fluids. The latter uses have recently been reviewed by J.K. Paul (7). Both thermal collectors and photovoltaic devices have been discussed in some detail in studies by the U.S. Office of Technology Assessment (3)(4). An earlier review on solar cells was prepared by Ranney (6).

It should further be noted that this book does not attempt to cover photovoltaic detector cells—the type of photoelectric cells which simply produce a low voltage adequate to cause a door to open, for example. The emphasis on the type of devices covered here is efficient energy production from incident sunlight not simply production of voltage in response to radiation of some wavelength.

Solar photovoltaic energy conversion (PV), the direct generation of electricity from sunlight, has a proven track record of about 10% conversion efficiency in satellite and other applications. The cost, however, is high. Interest in PV as an alternative terrestrial energy source is based not only on the fact that sunlight is inexhaustible, but also on the belief that eventually PV can be deployed on a scale sufficiently large to provide a significant fraction of the world's electricity economically (1).

As pointed out in one of the OTA reports cited above (3), photovoltaic devices, similar to the solar cells used to provide power for spacecraft, can convert sunlight directly into electricity with no moving parts. As a result, they can be extremely reliable and quiet. The cells are not as efficient as the best heat engines, but they can compete in efficiency with heat engines at lower temperatures (i.e., 400° to 500°F or lower). The main disadvantage of photovoltaic technology is its extremely high cost. While inexpensive heat engines may cost as little as \$100 to \$200 per kilowatt, photovoltaic systems currently sell for approximately \$11,000 per kilowatt. (The photovoltaic systems can provide this peak output only in bright sunlight.) Current Federal research programs have as their goal a cost for photovoltaic arrays of \$2,000 per kilowatt by 1982 and \$500 per kilowatt by 1986 (in 1975 dollars).

A simple photovoltaic system can consist of an array of cells connected with an inverter capable of converting the direct current produced by the cells into the 60-cycle alternating current which is compatible with electricity provided by electric utilities. It is possible to use direct current for most lighting, electric stoves, electric heating, and other purposes with little or no modification in the equipment, but a building would need special wiring and switching to use direct current, and this possibility will not be examined in detail.

Onsite storage can be provided using batteries, but it is usually preferable to sell excess electricity to the electric utility rather than storing it onsite. Utility storage tends to be less expensive, and excess onsite energy is typically available during periods when there is a large demand for utility electricity and the excess onsite electricity is particularly valuable to the utility.

A number of difficult legal, regulatory, and rate-setting problems will need to be overcome before onsite systems can be connected to utility grids in most areas. There should be no prohibitive technical problems (3).

## THE FUNDAMENTALS OF SOLAR CELLS

Photovoltaic cells (often called solar cells) are semiconductor devices which are capable of converting sunlight directly into electricity. All photovoltaic installations will consist of small, individual generating units—the photovoltaic cell. Individual cells will probably range in size from a few millimeters to a meter in linear dimension. A surprising variety of such cells is available or is in advanced development since it has only been since about 1972 that serious attention has been given to designing cells for use in anything other than spacecraft. Devices are available with a large range of efficiencies and voltages. Some cells are designed to withstand high solar intensities and high operating temperatures, while others are designed to minimize manufacturing costs.

The energy in light is transferred to electrons in a semiconductor material when a light photon collides with an atom in the material with enough energy to dislodge an electron from a fixed position in the material (i.e., from the valence band), giving it enough energy to move freely in the material (i.e., into the conduction band).

A vacant electron position or "hole" is left behind at the site of this collision; such holes can move if a neighboring electron leaves its site to fill the former hole site. A current is created if these pairs of electrons and holes (which act as positive charges) are separated by an intrinsic voltage in the cell material.

Creating and controlling this intrinsic voltage is the trick which has made semiconductor electronics possible. The most common technique for producing such a voltage is to create an abrupt discontinuity in the conductivity of the cell material (typically silicon in contemporary solid-state components) by adding small amounts of impurities or dopants to the pure material. This is called a homojunction cell. A typical homojunction device is shown in Figure 1.

An intrinsic voltage can also be created by joining two dissimilar semiconductor materials (such as CdS and $Cu_2S$), creating a heterojunction, or by joining a semiconductor to a metal (e.g., amorphous silicon to palladium) creating a Schottky barrier junction.

A fundamental limit on the performance of all of these devices results from the fact that: (A) light photons lacking the energy required to lift electrons from the valence to the conduction bands (the band gap energy) cannot contribute to photovoltaic current; and (B) the energy given to electrons which exceeds the minimum excitation threshold cannot be recovered as useful electrical current. Most of the unrecovered photon energy is dissipated by heating the cell.

**Figure 1: A Typical Photovoltaic Device**

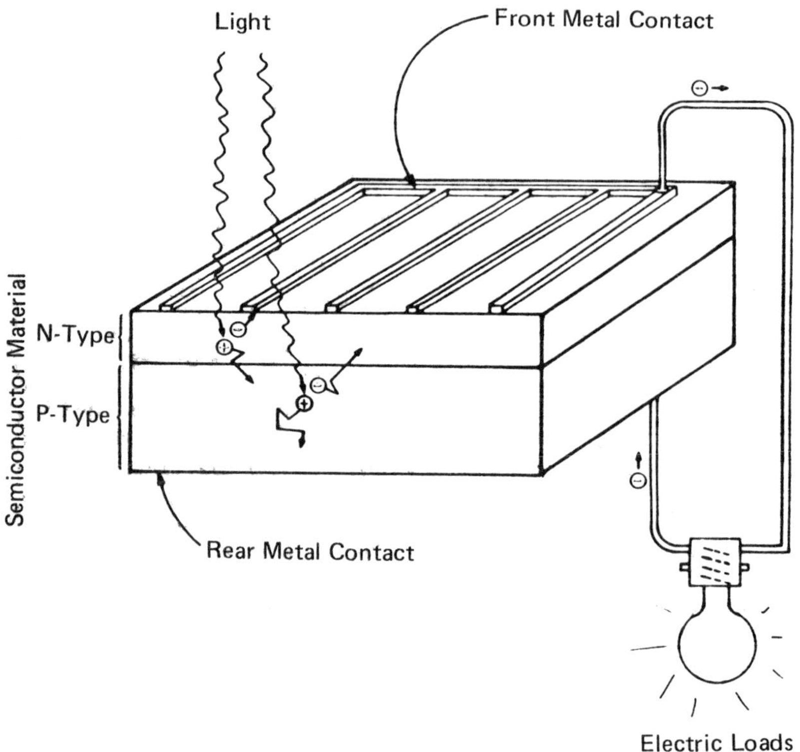


Source: Reference (4)

The bulk of the solar energy reaching the Earth's surface falls in the visible spectrum where photon energies vary from 1.8 eV (deep red) to 3 eV (violet). In silicon, only about 1.1 eV is required to produce a photovoltaic electron, and in GaAs about 1.4 eV. Choosing a material with a higher energy threshold results in capturing a larger fraction of the energy in higher energy photons but losing a larger fraction of lower energy photons. The theoretical efficiency peaks at approximately 1.5 eV, but the theoretical efficiency remains within 80% of this maximum for materials with band gap energies between 1 and 2.2 eV.

Electrons actually leave cells with energies below the excitation voltage because of losses attributable to internal resistance and other effects, not all of which are understood. An electron leaves a typical silicon cell with a useful energy of approximately 0.5 eV.

The same kinds of fundamental limits apply to photochemical reactions in which a light photon with energy above some fixed excitation threshold is able to produce a chemical reaction or a structural change which can be assigned a fixed energy. The theoretical limit to the performance of several types of cell designs is shown in Table 1.

**Table 1: Photovoltaic Cell Efficiencies***

| Device | Probable maximum achievable efficiency | Maximum measured efficiency | Performance of commercial cells |
|---|---|---|---|
| | *Silicon devices* | | |
| Single crystal homojunction | 20-22 | 19 | 10-15 |
| Polycrystalline homojunction | ? | 7-14(?) | — |
| Amorphous Schottky with platinum | 15 | 5.6 | — |
| | *Thin films* | | |
| $CdS/Cu_2S$ (chemical vapor deposit process)(heterojunction) | 15 | 8.6 | 2-3 |
| $CdS/Cu_2S$ (spray process) (heterojunction) | 8-10 | 5.6 | — |
| $(Cd/Zn)S/Cu_2S$ (heterojunction) | 15 | 6.3 | — |
| $CdS/CuInSe_2$ (single crystal) (heterojunction) | 24 | 12 | — |
| $CdS/CuInSe_2$ (thin film) (heterojunction) | 15 | 6.9 | — |
| GaAs (Schottky AMOS) | 25-28 | 14 | — |
| Single crystal Schottky with indium-tin oxide | 20 | 12 | — |
| | *Cells for use in concentrated sunlight* | | |
| Optimized silicon cell (single-crystal homojunction), 200 times concentration | 22 | 18 | 12.5 |
| Interdigitated back-contact silicon, single-crystal homojunction, 100 times concentration | 26-27 | 15(20?) | — |
| Thermophotovoltaic | 30-50 | 13 | — |
| $Ga_xAl_{1-x}As/GaAs$ (180 times) | 25-26 | 24.5 | — |
| $Ga_xAl_{1-x}As/GaAs$ (1,700 times) | | 19 | — |
| Multicolor cell (GaAs/Si/Ge) | 40 | — | — |
| Vertical multijunction (silicon) | 30 | 9.6 | — |

*Techniques for reporting efficiencies differ. Wherever possible, efficiencies were chosen which assume air mass 1 and include losses due to reflection and contact shading.

Source: Reference (3)

The performance of real cells (also shown in Table 1) falls below the theoretical maximum for a number of reasons. One obvious problem is reflection of light from the cell surface (which can be reduced with special coating and texturing) and reflection from the electrical contacts on the front surface of the cell (which can be reduced with careful contact design). Losses also result from the fact that the photo-generated electrons and holes, which fail to reach the region in the cell where they can be separated by the intrinsic voltage, cannot contribute to useful currents.

Photo-generated charges can be lost because of imperfections in the cell crystal structure, defects caused by impurities, surface effects, and other types of imperfections. Losses are minimized if a perfect crystal of a very pure semiconductor material is used, but producing such a crystal can be extremely expensive. Manufacturing costs can probably be greatly reduced if cells consisting of a number of small crystal "grains" can be made to operate with acceptable efficiencies. The size of the grains which can be tolerated depends on the light-absorbing properties of the cell material. If absorption is high, photovoltaic electron-hole pairs will be created close to the cell junction where the voltages exist; relatively small grain sizes can be tolerated since the charges need drift only a short distance before being sorted by the field.

As hinted at in Table 1, the two main avenues to solar cell improvement through research and development are:

(1) Improved materials for direct solar illumination; and

(2) Cell designs using similar or different materials but integrated with reflectors, lenses, or other devices to produce concentrated rather than "normal" incident sunlight.

High efficiencies are important for cells used in concentrated sunlight, since increased cell performance means a reduction in the area which must be covered with the magnifying optical equipment, which dominates the system's cost.

Modified silicon cells, designed to perform in concentrated sunlight are not inherently more expensive than ordinary cells, but are always likely to cost more per unit of cell area since production rates will be lower and since more care will be taken in their manufacture. However, since the cells cover only a fraction of the receiving area, much more can be spent on any individual cell. Cells such as the thermophotovoltaic device and the $Ga_xAl_{1-x}As/GaAs$ cells may be considerably more expensive per unit area than the silicon devices, but this cost may not be significant since the devices can be used with concentrations of 500 to 1,000 or more and can be much more efficient. A variety of approaches are being investigated for achieving high-efficiency performance in concentrated sunlight.

The current from a photovoltaic cell increases almost linearly with increasing sunlight intensity and the voltage increases slightly faster than the logarithm of the intensity. These effects would lead to an increase in overall cell efficiency, except that the increased current densities in the cell lead to increased resistive losses and other effects. The design of standard silicon cells can be optimized for operating in intense sunlight by carefully designing the wires used to draw current from the cells, optimizing the resistivity of the cell material, changing the thickness of the cell junction, and otherwise taking pains in cell manufacture (e.g., better antireflective coatings and surface texturing, higher quality silicon, precisely designed gridlines, etc.). Efficiencies as high as 17.9% have been reported for silicon cells operating at 100°C in sunlight concentrated 200 times (3).

An "interdigitated back-contact" cell exposes an unobstructed wafer of intrinsic (i.e., very pure) silicon crystal directly to concentrated sunlight. The junctions that produce the cell voltages, and that are attached to electrical leads, are entirely on the back of the cell. The name comes from the shape of the positive and negative electrical contacts on the back side of the cell. An efficiency of 15% has been reported for a preliminary version of this cell; it is believed that straightforward design improvements will result in cells which are at least 20% efficient.

Another approach to achieving high cell efficiencies involves the use of a number of different materials to form cells optimally designed for different colors of light. One basic approach involves the use of selective mirrors to separate colors and direct them to different cells.

Another ingenious scheme for separating colors has been proposed which uses a series of dyes capable of absorbing sunlight and reradiating the energy in a narrow frequency band matched to the band gap of each of a series of cell junctions. A considerable amount of development work will be required to design practical devices of this type, however, and the ultimate cost of fabricating the devices cannot be forecast with any confidence (3).

## FUTURE DEVELOPMENT PROSPECTS

The development of PV into an economically competitive and significant power source requires major scientific and/or technological advances. This statement applies both to the technology of crystal silicon cells, which forms a reference for the entire field, and the leading thin film candidate ($CdS/Cu_2S$), a cell whose active materials are sulfides of cadmium and copper. Nevertheless, the ultimate prospects are bright in view of the ferment and rapid rate of progress in both photovoltaic science and technology. Unexpected developments such as amorphous hydrogenated silicon cells and GaAs-liquid junction cells have occurred. Future developments of this kind may well advance solar photovoltaic energy conversion to the position of a dominant technology rather than one relatively minor component in the U.S. energy mix, or a form of insurance to be used in the event of crisis (1).

Utilization of PV as a major source of electricity in the U.S., with a market penetration exceeding 10% of the total consumption, will, in all probability, be a long-term venture requiring perhaps 30 to 50 years. The R&D effort required to identify the right technology will be extensive. Once identified, that technology will have to be developed and scaled up into a PV industry that is about a thousand times bigger than today's.

A ten-year build-up of PV to the significant level of generation of 5 x $10^{10}$ kilowatt hours per year (kWh/yr), or about 1% of U.S. annual electricity demand projected for the year 2000, would require an annual production of PV capacity at a rate of 2,000 peak megawatts per year ($MW_p$/yr). The annual cost would range between 1 and 2 billion dollars, depending on whether the PV system capital investment cost is \$0.50 or \$1.00/$W_p$. Assuming that technical success is achieved, the major factors leading to this modest rate of penetration are large capital risks and the time required to install the necessary facilities.

In the short term there might well be a significant intermediate market that includes foreign sales and remote applications, such as road lighting. Foreign sales might be particularly important in high insolation areas that do not have existing power grids, such as many Third World countries or the Middle East.

The cost which is economically competitive for PV systems operating in the high-insolation Sunbelt region of the U.S. around the year 2000 will be \$0.50 to \$1.00/$W_p$ (in 1975 dollars). In areas having average insolation, the permissible cost range is \$0.35 to \$0.75/$W_p$. The uncertainty is largely determined by that of future electricity generation costs. The permissible factory module price that is necessary to meet these system-cost requirements is \$0.10 to \$0.40/$W_p$ (1).

There are four basic approaches to achieving a cost reduction for solar cells (3):

(1) Reducing the cost of manufacturing silicon cells, which are the most common photovoltaic devices. This requires developing mass-production techniques to replace the inefficient processes used to fabricate cell arrays, and it will require developing inexpensive techniques for producing and slicing silicon crystals. Silicon is an attractive material because it is plentiful and nontoxic.

(2) Developing cells based on "thin films" of materials, such as cadmium sulfide or amorphous silicon, which can be applied directly to glass or other supporting material at very low cost. The main difficulty with the thin film cells is their relatively low efficiencies. Low efficiencies mean that relatively large areas are required and the cost of supporting these large areas of cells may exceed the cost of the cells themselves. Competitive thin film cells probably will require efficiencies greater than 10%.

(3) Using concentrating collectors to focus sunlight on photovoltaic cells, thereby reducing the area of cells required for a given energy output. A number of cell designs are being developed which are able to operate in a wide range of solar intensities. Some of these designs are variants of silicon designs, while others are based on gallium arsenide or other materials. The use of concentrating collectors replaces the problem of reducing cell costs

with the mechanical problem of designing a focusing collector which can be manufactured at low cost. One feature of the concentrating systems is that it may be economically attractive to cool the cells with a fluid and use the heated fluid for space-heating or other processes, thereby taking maximum advantage of the investment in the collector.

(4) Using properly designed plastic or glass sheets imbedded with a fluorescent dye to concentrate sunlight reaching the face of the sheet on the thin edge of the sheet. (Anyone holding a sheet or rod of clear plastic may have noticed how the edge or end sometimes seems to glow.) The use of such a concentrator would eliminate the need to develop a low-cost focusing and tracking system, but there would be a need to find a low-cost dye with the proper optical properties capable of surviving bright sunlight for many years.

One approach which has received much publicity but concerning which there is very little hard information available at the time of publication of this volume concerns the use of amorphous semiconductor alloys of silicon, fluorine and hydrogen. These materials have been developed by Energy Conversion Devices, Inc., headed by the inventor, Stanford R. Ovshinsky. In May 1979 Energy Conversion Devices, Inc. entered into a joint product development and license agreement with Arco Solar, Inc., an Atlantic Richfield Co. subsidiary, to develop and commercialize these "ovionic" cells. It is claimed that such cells would produce electricity from sunlight at costs comparable to power from conventional sources such as oil, gas or coal.

The APS (1) has concluded that a long-term and innovative R&D program is needed, which must include:

(a) the search for and development of new photosensitive materials;

(b) basic research on the interfacial phenomena that control photovoltaic conversion;

(c) investigation of nonbiological methods for direct production of fuels from sunlight;

(d) development of novel photovoltaic technologies or devices (for example, bubble concentrators, encapsulants for flat-plate cells, and noncell related structural materials) to aid in identifying the critical materials problems limiting performance and cost.

The various research programs funded by the U.S. Department of Energy are catalogued in Report (5) which lists: the title, organization performing the work, dollar amount of support, name of principal investigator, work location, dates of program award and duration, contract number, objective of project, and brief description of experimental approach.

# Application Areas for Solar Cells

Application of solar cells initially was in space vehicles where need was paramount and cost was secondary. Now, the big potential area is in terrestial applications where, unlike space vehicles, solar cells do not provide the only answer to the need and where cost competition is fierce.

## SPACECRAFT

The original application area for solar cells was in spacecraft, to provide power for transmission to earth of scientific data and photographs and for other onboard power requirements.

The performance of practically all spacecraft is limited to a great degree by the amount of electrical power that is available during flight. The most efficient power source has been found to be the conversion of sunlight into an electric current by an array of photovoltaic or solar cells. In early models, the exterior surface of a spacecraft was usually covered with solar cell array sections which were size-limited by the spacecraft dimensions and efficiency-limited by shadowing and the effects of angular solar incidence.

As the size of spacecraft increased, the volume available for power-consuming pay loads increased as the cube of the diameter ($V = 4/3\pi r^3$), while the surface area went up only as the square ($A = 4\pi r^2$). Accordingly, electrical power availability from surface-mounted solar cell arrays became less and less for each unit of pay load volume as the size of the craft increased.

A solution to this problem of improving the solar cell array electrical output was to increase the array area by the use of a folded or rolled-up array which is mechanically deployed or unfurled after the spacecraft was placed in its operational orbit or trajectory. This scheme has proved to be practical in the zero-gravity environment of space since such an extendable array can be designed with low weight and small power requirements. It is also desirable that the array be retractable for periods of power flight associated with course correction maneuvers, docking operations, and during the disposal of spacecraft waste to prevent solar cell contamination.

Since space travel was an area where need transcended cost, it provided an ideal trial area for the early high-cost developmental devices. Thus, a number of patents relating to construction and installation of solar cell devices for spacecraft will be reviewed here.

The patents in this section are not by any means all-inclusive but are representative of solutions to typical spacecraft solar cell system problems which include: means of deployment of cell arrays; means for handling damaged solar cells in space; and techniques for handling variations in solar radiation during space flights.

*C.E. Williamson and D.R. Baker; U.S. Patent 3,532,299; October 6, 1970; assigned to TRW Inc.* describe a deployable solar cell array for spacecraft having a foldable solar panel assembly composed of a number of thin film solar panels hinged edge-to-edge which are folded in accordion fashion into face-to-face relation flat against the spacecraft body during launch.

The apparatus includes extendable booms connected to the panel assembly for deploying the latter to an extended position of operation wherein the solar panels are disposed in generally coplanar edge-to-edge relation to provide a lightweight large area solar array. A spacecraft utilizes a pair of the deployable solar arrays for supplying electrical power to on-board electrical equipment, such as thrust and attitude control ion engines.

A somewhat similar device has been described by *P.A. Dillard; U.S. Patent 3,756,858; September 4, 1973; assigned to TRW Inc.*

*W.J. Billerbeck, Jr. and J.R. Owens; U.S. Patent 3,544,041; December 1, 1970; assigned to Communications Satellite Corporation* have designed a deployable solar array for a spin-oriented and stabilized spacecraft comprising a plurality of flexible solar cell panels, each panel having one end thereof rigidly coupled to the sidewall of the spacecraft along a line parallel to the spin axis with the panels initially wrapped about the periphery of the spacecraft. The apparatus includes means for releasing and guiding the panels, whereby centrifugal force, acting on the free end of the panels, causes them to be deployed under tension radially of the rotating spacecraft.

While solar cells have found widespread use in space and other applications their operation has not proven to be entirely satisfactory. Specifically, when a solar cell is placed in a space environment and subjected to bombardment by the omnipresent high energy radiation, defects occur in the cell's crystal lattice structure.

For example, defects are created when a solar cell is bombarded by high energy electrons, protons or neutrons. A defect occurs when one of the foregoing particles collides with an atom bonded in the crystal lattice structure of the solar cell. The collision displaces the atom from the crystal lattice creating a defect or vacancy therein. In some cases two adjacent atoms are displaced creating a defect known as a divacancy. These vacancies will then act as recombination centers for electrons and holes. This recombination traps the charge carriers and thereby reduces the possible current output from the solar cell. In this manner the cell's ability to generate electrical power can be deteriorated by high energy radiation.

*P.-H. Fang, G. Meszaros and W.G. Gdula; U.S. Patent 3,597,281; August 3, 1971; assigned to U.S. National Aeronautics and Space Administration* have discovered that silicon solar cells damaged by radiation can be completely recovered an indefinite number of times by subjecting such cells to high temperatures for a period of time.

More specifically, they have proposed the use of a "greenhouse effect" in which a plastic film which is resistant to high temperature and space radiation and is stable in vacuum is positioned between the sun and the damaged solar cells. This allows transmission of short wave light while prohibiting the escape of long wave light which is absorbed as heat by the solar cells. Thus the array is brought to annealing temperature. Under actual conditions this operation should take approximately one hour and can be repeated as often as necessary to maximize total quantum yield from the solar array.

A device developed by *H.P. Valentijn; U.S. Patent 3,620,846; November 16, 1971; assigned to U.S. National Aeronautics and Space Administration* provides an improved, deployable cantilever support having a multiplicity of panel segments deployed in a manner which minimizes the effects of the shifting of the instantaneous center of mass of the panels as the panels are deployed, whereby transient loading of a spacecraft, during deployment of the panel segments, substantially is avoided.

*A.A. Dollery, N.S. Reed and F.C. Treble; U.S. Patent 3,627,585; December 14, 1971; assigned to U.K. Ministry of Technology, England* have described a stowable solar cell array which includes solar cells mounted on a thin flexible substrate which is supported on an erectable frame. The frame and substrate are arranged so that in the stowed condition with the frame collapsed the substrate is held in flat concertinalike folds. The system includes frame erection means whereby the frame is capable of being erected to unfold the substrate and support it in a fully deployed condition.

One difficulty with using solar cells on spacecraft is that their performance is not constant under varying environmental conditions. For example, if a spacecraft's solar cells cool, the associated optimum solar cell output voltage increases; if a spacecraft's solar cells receive less sun illumination, their current output decreases.

In a practical situation, where it is intended that the destination of a spacecraft is Mars, then, as the distance from the sun increases, a spacecraft's solar cells cool and, as a result, an associated optimum solar cell output voltage increases. Simultaneously, sun illumination decreases and the optimum solar cell output current decreases.

Prior art solar cell arrays overcome the abovementioned variations in performance by being sized to accommodate all of the worst operating conditions for particular flights. For example, in an Earth to Mars mission, enough cells are connected in series to accommodate the relatively high temperatures that occur when the spacecraft is near Earth and enough solar cells are connected in parallel to accommodate the relatively low illumination that occurs when the spacecraft is near Mars. This results in an oversized array with attendant high cost and undue weight.

A scheme to provide optimum performance without using an excessive number of solar cells has been proposed by *E.M. Gaddy; U.S. Patent 3,636,539; Jan. 18, 1972; assigned to U.S. National Aeronautics and Space Administration*. It provides a spacecraft solar cell system including a switching circuit which comprises relay operated switches for changing a plurality of solar cells from a first series-parallel interconnection to a second series-parallel interconnection. The relays are actuated by a command device which may be a telemetry receiver. A protection circuit comprising a photodiode is connected between the command device and the relays to ensure appropriate solar cell orientation when switching occurs. This prevents arcing across the relay switches.

The most important consideration in space vehicles after reliability is weight. In the field of solar-electrical converters, known devices consist of a sturdy, rigid platform of relatively heavy construction to prevent damage to the individual cells. These cells in turn have heavy metal substrates and are adhesively attached to the supports. The cells have a fused silica (glass-like) cover bonded to the top of the solar cell with a specially formulated optical adhesive. The silica is thin and quite fragile and easily damaged if struck by an alien object or if bent or twisted along any one of its axes.

These known solar cells are not only heavy and have limited reliability but they are difficult to manufacture and require considerable time to assemble as the adhesives set.

An improved type of construction has been proposed by *B.D. Osborne; U.S. Patent 3,658,596; April 25, 1972; assigned to Lockheed Missiles and Space Co.* The cells are sandwiched between two sheets of a transparent thermoplastic such as FEP Teflon and suspended from lightweight supporting frames utilizing the tension capabilities of the plastic to secure them in relative position and further protect the cells from radiational damage.

A device developed by *J.M. Weingart and R.K. Yasui; U.S. Patent 3,715,600; Feb. 6, 1973; assigned to U.S. National Aeronautics and Space Administration* provides a solar cell stack construction for a spacecraft. In the stack only the top solar cell panel is exposed to solar energy to provide electrical power. Except for the top panel each panel is covered from exposure to damaging bombarding particles by the preceding panels in the stack toward the stack top. When the exposed top panel's performance is degraded, due to bombardment by particles or other unexpected failure, the top panel is ejected, thereby exposing the underlying panel in the stack, which now becomes the power supplying panel. Each panel in the stack is successively exposed when the panel above it is ejected.

It should be pointed out that the ejection of the top panel may affect the spacecraft direction due to undesired yawing or rolling. This can be avoided by placing several stacks symmetrically about the spacecraft and simultaneously ejecting the top panels of all the stacks. Such an arrangement is shown in Figure 2 wherein four stacks symmetrically arranged about the axis of a spacecraft **40** are designated by numerals **41** through **44**. The ejected top panel of each stack is numbered **45** and the underlying panel which is being exposed is numbered **46**. The directions of ejection of the four panels are designated by arrows **47**.

**Figure 2: Stacked Solar Cell Array Construction for Spacecraft**

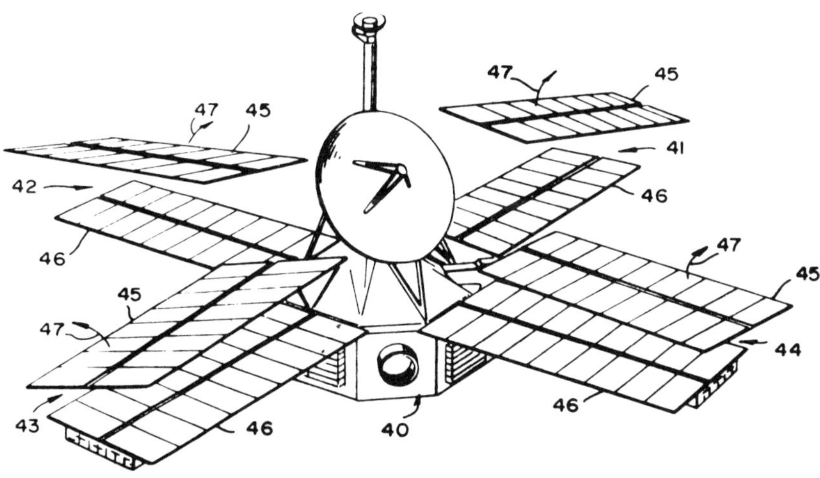


Source: U.S. Patent 3,715,600

A device described by *S. Karius; U.S. Patent 3,778,312; December 11, 1973; assigned to Licentia Patent-Verwaltungs GmbH, Germany* is a solar cell generator for use in spacecraft for missions in the vicinity of the sun. The generator includes a support carrying a field of solar cells which are bounded by a leading end region and a trailing end region. The support is movable relative to the spacecraft between a retracted position, in which the field of solar cells is out of the path of solar radiation, and an exposed position wherein the field of solar cells is exposed to the solar radiation. The support is capable of occupying intermediate positions, in each of which the leading end region of the field of solar cells and a portion of the field are exposed. The field of solar cells is covered by means for filtering the solar radiation. This filtering means has its minimum radiation permeability at the leading end region of the field and its maximum radiation permeability at the trailing end region, the radiation permeability thus increasing from the leading end to the trailing end region. The solar cell generator finally includes a device responsive to the amount of radiation striking the solar cells for moving the support between the retracted and the exposed positions to maintain a constant output from the field of solar cells.

A satellite structure incorporating solar cells has been described by *P.A. Dillard and C.E. Williamson; U.S. Patent 3,783,029; January 1, 1974; assigned to TRW Inc.* incorporates a deployable panel structure embodying a pair of the panels and an intervening deployment boom for extending the panels from folded to unfolded configuration.

**Figure 3: Space Vehicle with Solar Cell Structure in Stowed Position at Launch**

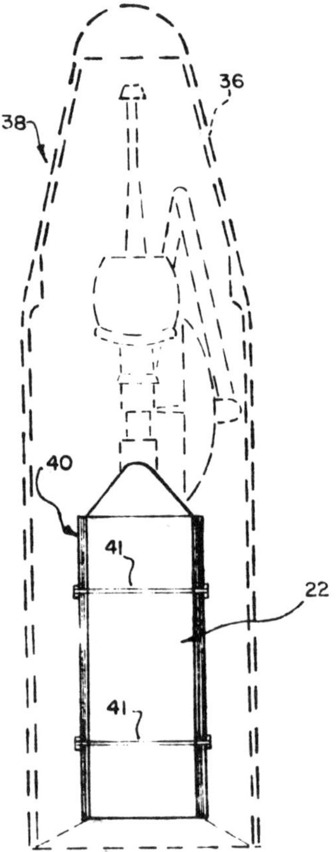


Source: U.S. Patent 3,783,029

Figure 3 shows the spacecraft in stowed configuration in the launch vehicle. The body of the spacecraft is indicated by **22**. Numerals **36** and **38** indicate the shroud and the launch vehicle proper. Numeral **40** indicates the area of the folded solar panels and **41** the retaining bands for the panels which are sheared by pyrotechnic shear devices when an orbit of the space vehicle is achieved.

For contrast with Figure 3, Figure 4 shows a similar spacecraft with its solar array deployed as described by *D.L. Wentworth; U.S. Patent 3,785,590; January 15, 1974; assigned to Cummunications Satellite Corporation.*

In the particular design, rotatable, roller booms are carried by an extendable member which moves outward from a spacecraft body with one end of each flexible solar cell panel rigidly coupled to the spacecraft body and the other end rigidly coupled to the periphery of a rotatable roller boom. During deployment, the solar cell panels which are rolled up on the roller booms unwind. A spring-operated negator motor coupled to the rotating boom at one end and to the fixed housing at the end of the deployment member, winds up during deployment of the solar cell panel to facilitate retracting of the thin film solar array by rewinding the flexible solar panels about the periphery of the roller booms.

There is shown in Figure 4, a spacecraft **10** having a cylindrical body **12** with communication system components **14** positioned about the longitudinal axis of the spacecraft. The solar cell array comprises three identical flexible solar cell panel assemblies **18, 18′** and **18″** which are extendable, that is deployable, from within the spacecraft through narrow longitudinal slots **20** extending the length of the cylindrical body and generally parallel to the axis of the spacecraft.

**Figure 4: Spacecraft with Solar Cell Structure in Deployed Position in Space**

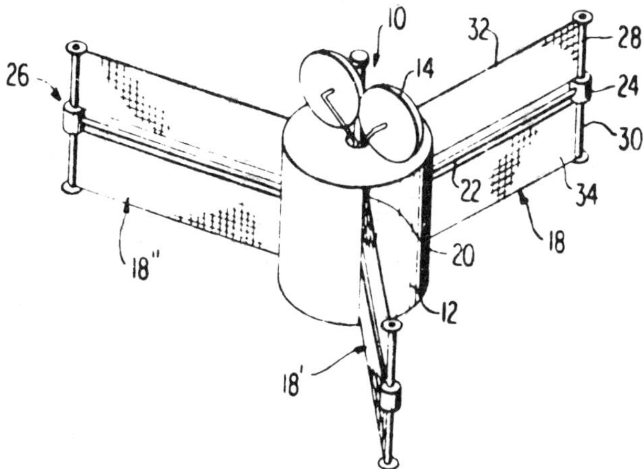


Source: U.S. Patent 3,785,590

Each assembly comprises a radially extensible member **22** in the form of an extensible hollow shaft which may constitute for instance a series of telescoping light weight metal tubes and carries at the outer end, a cylindrical metal housing **24** which in turn rotatably supports a roller boom assembly **26**. The assembly constitutes paired roller booms **28** and **30** extending outwardly from cylindrical metal housing **24** on each side thereof for rotation about an axis passing through the cylindrical metal housing, and at right angles to member **22**.

Each solar cell assembly further comprises left and right hand solar array panels **32** and **34**. The outer edges of these panels are coupled directly to the periphery of respective roller booms **28** and **30** by suitable coupling means such as an adhesive or the like. The inner edge of the rectangular flexible solar cell panels **32** and **34** are rigidly and electrically coupled to the spacecraft power system internally of the cylindrical side wall **12** (by means not shown).

In a configuration described by *W.I. Berks and H.F. Meissinger; U.S. Patent 3,817,481; June 18, 1974; assigned to TRW Inc.* the ends of the solar panels are wound on rotary drums from which the panels unwind during deployment in such a way that the deployed panels have lobe-like or wing-like configurations and the panels may be retracted to stowed configuration and deployed any number of times in flight.

A configuration that was developed by *G.J. Andrews and H.A. Rosen; U.S. Patent 3,863,870; February 4, 1975; assigned to Hughes Aircraft Company* consists of a spin-stabilized vehicle comprising (1) a body and (2) three deployable panels, arranged to present body and panel surfaces covered with photovoltaic or solar cells for increasing the stability and the sun derived power of the craft.

The panels are stowed within or wrapped about the craft during ascent towards its utilizable position. When deployed, the panels extend radially from the craft. When extended, the panels are at least as long as the radius of the body. The use of three panels results in a relatively small ripple of the power supplied by the cells as craft rotates and therefore a relatively high efficiency of cell utilization is provided.

A similar device has been described by *R.E. Coltrin, T.C. Eakins and J.E. McIntyre; U.S. Patent 3,973,745; August 10, 1976; assigned to Hughes Aircraft Company.*

A scheme described by *A. Anchutin; U.S. Patent 3,948,468; April 6, 1976; assigned to RCA Corporation* is one in which a replaceable solar array panel for spacecraft is formed of two or more panels one surface being visible or exposed to the sun and other surfaces being shielded from the sun. The exposed and shielded surfaces are arranged to be replaced by moving or rotating a shielded surface into an exposed surface orientation and to eject, move or rotate an exposed panel surface. The activation of the movements required may be automatic or remotely controlled from a ground-based station upon appropriate indications of the need for replacement of damaged or otherwise deteriorated exposed surfaces.

The solar cells in the abovedescribed deployable arrays and also in fixed arrays are typically mounted in panel systems wherein two or three adjacent rows of a plurality of series-connected solar cells are connected in parallel. The series-parallel groups are then connected via appropriate bus bars to the spacecraft power utilization and battery charging systems. Nearly all of these solar cell panel systems use diodes connected in series between each solar cell group and a solar cell bus bar to prevent total panel failure in event of shorts developing in a single cell string in a group, and to prevent major power loss in case of partial shadowing of the panel. Also, these diodes help prevent battery discharge through the panel when it is in eclipse, and help prevent local heating effects due to shadows or cell output anomalies.

Conventional diodes have been used for this "blocking" purpose and are generally cylindrical in shape and, on rigid flat solar cell panels and cylindrical arrays, the large diameter of the diodes compared to the thickness of the solar cell cover glass has been accommodated by installing the diodes in holes drilled in the substrates.

In fixed cylindrical arrays comprising the craft's exterior surface, the mounting of the diodes in holes presents no serious problem. However, in the more desirable-larger surface-rigid fixed and foldable panel arrays, these holes present a serious structural weakness since they are usually aligned in a row at the end of the solar cell group adjacent the junction of the panels and the main body of the spacecraft where structural loads are concentrated.

On a flexible roll-up solar array, which typically may have a 2 mil substrate and a 13 to 14 mil thick solar cell/cover glass, the large diameter diodes present major design problems. One solution has been to mount the diodes in holes drilled in the drum or take-up roller of the roll-up array and run isolated bus bars from each panel group to the drum where they are connected to diodes mounted in a heatsink.

Blocking diodes are especially necessary in a roll-up array where part of the array is retracted (i.e., rolled up on the drum) and part of the array is illuminated. The retracted section will act as an electrical load to the illuminated part unless the retracted section is blocked off by diodes. If it is not blocked off, current leaks through the retracted, roll-up sections will cause a temperature rise therein and they will accept more current, which further raises their temperature, permitting still further increase in leakage current. These phenomena could continue until cell damage occurs and/or the retracted sections act as a dead short to the rest of the current-producing panel. It can thus be seen that without the blocking diodes, the output of the entire panel might be lost.

As roll-up or thin solar panels become larger and more complex by the addition of panel regulators, for example, the prior art techniques of using conventional diodes become cumbersome, the number of isolated bus bars increases, and the area available for heatsinking diodes is restricted. Accordingly, it should be evident that a technique which would provide the necessary diode protection while eliminating the need for heatsinks, allow maximum usage of diode isolation, and minimize the need for numerous long isolated bus bars would constitute a significant advancement of the art.

A type of construction developed by *G. Wolff, G.R. Brooks and T.C. Eakins; U.S. Patent 3,952,324; April 20, 1976; assigned to Hughes Aircraft Co.* provides a type of isolation or blocking diode which has the thickness and shape of adjacent solar cells.

A type of construction developed by *C.J. Pentlicki; U.S. Patent 4,133,501; Jan. 9, 1979; assigned to Communications Satellite Corporation* is one in which a pair of semirigid solar cell panels are curved under stress and latched in arcuate edge abutting fashion about the exterior of a spacecraft body. Upon release of the latched ends they flatten into tangency at opposite sides of the spacecraft body. Spring hinges at the ends of a deployment arm cause each panel to automatically move radially outward of the spacecraft body, and a motor mounted to the spacecraft body and to one end of the deployment arm rotates the panel to orient the solar cells relative to the source of solar energy.

## HOUSEHOLD HEATING AND COOLING

The concept of heating and cooling households by individually owned systems is obviously an intriguing one. In some climates at least, this might make the owner independent of outside suppliers of oil and electricity and could result in lower costs to the householder. This has of course been the goal of many installations of collector systems (nonphotovoltaic systems where fluids are heated in rooftop collectors).

Some of the economics of possible applications of photovoltaic systems to single family houses have been reported (3) and are covered in the last chapter of this volume.

One interesting concept, touched upon elsewhere in this volume as well, involves the use of roof shingles which double as solar cells. In a process described by *A.F. Forestieri, A.F. Ratajczak and L.G. Sidorak; U.S. Patent 4,040,867; August 9, 1977; assigned to U.S. National Aeronautics and Space Administration,* a solar cell shingle may be made of an array of solar cells on a lower portion of a substantially rectangular shingle substrate made of fiberglass cloth or the like. The solar cells may be encapsulated in fluorinated ethylene propylene (FEP) or some other weatherproof translucent or transparent encapsulant to form a combined electrical module and a roof shingle. The interconnected solar cells are connected to connectors at the edge of the substrate through a connection to a common electrical bus or busses.

An overlap area is arranged to receive the overlap of a cooperating similar shingle so that the cell portion of the cooperating shingle may overlie the overlap area of the roof shingle. Accordingly the same shingle serves the double function of an ordinary roof shingle which may be applied in the usual way and an array of co-operating solar cells from which electrical energy may be collected.

A type of construction developed by *A.J. Manelas; U.S. Patent 4,080,221; March 21, 1978* shows a system for household electrical power and heating.

It makes use of an array of light sensitive, voltage producing solar cells of the flat disc silicon type. To increase power, while using fewer costly cells, each cell of the array has a truncated conical shell mounted on legs at a spaced distance thereover, the shell having a mirror-like reflective inner surface. Thus, sunlight is received in the large end and reflected through the small end to the cell. A sealed, weather-tight enclosure for the array has fluid inlets and outlets for producing heat, the heat conductive shells absorbing and radiating heat.

As shown in Figure 5, a typical building such as the residence **30**, may have a breezeway **31**, an ell **32**, cellar **33**, hot air heating system **34**, main tap water heating system **35**, electric appliances **36** and a television **37**. In the ambient air space **38** of the cellar there is a heat storage member **39** which may be constructed of about five tons of apertured fire bricks and which is arranged to store solar heat for use when needed on sunless days.

The power pack panel **61** is about 8 feet wide, 16 feet long and 12 inches thick and is preferably mounted at an angle of 45° from the horizontal facing south for example on the roof of the ell. The angle of inclination will depend on the geographical location, 45° having been found best for the Commonwealth of Massachusetts.

The power pack unit **41**, includes a substantially sealed, weather-tight enclosure **42** having a rigid, planar bottom **43**, which is preferably black, side walls **44, 45, 46** and **47**, upstanding from bottom **43** and preferably of rectangular configuration, and a top, or cover, **48** of transparent material such as plastic of the Plexiglass, or glass, type. The bottom and the side walls **44, 45, 46** and **47** are of heat insulation sheet material, or include a central layer **49** of heat insulation such as Fiberglas to prevent losing heat to the ambient atmosphere.

Within the sealed enclosure **42** a plurality of light-sensitive, voltage producing, solar cells **51** are arrayed in a common plane, at predetermined spaced distances apart, on the rigid planar bottom. The cells may be of any suitable type, but preferably are silicon cells of flat, thin disc, or wafer, configuration, commercially available, for example, from Solar Power Corporation.

Each solar cell **51** is of predetermined area with a diameter of about 3½" and includes an electrically conductive backbone, with spaced rods, or wires, normal thereto and in parallelism with each other, all embedded in a black backing with one terminal on the backbone on the exposed face of the silicon cell and the other terminal on the foil on the underface of the cell. Each cell is capable of producing 0.6 watt of energy in strong sunlight.

Figure 5: Solar Heated House (Above) and Detail of PV Panel (Below)

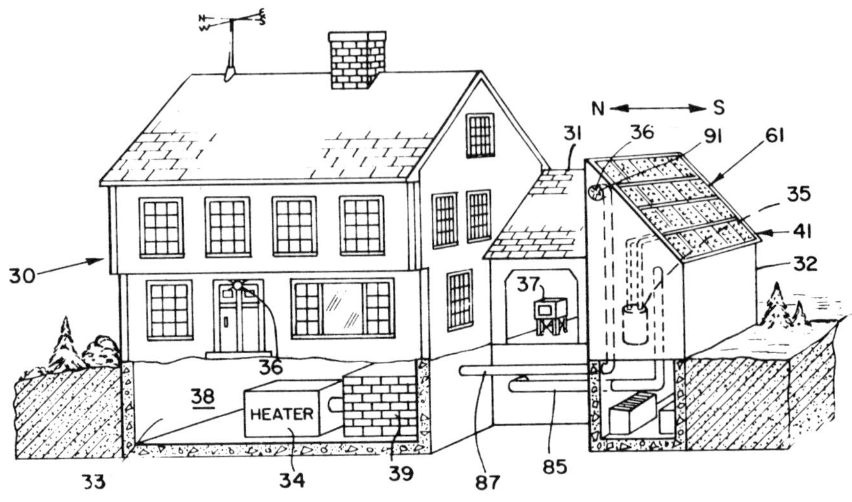


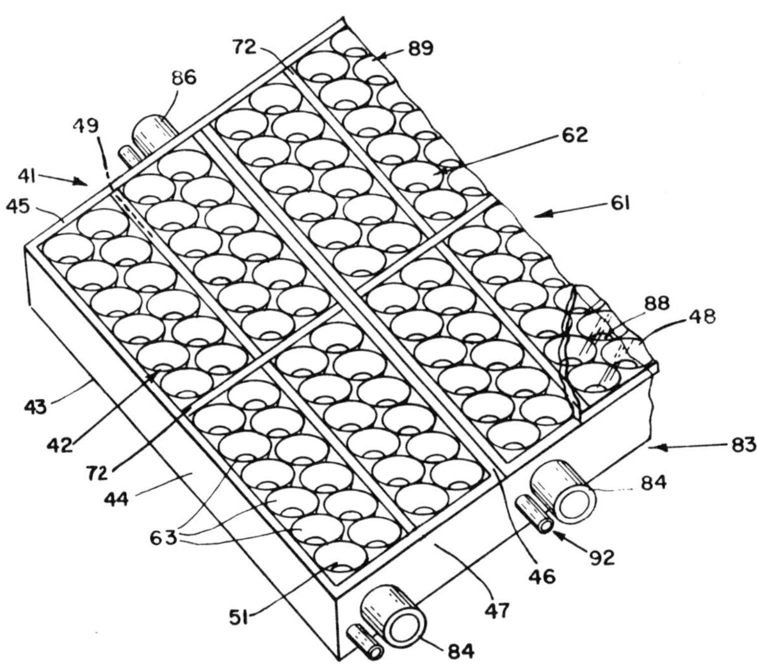


Source: U.S. Patent 4,080,221

To permit the use of fewer costly solar cells, and thereby bring the cost of the solar energy system **61** down to a reasonable amount, flared light reflector means **62** is provided in each enclosure **42**, preferably one for each cell **51** therein. The flared light reflector means **62** is mounted within the enclosure **42** between the array of spaced apart, common-planar cells **51** and the transparent top **48** for collecting light from the sun over a larger total area than the total area of the cells in the enclosure and reflecting the light onto the cells as the sun rises and sets during the day.

Flared light reflector means **62** preferably comprises a plurality of truncated conical shells **63** of aluminum, or other heat conductive material. Each shell has a large open end proximate the top, and of substantially greater area than the area of a cell **51**, a truncated-conical side wall with a mirror-like polished, light-reflective, inner, flared surface and a small open end, encircling a cell **51** at a spaced distance thereabove by support of a pair of integral legs affixed to the planar bottom **43**.

In the illustrated embodiment wherein the cells **51** are each about 3½" in diameter, the small open end **67** is about 3¾" in diameter and supported about ½" above the cell to provide air circulation space. The large open end is about 7½" in diameter and the side wall is about 4" in vertical height.

Preferably the truncated conical side wall of each shell **63** is inclined at about 15° from the vertical, or 75° from the plane of enclosure bottom **43**, since this degree of inclination in the Massachusetts area provides the optimum angle for collecting useful sunlight in the open ends and reflecting the light out through the small ends onto the cells **51**.

Each power pack unit **41**, of the process preferably comprises 48 solar cells **51** with shells **63** enclosed in a weatherproof enclosure **42**, about 40" wide by 96" long and 7½" deep, there being cruciform dividers **72** to form groups of twelve. The 48 cells of each unit are connected in series to conductors. The approximate cost of each unit **41** is $600.00 and it produces 150 watts of electrical power, each cell **51** producing 3.5 watts because of the truncated-conical shells **63** rather than 48 x 0.6 or 28.8 watts without the shells, an increase by a factor of six.

The solar system **61** includes about ten power pack units **41**, which may be arranged in a string, or row, or in multiple in a frame, the units **41** being electrically connected in series by conductors into a power circuit including about nine storage batteries, and a one way diode, to energize the batteries from the light of the sun.

The household appliances **36** are powered by batteries, or by a backup unit, if five or six successive sunless days occur. A separate large battery is connected through a DC/AC converter to the television **37** which requires alternating current. The batteries are preferably of the lead acid type such as an Eastern 375 amp hour battery having a storage capacity of 4,500 watts. Nine such batteries in series produce 110 volts DC at 47 amperes for 8 hours and can supply the building **30** for 7 days of energy with 25% generated from the solar panels **61**. Such batteries cost in total $847, and weigh about 2,000 pounds.

A backup unit is an electric starting gasoline engine motor generator, which is automatically connected into the system by the sensing of dull days, rain, etc. Low voltage circuit breakers, voltage regulators and DC to AC inverters, are provided to prevent over-discharge of the batteries, provide a signal to the circuit breakers to open when the energy stored is at its minimum (100 volts) and to close when the batteries are at 110 volts. The DC to AC inverter is a 1,200 watt square wave, 60 cycle generator and it is powered with 12 volts DC and turned on by a command from the small appliances. DC to AC Invert Synchronous Convert is a generator powered from 110 volt batteries to operate the heavy appliances, this device consists of a 1.2 horsepower, 110 volt DC motor permanent magnet type driving direct a 3,000 watt alternator at a speed of 2,500 rpm. A transducer relay supplies the power for on and off as commanded by the appliances.

Air heating means **83** includes the sealed enclosures **42**, the heat conductive aluminum shells **63**, a cold air inlet **84**, connected by air ducts **85** to the space **38** in the cellar **33**, or elsewhere in residence **30**, and a hot air outlet **86** connected by hot air duct **87** to the heat storage member **39** and thence to the house air heating system **34**. The air within the enclosures **42** rises to 200°F or more due to the heat radiation of the sun and the inner walls of the enclosure and the bottom are black to absorb heat. The shells **63** serve as heat absorbers and heat radiators to heat the cold air entering in the lower portions **88** of the enclosures and exiting from the upper portions **89** as hot air. The heat storage member **39** preferably comprises a maze of apertured fire brick, but it could be any well-known type of rock storage device. A suction fan **91**, powered by the batteries, is provided in hot air duct **87**.

## COMMUNITY ENERGY SYSTEMS

The provision of solar energy to entire communities similar to the function performed now by public utilities is obviously a concept of great interest.

The primary source of energy is now fossil fuels, i.e., coal, oil and natural gas. In recent years nuclear energy generated power has received considerable attention and is now being used to supplement the power derived from the fossil fuels. Although no one knows the extent of fossil fuel reserves, it appears that their availability may begin to drop off while the demand for energy and power continues to rise. There are at present serious doubts that nuclear energy can be relied upon to fill a continuously larger percentage of energy requirements, since there are no known ways as yet to provide complete insurance against environmental pollution by nuclear energy plants.

Increases in the level of energy consumption in the United States can be directly related to increases in GNP. Consumption of energy has increased from 3%

a year between 1947 and 1965 to 5% a year since 1965. The United States in the early 1970s was burning annually 500 million tons of coal, 20 trillion ft$^3$ of gas and 5 billion barrels of oil. It is estimated that in order to meet an ever-increasing power demand, the United States will have to increase its power-generating capacity to six times its present level by the year 2000. This means an increase from 330 million kilowatts to nearly 2 billion. Therefore, in addition to raw material supply considerations, these tremendous requirements for energy raise equally tremendous problems in ecology.

All available energy sources now in use (coal, oil, gas and nuclear fuel) have a major drawback, they produce by-products that pollute the air, water and land. The resulting problems of pollution control and waste disposal are fast becoming staggering both in magnitude and cost.

In contrast, use of radiation energy from the sun in an effective and efficient manner offers the possibility of providing electrical power without the problem of pollution control and waste disposal.

The practicality of generating electrical power on earth from solar energy has previously been questionable because of the absorption of solar energy by the atmosphere, obscuration by clouds, dust deposition, wind effects on structure, the limited availability of the sun's radiation at low sun angles and the lack of energy at night. Even in areas where solar energy is received throughout most of the year, a sudden interruption by a cloud could require spinning standby power generators of equivalent capacity or an energy-storage system which would be orders of magnitude more effective than batteries.

Another factor to be considered in the direct collection of solar radiation on earth is the solar collector surface area that would be required to supply the projected electrical power demand. Assuming for the sake of argument that the conversion efficiency of solar radiation collected on earth is 100%, the estimated total solar collector area required by the year 2000 for the United States would be about 284 square miles.

A proposed system which could provide power in large amounts and which is free from such factors as cloud cover has been described by *P.E. Glaser; U.S. Patent 3,781,647; December 25, 1973; assigned to Arthur D. Little, Inc.* In this system solar radiation is collected and converted to microwave energy and is then transmitted to earth and converted to electrical power for distribution.

Such a system is shown in Figure 6. Two geostationary satellites **10** and **11** are shown positioned with respect to the earth **12** so that at least one is illuminated by the sun at all times. As an example, at an altitude of 22,300 miles in an orbit parallel to the earth's equatorial plane, a satellite moving east to west would be stationary with respect to any point on earth. However, at times the satellite would pass through the earth's shadow. Thus having two satellites in the same orbit but out of phase permits one to be illuminated while the other is in shadow. For example, at an altitude of 22,300 miles, the two satellites could be

**Figure 6: Satellite Solar Energy Collector with Microwave Energy Transmission to Earth Stations**

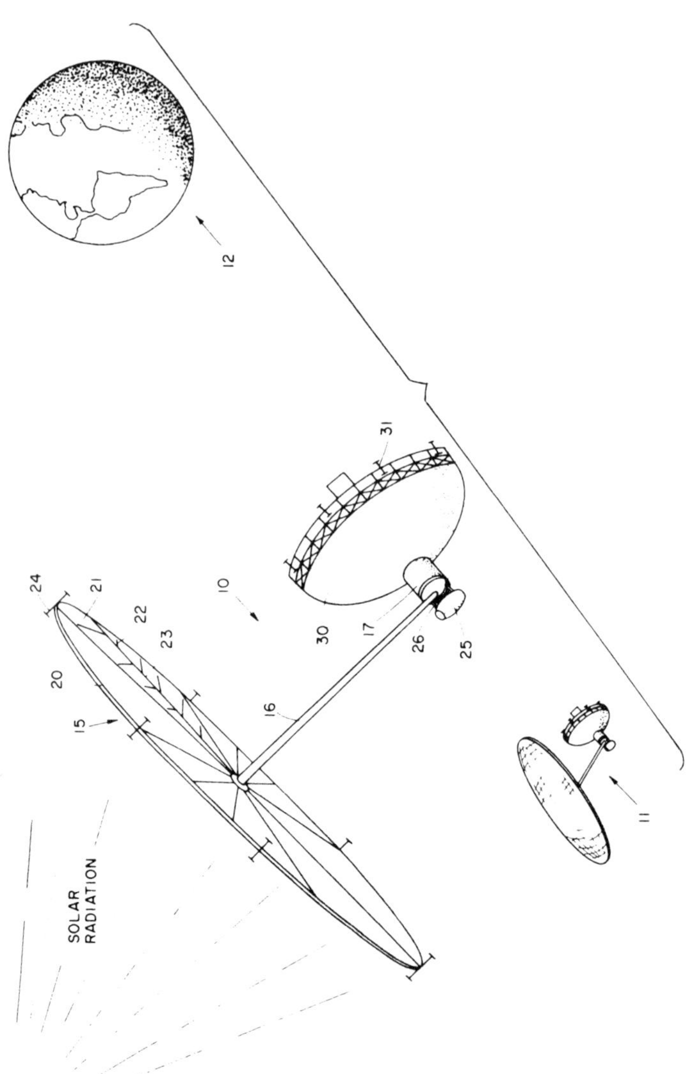


Source: U.S. Patent 3,781,647

placed about 21° out of phase or about 7,900 miles apart. Such a phase difference would keep the satellites above the horizon and both would have a direct line of sight to the same point on earth. A network of satellite stations, such as those shown could be employed to position stations in orbit to achieve the most effective system operation and to supply widely dispersed points on earth either continuously or as required to meet peak power demands.

If, on the other hand, it is not necessary to provide a continuous supply of electric power, then a single satellite may be used, taking power from it at will or whenever it was positioned in solar energy receiving relationship with the sun.

Satellite **10** has a solar collector/converter **15** pivotally mounted on a support member **16** which is pivotally attached to an equipment housing **17** of the satellite. The embodiment of the solar collector/converter comprises a large-surface area platform such as a disc or dish-shaped member **20** which is divided into sectors such as sector **21**.

Associated with each sector are current collecting and transmitting means such as branch wires **22** and main lead wire **23**. The main lead wires from the sectors are then connected to a main cable (not shown) which extends down through the support to microwave energy generating equipment located in the equipment housing.

Means, such as small attitude control rockets **24**, are provided to position the solar collector/converter in a continuously optimum orientation. It is, of course, within the scope of this process to form the solar energy collector/converter in any configuration, the circular one shown representing but one illustrative embodiment.

If the satellite is a manned space station then personnel quarters **25** are provided with proper communication means being included such as passage **26** connecting quarters **25** with the equipment housing and means for docking earth-to-satellite supply spacecraft. An antenna **30** with appropriate guidance and attitude control means **31** is affixed to the equipment housing **30**. The positioning and guiding of the antenna must be very accurate in order to direct the beam of microwave energy to the receiving stations on the earth.

## AUTOMOBILE PROPULSION

Since the world has become increasingly dependent on the automobile and since liquid fuels are in increasingly short supply at increased unit costs, alternative energy sources for vehicles are of obvious interest.

A solar-celled hybrid vehicle has been proposed by *W.H. Moore; U.S. Patent 4,090,577, May 23, 1978.* A front wheel driven, gas powered vehicle is converted to include a rear differential connected for power input to a first and second electrical motor, or a single electric motor if desired. When first and second electrical motors are used they are connected in parallel, the power input thereto being brought across a current limiting series of resistors to protect and control the current level thereto. A switching circuit connects in various series and parallel combinations a plurality of batteries and concurrently switches the necessary current limiting resistance. Thus a control combination is provided including a manual selector for the desired forward and reverse directions and the low and high current ranges which is further multiplied by the various resistances. In this configuration, the normally available gasoline power plant is retained in the vehicle and is augmented during periods of nonoptimal use by the above electric motor provisions. This electric power can be periodically replenished either by way of a charger or a set of solar panels placed on the roof of the vehicle.

Such a vehicle design is shown in Figure 7. The vehicle **10** includes a greenhouse or passenger compartment **11** disposed between an engine housing **12** and a cargo compartment **13**. In the more recent configurations in the prior art, the vehicle is most often powered or driven by front wheels **14**, a typical example of a configuration of this type being a 1976 Honda station wagon model. In this form the high density of the engine contained in the engine compartment **12** and the immediately adjacent differential and transmission connected to the wheels, provide an inordinate forward weight bias, particularly in cases where the rear seats of the passenger compartment are unoccupied. While conventionally this weight bias is accomodated by sophisticated suspension arrangements, these same arrangements are designed to handle large cargo loads and therefore may be modified to convert this prior art vehicle to a hybrid drive.

**Figure 7: Solar Celled Hybrid Vehicle Design**

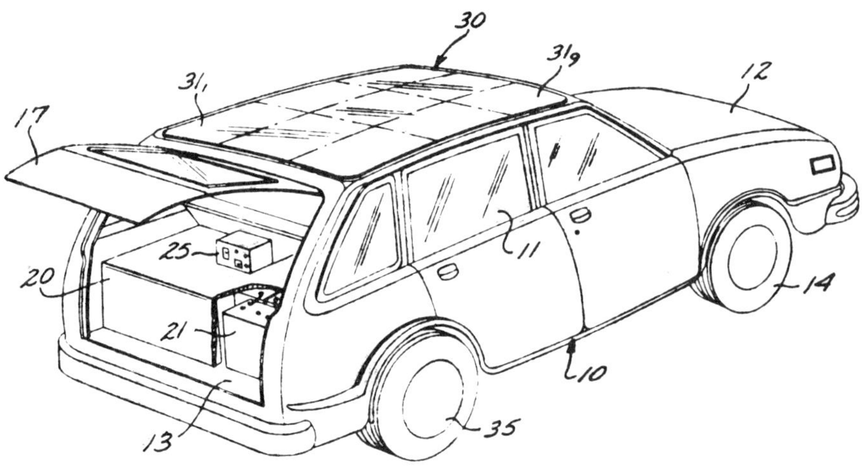


Source: U.S. Patent 4,090,577

More specifically, shown disposed in the cargo compartment, and accessible through a cargo door **17**, is an electric drive enclosure **20** storing a plurality of batteries **21** and supporting a charger **25** which may be connected to any conventional home outlet to periodically bring the batteries up to full charge.

Included further on the upper surface or roof of the vehicle is an array of solar cells generally designated by the numeral **30** comprising nine rectangular array segments $\mathbf{31}_1$ through $\mathbf{31}_9$ such as those produced by Sensor Technology, each solar panel $\mathbf{31}_1$ through $\mathbf{31}_9$ including a plurality of silicon wafers. This combination of the silicon wafers provides approximately 100 watts of electric power when exposed to full radiation from the sun. It is the power supplied by way of batteries **21** which may be utilized to drive a rear axle **35**.

## AUTOMOBILE COOLING

A design by *J.H. Miller; U.S. Patent 3,943,726; March 16, 1976; assigned to Lawrence Peska Associates, Inc.* is one in which a solar energy cell system is used as an electric supply source to operate an air conditioner or a fan ventilation system contained within the interior chamber of an automobile. The electrical circuit comprises a series circuit consisting of: the solar energy cells, a voltage regulator, a storage battery, a thermostatic temperature control and a ventilation fan or an automobile air conditioner.

Such a system is shown schematically in Figure 8. It shows the solar cooling system **10** for an automobile **11**, wherein the cooling system is contained within an interior chamber **13** of an automobile **11**. The solar cooling system comprises a standard automobile **11** with a plurality of solar energy cells **14** embedded in the roof **15**, the trunk lid **16**, the front hood **17** and along the periphery of the rear window **18** of the automobile.

A plurality of cool air ducts **19** and hot air ducts **20** communicate between the interior chamber and the outside surface **21** of the automobile. The plurality of the solar energy cells are wired in series to a voltage regulator **22**, a storage battery **23**, and a thermostatic temperature control unit **24** as shown. A ventilation fan **25** is positioned in the forward section **26** of chamber **13** as well as second ventilation fan **27** in the rear section **28** of chamber **13**, wherein fans **25, 27** are wired in series circuit to the thermostatic temperature control unit. Alternatively, in place of the fans **25, 27,** a standard automobile air conditioning unit **29** is wired in series circuit to control unit **24**.

The radiant energy of the sun is absorbed by the plurality of the solar energy cells. The generated electrical energy from the solar energy cells is stored by the storage battery until control unit **24** activates electrical energy flow from the storage battery to the cooling unit. Consequently, anytime the temperature of chamber **13** exceeds the control unit setting 24, the cooling unit is activated. The solar energy cells are constructed from well-known standard materials such as a layer composite of silicon and boron.

**Figure 8: Application of Solar Cells to Automotive Cooling Shown in Block Diagram Form and Elevations of Vehicle**

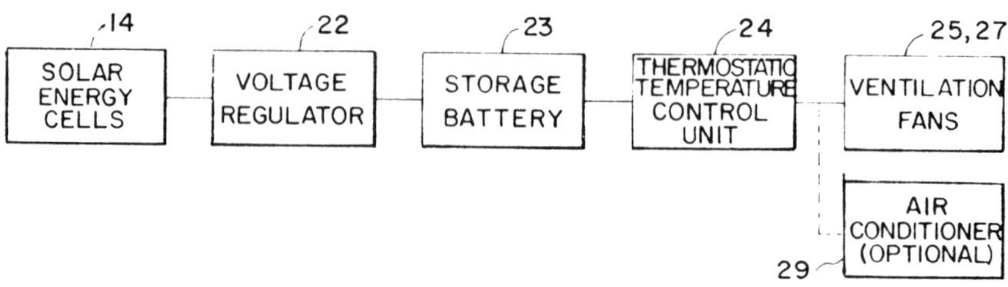


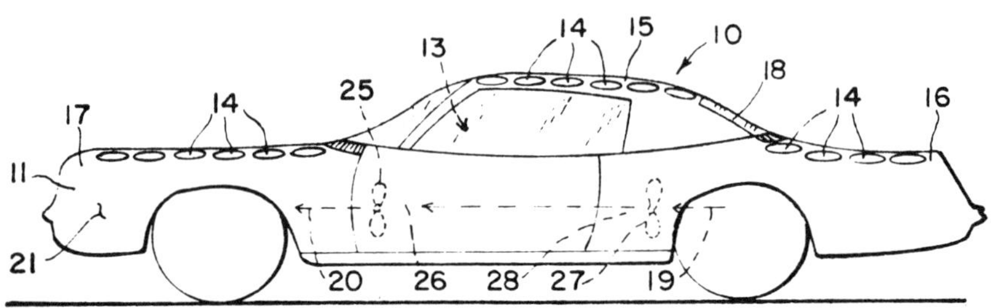


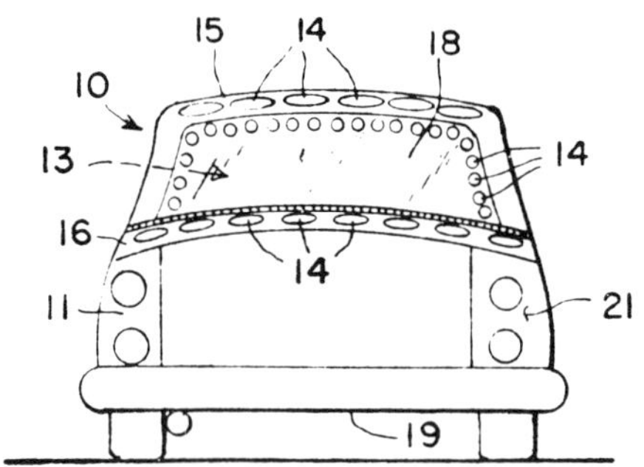


Source: U.S. Patent 3,943,726

## WRIST WATCH

Somewhat surprisingly, perhaps, a fair bit of developmental effort has resulted in the commercialization of solar-powered wrist watches. They are commercially available in both Europe and the U.S.

Several patents have been issued on such devices as well. One such development is described by *T.S. Grohoski; U.S. Patent 3,509,712; May 5, 1970; assigned to Timex Corporation.* It covers a watch in which a phototropic (photochromic) material is placed over the light sensitive surface of the solar cell. The phototropic material darkens in proportion to the amount of light incident upon it. As the amount of incident light increases, the phototropic material becomes darker and permits the passage of a decreasing percentage of the incident light through the material. In this way the amount of light reaching the solar cell is limited to a relatively constant level or a level below some predetermined maximum level. Since the quantity of light reaching the cell is limited, the cell's electrical output is also limited to a level below the current which would injure the battery.

Such a watch design is shown in Figure 9. It contains a solar cell **4a**, the dial plate of the watch. It is protected from the atmosphere by the watch crystal **8.** Alternatively, the dial is an uncovered solar cell, i.e., consisting of P-layer and N-layer, and the watch crystal **8** is of a phototropic glass. The crystal is mounted in a bezel 9. Wires from the solar cell **4a** lead to a rechargeable miniature battery **10.** The wire **11** from the battery leads to a miniature electric motor **12** whose output shaft **13** drives the gear mechanism of the watch.

**Figure 9: Application of Solar Cell to Electric Wrist Watch**

Source: U.S..Patent 3,509,712

In an array of solar cells used as the power source for a device such as a clock, a wrist watch, an electronic calculator or the like, it is generally the case that only a limited area is available in which solar cells can be exposed to the light. Thus, it is essential that the active area of the cells utilize the available area as nearly completely as possible. Consequently, the space consumed by insulative materials separating adjoining cells must be minimized.

Conventionally, solar cells are placed together as closely as possible without incurring substantial danger of making contact between adjoining cells and thus developing a short which will limit the output of the array. Moreover, in order to make maximum utilization of the space available, the cells must be placed in some sort of orderly configuration. With rectangular cells, the configuration will be essentially rectangular. For a circular array, the cells are conveniently sectorial in shape.

Up to now, conventional techniques have made it essentially impossible to decrease the separation between cells below about 0.3 mm, even where the individual carrying out the fabrication is of the highest skill. A separation of this magnitude, as indicated by the discussion up to this point, involves a serious waste of the available area. In addition, the gap between the solar cells is relatively conspicuous and where the cells are to be used in a device such as a lady's wrist watch, the gap between the cells constitutes an eyesore. The same holds true, though perhaps to a lesser extent, whenever a solar battery is to be used in ornamental goods such as clocks and wrist watches for men.

As is evident, it would be desirable to improve methods of fabrication so that solar batteries could be made available which utilize virtually all of the space provided and which do not suffer from the defect of conspicuous gaps between adjoining cells.

In a scheme devised by *O. Tsutomu; U.S. Patent 4,038,104; July 26, 1977; assigned Suwa Seikosha KK, Japan* solar cells are formed in a wafer, preferably of silicon, with a minimum spacing between the cells. The wafer is mounted on a substrate of glass, ceramic or another silicon wafer. Channels are cut between adjoining solar cells using a dicing saw, the channel being cut to a depth such that it completely separates the cells. Using a dicing saw having a thickness of 25 $\mu$, the width of the channel will be between 30 and 40 microns.

If desired, where a silicon wafer is used as the substrate, the silicon substrate may be oxidized to form an insulating layer of silicon dioxide thereon prior to mounting the solar cell array thereon. After cutting the channels, it is desirable to fill them with an insulating material such as a wax or a resin.

A process developed by *O. Koichi; U.S. Patent 4,087,960; May 9, 1978; assigned to Suwa Seikosha KK, Japan* is one in which a solar-battery wrist watch makes use of a flexible printed-circuit board for connecting the cells of the battery in series and for connecting the battery with a storage battery. The arrangement eliminates the need for connecting wires and extraneous supports in that the flexible printed-circuit board holds the solar battery against the dial plate of the watch.

A type of construction developed by *H. Nagao and K. Kawamura; U.S. Patent 4,140,545; February 20, 1979; assigned to Sharp KK, Japan* is one in which the electrodes formed on the light receiving surfaces of the plural solar cell elements are electrically connected to each other by light transmitting, electrically conductive wiring means formed on the light receiving surfaces.

A design developed by *S. Wada and Y. Isobe; U.S. Patent 4,144,096; March 13, 1979; assigned to Suwa Seikosha KK, Japan* provides a monolithic solar battery including a plurality of unit solar cells wherein each unit solar cell is electrically isolated by buffer regions and an active photovoltaic diffusion region in each unit solar cell is electrically connected with one of the two buffer regions defining the cell. In one embodiment wiring between unit solar cells of the battery and battery electrodes is provided on the upper light receiving surface of the solar battery. In a second embodiment the electrical connection between unit solar cells of the battery and electrodes is provided on the surface opposed to the light receiving surface of the solar battery. When electrical connection between the unit cells of the solar battery and lead-out electrodes is provided on the lower (opposed) surface of the substrate, a substantial increase in the active photovoltaic area is obtained, permitting an increased electrical output of the solar battery having a fixed surface area.

## BEACONS AT REMOTE LOCATIONS

In remote areas where no primary source of electrical power is available, it is necessary to have a standby source of electrical power to operate equipment, particularly emergency equipment. Thus, for example, in order to summon assistance in the event of an emergency, ships and aircraft are usually provided with an emergency position indicating radio beacon which is powered by a battery pack.

In such times of distress, the single-use battery pack, is usually the sole source of power and when the battery's power has been depleted after one use, the emergency equipment is rendered inoperative for want of a power source. Furthermore, since the available power of the battery deteriorates rapidly under load, particularly at elevated temperatures, an inordinately large number of batteries must be available for emergency power requirements. Thus the use of solar energy to power such remote device is very attractive.

An apparatus described by *E.W. Kazis, R. Mark and T.J. Wetherell; U.S. Patent 4,009,051; February 22, 1977; assigned to General Solar Power Corporation* is a solar power pack apparatus comprising photovoltaic cell means, first electrochemical cell battery means in circuit with the photovoltaic cell means and second electrochemical cell battery means. The second electrochemical cell battery means has a watt-hour capacity sufficient to increase the watt-hour capacity of the first electrochemical cell battery means when the first and second electrochemical battery means are intercoupled. Further provided are electrochemical coupling means operative to selectively couple the first and second electrochemical cell battery means and electrical load terminal means adaptable for connection thereto of an electrical load in circuit with the electrical coupling means.

Such a system provides a source of continuous electrical power to power electrical apparatus under adverse environmental conditions. The system has a long shelf-life capability in addition to a long operating life under load. These characteristics make such a system particularly suitable for powering emergency equipment under adverse environmental conditions at remote locations, such as an emergency position indicating radio beacon.

## MISCELLANEOUS APPLICATIONS

To illustrate some varied and unusual potential applications of solar cells, the following examples are cited.

According to *R.C. Bender; U.S. Patent 3,844,840; October 29, 1974* the substantially hemispherical portion of a cycle rider's helmet may be covered with a plurality of solar energy electric current generating cells. Conductors, connected with the cells, are in turn connected to a junction plate for energizing a hearing aid or a small transistorized radio carried by the user.

A device developed by *E.J. Schiel and R.R. Gammarino; U.S. Patent 3,950,862; April 20, 1976; assigned to U.S. Secretary of the Army* relates to a multisolar cell detection arrangement for detection of pulsed laser radiation inside a military helmet. Each solar cell is coupled to a ferrite core which allows operation in full sunlight without the need of optical filters.

The solar cells are activated by short-pulsed laser radiation, directed from a laser transmitter such as an injection laser transmitter mounted on an M-16 rifle and bore-sighted with a rifle sight so as to simulate rifle fire. The device permits accurate scorekeeping during military field exercises.

# Materials for Solar Cells

To summarize the present status of solar cell materials, the entire U.S. Dept. of Energy research and development program is focussed on silicon. It is the recipient of about 60% of U.S. Federal funding.

Secondary U.S. Dept. of Energy support emphasis is given to concentrator systems at a level of about 10% of the funding allotted to R&D on silicon solar arrays. The concentrator systems may use a variety of materials including

cadmium and copper sulfide
amorphous silicon
indium phosphide and gallium arsenide
semiconductor oxides on silicon

Thus, in the short term, mass-produced silicon cells seem most likely to dominate whatever markets exist. Then, as the industry grows, becomes healthy and able to feed itself, thin film devices using other materials than crystalline silicon—probably in optical concentrator systems—may provide the second generation of practical solar cell devices.

## SILICON

Silicon-based PV is the reference technology for both flat plate and concentrator systems. It is clear that cells with good electrical characteristics can be produced. Field experience indicates that useful cell life under operating conditions can be adequate. This technology, therefore, provides a vehicle for evaluating systems, with known bounds on performance and manufacturing costs.

Commercial markets for applications such as cathodic protection, radio repeaters, and navigational aids are developing for flat plate cells even at the current high prices of \$5 to \$10/$W_p$.

> Flat plate silicon module costs using technology now foreseeable will approach the price range 50-75¢/$W_p$. This price range will not produce electricity competitive with projected coal central power generation, but it does offer important insurance value against unexpected energy price rises and would permit the growth of a significant export market to developing countries (1).

Three main directions for low-cost silicon PV production are now being followed. One is based on the use of single crystal material sliced from large diameter ingots. Another approach employs directly fabricated sheet material of reduced quality that offers the potential for lower area costs of silicon. The third alternative is based on wafers sliced from cast polycrystalline blocks. These last alternatives lead to less efficient cells than the first.

Current solar cell technology is based on the use of Si wafers obtained by slicing large Czochralski or float-zone ingots (up to 12.5 cm in diameter), using single-blade inner diameter (i.d.) diamond saws. This method of obtaining single crystalline Si wafers is tailored to the needs of large volume semiconductor products (i.e., integrated circuits plus discrete power and control devices other than solar cells). Indeed, the small market offered by present solar cell users does not justify the development of Si high-volume production techniques which would result in low-cost electrical energy.

Growth of Si crystalline material in a geometry that does not require cutting to achieve proper thickness is an obvious way to eliminate costly processing and material waste. Growth techniques such as edge-defined film-fed growth (EFG), web-dendritic growth, chemical vapor deposition (CVD), etc., are possible candidates for the growing of solar cell material. The growing of large ingots with optimum shapes for solar cell needs (e.g., hexagonal cross sections) requiring very little manpower and machinery would also appear plausible. However, it appears that the cutting of the large ingots into wafers must be done using multiple rather than single blades in order to be cost-effective (8).

> Efforts to reduce the cost of silicon sheet from sliced single-crystal material should be emphasized because of the importance of high efficiency in minimizing systems costs in both flat plate and concentrator applications. With current technology the cost per unit area of cells fabricated from cast semicrystalline silicon or ribbon growth is slightly lower than that of cells made from Czochralski material. This does not offset the higher systems costs associated with lower efficiency. Current ribbon growth efforts should be continued because of the potential for possible advances, particularly in cell efficiency.

New developments in silicon cell design have been taking place with encouraging frequency in recent years and it is reasonable to expect further innovations in the future. These developments are exemplified by devices such as the violet cell, back-surface-field cell, vertical multijunction cell, and tandem junction cell. Yet

> silicon cell performance currently falls below theoretical limits for reasons that are not adequately understood. Innovations in cell design and processing technology that can lead to improved efficiency and lower cost should be supported (1).

A major effort of the U.S. Department of Energy lies in the Low-Cost Solar Array Project (LSSA) and its subsidiary program the Large-Area Silicon Sheet Task (8)(9).

The objective of the Large-Area Silicon Sheet Task is to develop and demonstrate the feasibility of several alternative processes for producing large areas of Si sheet material suitable for low-cost, high efficiency solar photovoltaic energy conversion. To meet the objective of the LSSA Project, sufficient research and development must be performed on a number of processes to determine the capability of each for producing large areas of crystallized Si. The final sheet growth configurations must be suitable for direct incorporation into an automated solar-array processing scheme.

At the time the LSSA Project was initiated (January 1975) a number of methods potentially suitable for growing Si crystals for solar cell manufacture were known. Some of these were under development; others existed only in concept. Development work on the most promising methods is now being funded. After a period of accelerated development, the various methods will be evaluated and the best selected for advanced development. As the growth methods are refined, manufacturing plants will be developed from which the most cost-effective solar cells can be manufactured. The Large-Area Silicon Sheet Task effort is organized into four phases: research and development on sheet growth methods (1975-77); advanced development of selected growth methods (1977-80); prototype production development (1981-82); development, fabrication, and operation of production growth plants (1983-86).

Silicon is a relatively poor absorber of light and, as a result, cells must be 50 to 200 microns thick to capture an acceptable fraction of the incident light. This places rather rigorous standards on the sizes of crystal grains which can be tolerated, and all commercial silicon cells are now manufactured from single crystals of silicon.

It is believed that if polycrystalline silicon is to be used, individual crystal grains must be at least 100 microns on a side if efficiencies as high as 10% are to be achieved. It is important that the grains be oriented with the grain boundaries perpendicular to the cell junction so that charge carriers can reach the junction without crossing a grain boundary. A number of research projects are underway to develop inexpensive techniques for growing such polycrystalline materials and for minimizing the impact of the grain boundaries. Efficiencies as high as 6.7% have been reported for vapor-deposited polycrystalline silicon cells with grains about 20 to 30 microns on a side, and a proprietary process capable of producing grains nearly a millimeter on a side reportedly can be used to produce cells with efficiencies as high as 14%. Work is underway to improve crystal growing techniques and to enlarge grains with lasers and electron beams (3).

As summarized by Hovel (2), there are three main efforts in Si solar cell development: reducing the cost of the starting Si, reducing the cost of fabricating cells, and improving the conversion efficiency. He goes on to describe these in some detail and includes an extensive bibliography.

One aspect of the overall silicon development program of the Department of Energy is the definition of the optimum material for contact with the molten or solidifying silicon.

In the crystal-growing processes, a refractory crucible is required to hold the molten silicon, while in the ribbon processes an additional refractory shaping die is needed. The objective of these contracts is to develop and evaluate cost effective refractory die and container materials. The material must be mechanically stable to temperatures above the melting point of silicon, must not excessively contaminate the silicon processed through it, be amenable to the fabrication of dies and containers with close tolerances and of varying geometries, and be cost effective.

Table 2 shows the die and container materials study projects supported by DOE through JPL. Two of the contracts in this area, RCA and Tylan, are to develop a substrate material for supported film growth and a coating for substrates, dies, and containers.

**Table 2: Die and Container Materials Study Projects Supported by DOE Through JPL**

| | |
|---|---|
| Battelle Labs<br>Columbus, Ohio<br>(JPL Contract No. 954876) | Silicon nitride for dies |
| Coors Porcelain<br>Golden, Colorado<br>(JPL Contract No. 954878) | Mullite for container and substrates |
| Eagle Picher<br>Miami, Oklahoma<br>(JPL Contract No. 954877) | CVD silicon nitride and carbide |
| RCA Labs<br>Princeton, New Jersey<br>(JPL Contract No. 954901) | CVD silicon nitride |
| Tylan<br>Torrance, California<br>(JPL Contract No. 954896) | Vitreous carbon |

Source: Reference (9)

**Slicing or Sawing Crystals**

All of the silicon photovoltaic devices now sold are manufactured from wafers sawn from single-crystal boules (2- to 3-inch-diameter cylinders of silicon) (3). These wafers represented about 35% of the cost of photovoltaic arrays in 1976 (see Table 4 at the end of this volume).

The crystals are commonly grown by dipping a seed crystal into a silicon melt in a quartz crucible which is a few degrees above silicon's melting point and by slowly withdrawing the growing crystal from the melt. The crucible and the crystal are counterrotated to grow a straight crystal of uniform, circular cross section. The crystal is pulled from the crucible until most of the molten silicon has been withdrawn and removed from the melt. The entire apparatus is then allowed to cool so that the crystal can be taken out of the airtight chamber.

As the remaining molten silicon solidifies in the crucible, the crucible usually breaks, adding about $50 to the cost of the boule. This procedure is called the "Czochralski" (CZ) or "Teal-Little" method. The boule is sawed into thin wafers which are sent to the next stage of the cell fabrication process. A variety of new concepts have been proposed for reducing the cost of the crystal-growing processes:

- Improved sawing techniques might reduce the material lost in the sawing process; currently, nearly 50% of the crystal grown is lost as silicon "sawdust." Techniques are being developed which use multiple saws with thin sawblades, sawing wires, and other advanced processes to decrease the material lost in sawing and increase the number of silicon wafers produced from a single crystal by producing thinner wafers.
- The molten silicon in the quartz crucible from which the crystal boule is withdrawn can be continuously replenished, leading to longer crystal draws and a lower requirement for crucibles. (This process could also save a number of processing steps.)
- The silicon material lost in sawing could be recycled for further use if it is not contaminated in the cutting process.
- Techniques can be used to increase the diameter of the crystals grown by using the standard Czochralski process. Crystals now in use are typically 3" in diameter, but crystals 10" in diameter have been grown in laboratories. An experiment is now underway to grow three 12" diameter crystals in one continuous heating cycle. Choice of an optimum diameter will depend on a detailed study of the manufacturing

process. (Problems in cutting the crystals increase with crystal diameter.)

- The rate of crystal growth can be increased if columnar grains can be tolerated in cell wafers. (Research is required to determine the extent to which such grains can be tolerated or their effects minimized.)
- Considerable energy and capital equipment could be saved if a process could be developed for growing a crystal in a mold. The Crystal Systems Company is currently examining a concept in which a seed crystal is placed in one end of an insulated mold filled with molten silicon. A temperature gradient is maintained along the mold so that the crystal grows from the end with the seed crystal to fill the mold. This technique is commonly used to grow metal crystals, but has not yet been successfully adapted to the growth of semiconductor-grade silicon crystals.
- Improvements in cutting circular wafers into the hexagons required for close packing are possible through the use of a laser-slicing technique being developed by Texas Instruments.

In solar cells the active area is the junction, which is very close to the surface of the device and it extends only several thousand Angstroms downward from the surface. This junction is the vital zone where the efficiency and usefulness of the device are characterized. If this junction is damaged, the cell efficiency is drastically reduced. It has, therefore, been required that solar cells be formed into the desired size and shape before the junction is formed.

It has been proposed, however, by *J. Lindmayer; U.S. Patent 4,097,310; June 27, 1978* that the silicon be treated so as to form a junction for generating electrons by light before the wafer has been reduced to its desired shape. Thereafter, the wafer is formed with the desired shape by using a saw having a diamond blade and cutting through the wafer and the electron-generating junction to produce a solar energy cell of at least one dimension smaller than that of the wafer prior to cutting.

The saw may be of the rotary blade or oscillating type. Also, the electron-generating junction may be of the P-N or N-P type. In any case, there is a substantial and unexpected improvement in the electron-generating efficiency of the cell when a diamond saw is used to cut directly through the junction.

The mode of operation favored by Lindmayer comprises taking a substantially single crystal ingot of silicon (Monsanto Company) and cutting it into substantially uniform slices. After slicing by known means, it was found that each sliced, silicon wafer had a uniform thickness of 12 mils and a radius of approximately one inch.

The silicon wafers so produced were treated in a known manner to produce a P-N junction therein. After the P-N junction had been formed, the approximately 2-inch-diameter wafer was cut to the form of a square having sides 0.8 inch by 0.8 inch with a diamond-impregnated peripheral saw, (manufactured by Lindberg Tempress Company, Model No. 602). The saw had a rotary blade 0.005 inch thick, having diamond particles embedded therein. The blade was Model No. 12-361.

The speed of rotation of the blade was about 18,000 rpm. Utilizing such speed, the wafer was cutting entirely through its thickness, including the P-N junction, to form a rectangular cell 0.8 inch by 0.8 inch at a speed of one inch per second. However, blade speeds of about 5,000 to 20,000 rpm can also be used with success, as were cutting speeds of 0.05 to 10 inches per second.

As noted above, the growing of large silicon ingots with optimum shapes for solar cell needs (e.g, hexagonal cross sections) by techniques which require very little manpower and machinery seem plausible (8)(9). However, also as noted above, the cutting of such large ingots into wafers will require multiple-blade rather than single-blade sawing in order to be cost-effective.

As part of the Low Cost Solar Array Project and the Large Area Silicon Sheet Task under that project, a number of ingot technology projects are now supported by the U.S. Department of Energy through the Jet Propulsion Laboratory (JPL) of the California Institute of Technology. These ingot technology projects are listed in Table 3.

**Table 3: Current Ingot Technology Development Projects Supported by DOE Through JPL**

| | |
|---|---|
| Crystal Systems, Inc.<br>Salem, Massachusetts<br>(JPL Contract No. 954373) | Heat exchanger method (HEM), cast ingot, and multiwire fixed abrasive slicing |
| Kayex Corp.<br>Rochester, New York<br>(JPL Contract No. 954888) | Advanced CZ growth |
| Siltec Corp.<br>Menlo Park, California<br>(JPL Contract No. 954886) | Advanced CZ growth |
| Texas Instruments<br>Dallas, Texas<br>(JPL Contract No. 954887) | Advanced CZ growth |
| Varian Vacuum Division<br>Lexington, Massachusetts<br>(JPL Contract No. 954374) | Multiblade slurry sawing |
| Varian Vacuum Division<br>Lexington, Massachusetts<br>(JPL Contract No. 954884) | Advanced CZ growth |

Source: Reference (9)

Details of progress in some of these ingot technology programs are as follows (9):

*Heat Exchanger Method–Crystal Systems:* The Schmid-Vicchnicki technique (heat exchanger method) has been developed to grow large single-crystal sapphire (Figure 10). Heat is removed from the crystal by means of a high-temperature heat exchanger. The heat removal is controlled by the flow of helium gas (the cooling medium) through the heat exchanger. This eliminates the need for motion of the crystal, crucible, or heat zone. In essence this method involves directional solidification from the melt where the temperature gradient in the solid might be controlled by the heat exchanger and the gradient in the liquid controlled by the furnace temperature.

The overall goal of this program is to determine if the heat exchanger ingot casting method can be applied to the growth of large-shaped Si crystals (>8 in cube dimensions) in a form suitable for the eventual fabrication of solar cells. This goal is to be accomplished by the transfer of sapphire growth technology (50-lb ingots have already been grown), and theoretical considerations of seeding, crystallization kinetics, fluid dynamics, and heat flow for Si.

**Figure 10: Crystal Growth Using the Heat Exchanger Method–Crystal Systems**

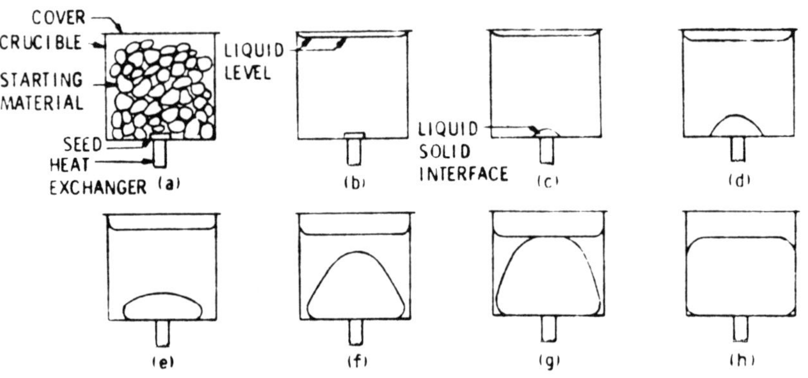


Growth of a crystal by the heat exchanger method:

(a) Crucible, cover, starting material, and seed prior to melting
(b) Starting material melted
(c) Seed partially melted to insure good nucleation
(d) Growth of crystal commences
(e) Growth of crystal covers crucible bottom
(f) Liquid-solid interface expands in nearly ellipsoidal fashion
(g) Liquid-solid interface breaks liquid surface
(h) Crystal growth completed

Source: Reference (9)

*Advanced CZO–Varian, Texas Instruments, Siltec, and Kayex Corp:* In the advanced CZO contracts, efforts are geared toward developing equipment and a process in order to achieve the cost goals and demonstrate the feasibility of continuous CZ solar-grade crystal production (Figure 11). Varian will modify an existing furnace for continuous growth using granular silicon for recharging (molten silicon will also be considered), and a new puller is to be designed. Texas Instruments' technique is based on an incoming flow of solid granular or nugget polysilicon, premelted in a small auxiliary crucible from which liquid silicon will be introduced into the primary crucible. Siltec's approach is to develop a furnace with continuous liquid replenishment of the growth crucible accomplished by a meltdown system and a liquid transfer mechanism with associated automatic feedback controls. Kayex will demonstrate the growth of 100 kg of single crystal material using only one crucible by periodic melt replenishment.

**Figure 11: Continuous CZ Crystal Growth Machines**

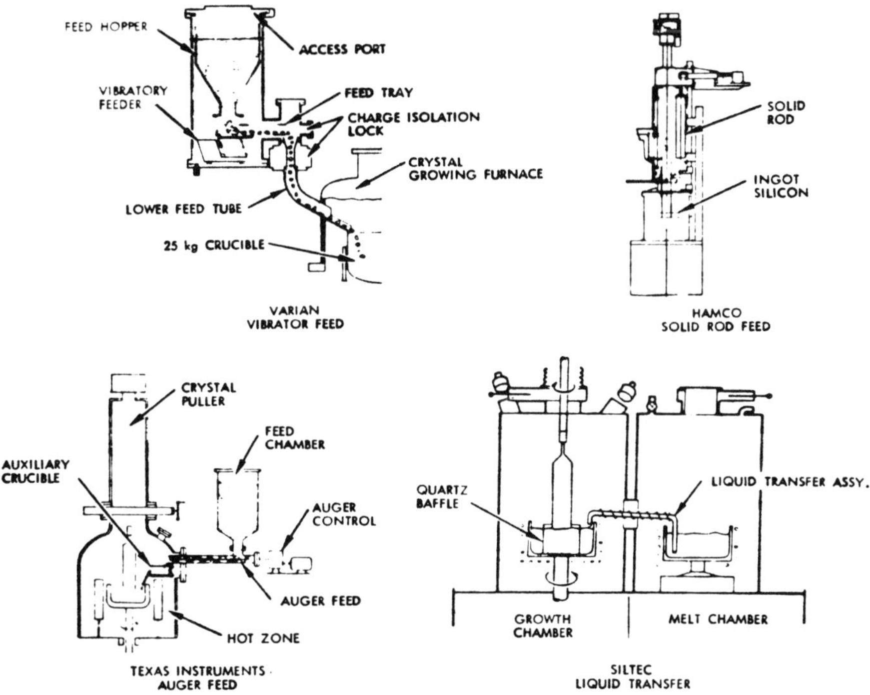


Source: Reference (9)

*Multiwire Sawing–Crystal Systems; Multiblade Sawing–Varian:* Today most Si is sliced into wafers with an inside diameter saw, one wafer at a time being cut from the crystal. This is a large cost factor in producing solar cells. The multiblade and multiwire slicing operations employ similar reciprocating blade head motion with a fixed workpiece.

Multiblade slicing is accomplished with a slurry suspension of cutting fluid and silicon carbide abrasive and tensioned steel blades of 6 mm height and 0.2 mm thickness. Multiwire slicing uses 0.5 mm steel wires surrounded by a 0.25 mm copper sheet, which is impregnated with diamond as an abrasive.

**Ribbons**

Research has been underway for at least 15 years to develop a process by which single-crystal silicon can be produced in the form of ribbons or sheets (3). Thin ribbons would be appropriate for use in solar cells without the crystal-slicing steps required for boules. This would eliminate the costs involved in slicing the crystals and could make more efficient use of silicon. This would not only eliminate expensive operating steps, but could reduce trimming losses when cutting round cells to hexagonal cells.

The Department of Energy has been funding Mobil-Tyco, Solar Energy Corp., IBM, Motorola, RCA, the University of South Carolina, and Westinghouse to assess the merits of a variety of processes for growing single-crystal ribbons and sheets.

One technique for doing this is the "edge-defined, film-fed growth" process (EFG); work began on this process about a decade ago. The EFG process utilizes a graphite die on top of a crucible of molten silicon. A narrow pool of molten silicon forms on the top of the die by capillary action. A thin ribbon of silicon is slowly withdrawn from this pool.

The edge-defined film-fed growth (EFG) technique as practiced by Mobil-Tyco Solar Energy Corp. of Waltham, Mass. under DOE through JPL Contract No. 954355 is based on feeding molten Si through a slotted die as illustrated in Figure 12. In this technique, the shape of the ribbon is determined by the contact of molten Si with the outer edge of the die. The die is constructed from material that is wetted by molten Si (e.g., graphite). Efforts under this contract are directed toward extending the capacity of the EFG process to a speed of 7.5 cm/min and a width of 7.5 cm.

In addition to the development of EFG machines and the growing of ribbons, the program includes economic analysis, characterization of the ribbon, production and analysis of solar cells, and theoretical analysis of thermal and stress conditions.

Photovoltaic cells made from EFG ribbons have shown efficiencies as high as 10%. Intensive work is proceeding on the EFG process but several difficulties remain: 1) contamination of the ribbon by impurities from the die results in efficiencies lower than other silicon cells; 2) the graphite die is attacked by the molten silicon; and 3) the process is currently quite slow. Ribbons 2- to 2½-cm wide can now be grown at rates of 2 to 7 cm per minute. The objective is a growth rate of approximtely 18 cm/min. A recent analysis by IBM indicated that a large-diameter Czochralski boule can grow silicon crystal areas at a rate equivalent to 20 to 40 simultaneously pulled ribbons.

**Figure 12: Edge-Defined Film-Fed Growth (EFG)–Mobil Tyco**

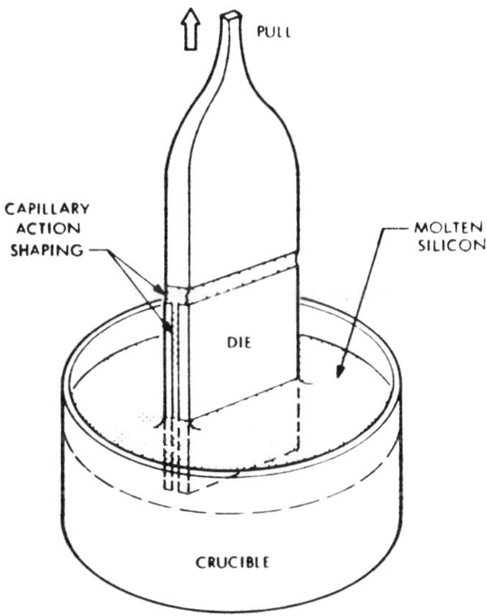


Source: Reference (9)

Other ribbon techniques under investigation include the IBM "Capillary Action Shaping Technique" (CAST) which uses a wetted die, and the RCA "inverted Stepanov" technique which uses a nonwetted die. Both techniques share many of the problems of the EFG approach (3).

Another process for manufacturing ribbons is the web-dendrite crystal growth process. Unlike EFG, this process requires no die, and the die contamination and die erosion problems are, therefore, avoided. Strips of silicon 2 meters long, 0.15 mm thick, and 22 mm wide have been grown at rates of 2 to 3 cm/min; the objective is a growth rate of 18 cm/min. Difficulties which remain with the process include: 1) the fact that careful temperature control is necessary, which probably precludes the simultaneous growth of several ribbons from one melt; and 2) relatively slow growth rates.

As shown in Figure 13, the dendritic web is a thin, wide, ribbon form of single crystal silicon. "Dendritic" refers to the two wire-like dendrites on either side of the ribbon, and "web" refers to the silicon sheet that results from the freezing of the liquid film supported by the bounding dendrites. Dendritic web is particularly suited for fabrication into PV convertors for a number of reasons, including the high efficiency of the cells that can be fabricated from it, the excellent packing factor of the cells into subsequent arrays, and the cost effective conversion of raw silicon into substrates.

Figure 13: Schematic Section of Web Growth–Westinghouse

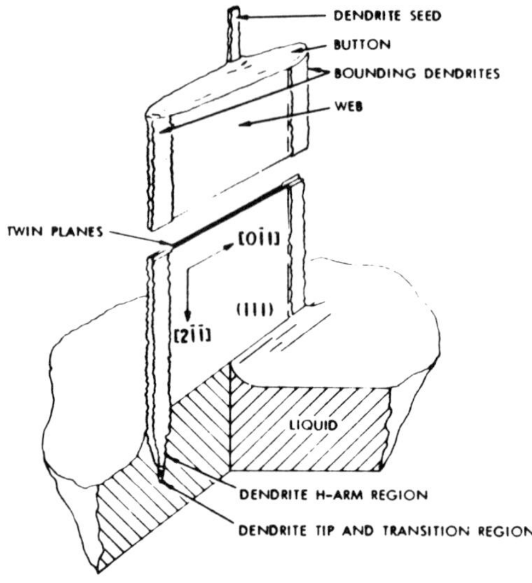


Figure 14: Laser Zone Regrowth–Motorola

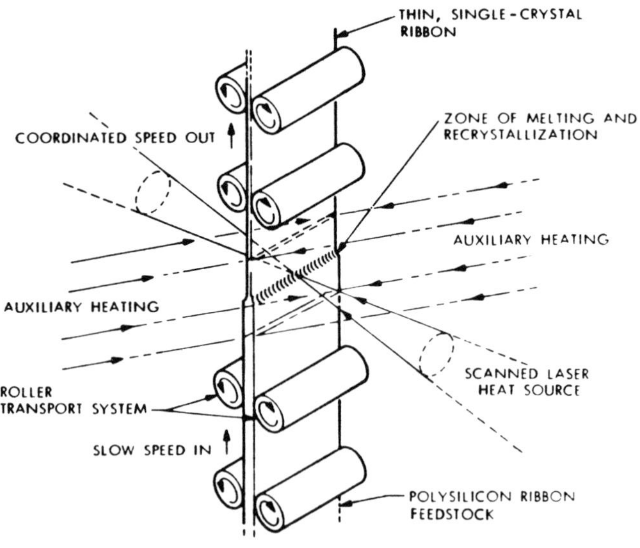


Source: Reference (9)

Still another technique for producing single crystal silicon ribbons is a ribbon-to-ribbon (or more precisely a polysilicon-ribbon-to-single-crystal-ribbon) laser zone growth process as developed by Motorola, Inc. of Phoenix, Arizona under DOE support through JPL under JPL Contract No. 954376.

The ribbon-to-ribbon process is basically a float-zone crystal growth method in which the feedstock is a polycrystalline Si ribbon (Figure 14). The polysilicon ribbon is fed into a preheated region that is additionally heated by a focused laser beam, melted, and crystallized. The liquid Si is held in place by its own surface tension. The shape of the resulting crystal is defined by the shape of the feedstock and the orientation is determined by that of a seed single-crystal ribbon.

A process developed by *H.E. LaBelle, Jr.; U.S. Patent 3,591,348; July 6, 1971; assigned to Tyco Laboratories, Inc.* provides a method of growing crystalline materials in the form of elongate bodies of predetermined constant cross section. The method involves provision of a shaping member having a surface with a gross configuration conforming to the desired cross-sectional shape of the body to be grown, establishing on the surface a liquid film of the material to be grown, and continuously growing the body from the liquid film while simultaneously feeding additional material to the surface to replenish the film.

A process developed by *J.S. Bailey; U.S. Patent 3,826,625; July 30, 1974; assigned to Tyco Laboratories, Inc.* is an improvement over the method described in U.S. Patent 3,591,348 for growing crystalline bodies from the melt. The improvement essentially consists of using a porous forming member (also called a die) that is characterized by an interconnecting network of pores or cells of capillary proportions.

This, then, is a refinement in the edge-defined, film-fed growth process as described by LaBelle in the earlier patent.

Silicon ribbons employed in solar cells must be substantially monocrystalline, uniform in size and shape and substantially free of crystal defects. It is not difficult to control the size and shape of substantially monocrystalline ribbons grown by the EFG process.

However, one problem which results during the production of flat elongate monocrystals grown from the melt is the formation of defects adjacent the ribbon edges. Although not known for certain, it is believed that such edge defects result from the shape of the liquid/solid interface at the ribbon edges or the accumulation adjacent the crystal edges of impurities present in the melt. These edge defects are objectionable and the ribbons must be processed further to remove the defects before they can be used.

Some of these problems have been overcome in a process described by *K.V. Ravi; U.S. Patent 4,095,329; June 20, 1978; assigned to Mobil Tyco Solar Energy Corp.* This process basically comprises first producing a substantially monocrystalline tubular body of silicon or other suitable semiconductor material and then cutting the tubular body along its length to produce a plurality of nearly flat, monocrystalline ribbons. Preferably the cutting of the tubular body is achieved by etching, as with an acid jet. In a preferred mode of operation to produce solar cells, the tubular body is treated to form an annular rectifying junction, then it is severed longitudinally to form a plurality of nearly flat ribbons, and finally the ribbons are modified to form solar cells.

Since a tubular body is continuous in cross section, it has no edge regions comparable to the long side edges of a ribbon. Accordingly, tubular bodies do not have the edge surface defects as normally found in flat ribbons or other shapes having two or more defined side edges. More precisely, a tubular body grown by the EFG process has better crystallinity than ribbons grown by same EFG process under the same conditions.

In addition, the absence of edges leads to better stability during growth, permitting greater growth flexibility and, hence, quality of the crystals. Furthermore tubes of silicon can be grown at quite high growth rates. Figure 15 shows the four steps in manufacturing a silicon ribbon by this process.

Turning first to the view labelled **1**, a tubular body **10** of a substantially monocrystalline P-type silicon is provided by growing it from a boron doped, semiconductor grade silicon melt under an inert atmosphere using the abovedescribed EFG process. The tubular body is grown from a melt contained in a quartz crucible (not shown) using a die (not shown) consisting of two graphite cylinders disposed concentrically one inside the other and locked together in the manner of U.S. Patent 3,687,633.

The gap between the two graphite cylinders is sized to serve as a capillary for molten silicon and the die assembly is disposed so that melt can enter the bottom end of the capillary and rise to its upper end by capillary action. This tubular body **10** is then introduced into a diffusion furnace where it is exposed to a gaseous mixture of oxygen and phosphorus oxychloride at a temperature of about 1000°C for a period of about 15 to 30 minutes.

As a consequence of this diffusion step, phosphorus is diffused into the outer and inner surfaces of the tube so as to form an N-P-N structure (see view **2** and **3A**) with relatively shallow outer and inner N regions **12** and **14** and thin layers **16** and **18** of silicon dioxide covering the outer and inner surfaces. The N regions each have a depth of about 0.5 micron and the diffusion oxide layers each have a thickness of about 3000 A. The formation of the diffusion oxide layers results from the presence of oxygen which is used as the transport medium for the phosphorus oxychloride.

Thereafter as shown in view **2**, the outer and inner surfaces of the tube are coated with a conventional polymethyl methacrylate positive resist material as represented at **20** and **22**. (For convenience of illustration the N regions **13** and **14** and the oxide layers **16** and **18** are not specifically shown in views **3** and

**Figure 15: Steps in Making Silicon Ribbon from Silicon Tubes by Mobil Tyco Process**

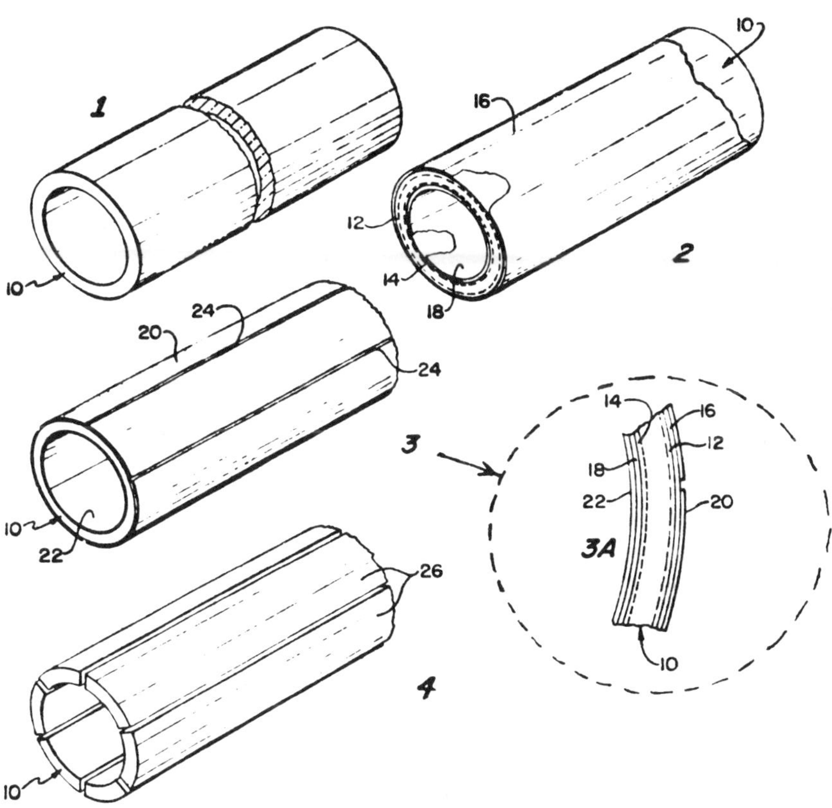


Source: U.S. Patent 4,095,329

4.) Then the outer photoresist layer **20** is exposed to a narrow light beam, so that a plurality of circumferentially spaced, straight and narrow longitudinally extending areas of the resist coating **20** are exposed to the beam and thereby altered to a lower molecular weight polymer. The tube is then immersed in a preferential solvent or etchant such as methyl isobutyl ketone, with the result that the unexposed portions of the resist coating **20** remain intact while the exposed areas are dissolved away as represented at **24** in view **3** to expose narrow line portions of the outer oxide layer **16**.

The next step involves etching the tube so as to subdivide it into a plurality of narrow strips **26** as shown in view **4**. This is achieved in two stages. In the first stage the tube is immersed in HF at room temperature for about 1-2 minutes so as to dissolve the exposed narrow portions of the outer oxide layer **16**. In the second stage the tube is immersed in KOH at room temperature for about 10 minutes (this time being determined by the tube thickness) or in a mixture of one part HF and three parts $HNO_3$, whereby the silicon tube is etched into precise-width ribbon-like sections **26**. Depending upon the tensile strength of the inner resist layer **22** and its adherence to the tube, the sections **26** may or may not detach themselves from that layer when the etchant has dissolved through the full wall thickness of the tube.

In any event, etch-cut sections **26** are removed from the etchant bath and trichloroethylene is applied so as to dissolve away the inner resist layer from each section. Then the ribbon-like sections **26** are immersed in HF followed by KOH (or a mixture of $HNO_3$ and HF) at room temperature for a period of about 2-3 minutes. This etch step serves to remove their inner oxide layers and their inner N conductivity regions **14**.

Thereafter trichloroethylene is applied to each ribbon-like section **26** to dissolve away its outer resist layer **20** and then the sections **26** are again immersed in HF at room temperature long enough (about 2-3 minutes) to remove the outer oxide layer **16** but not the outer N-conductivity region **12**.

A problem with the EFG process is the random occurrence of twinning and other effects as the silicon ribbon is grown. When twinning occurs in a direction parallel to the length of the ribbon, the ribbon is better for the manufacture of solar cells, but when twinning occurs randomly, and in an uncontrolled manner so that there are different orientations of the twinning defects along the length of the ribbon, that portion of the ribbon is poorer and is discarded.

The defect called twinning is evidenced by replication of crystal planes in mirror image formations. As much as 5 feet of ribbon may be crystallized before twinning stabilizes in the preferred direction parallel to the length of the ribbon and normal to the crystallographic plane. Since ribbon growth is a slow process, such loss of ribbon is costly because of the substantial loss of process time and the unnecessary wear on the walls of the die surrounding the opening imposed by the passage of the hot, corrosive liquid silicon. The latter affects both configuration and dimension of the opening in time so that the die must be replaced. A ribbon growing technique which would reduce or eliminate the uncontrolled orientation of the twinning effect would be of considerable advantage.

Such an improved process is claimed by *M.H. Leipold; U.S. Patent 4,121,965; October 24, 1978; assigned to U.S. National Aeronautics and Space Administration* in which the orientation of twinning and other effects in silicon crystal ribbon growth is controlled by use of a starting seed crystal having a specific {110} crystallographic plane and <112> crystallographic growth direction.

Figure 16 shows the operation of this process. The EFG process by which silicon crystals may be grown is generally practiced with a crucible **10** made of refractory material such as quartz (silicon dioxide) or silicon nitride heated, as by an RF coil **11**, to produce a melt (silicon solution). Two sheets of silicon carbide **12**

**Figure 16: Method of Controlling Defect Orientation in Silicon Crystal Ribbon Growth**

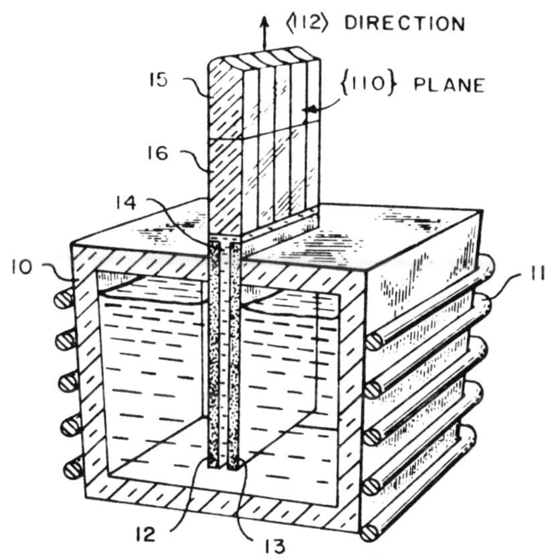


Source: U.S. Patent 4,121,965

and **13** are semi-immersed in the melt. The sheets are spaced sufficiently close to each other to make the liquid silicon rise by capillary force and wet a slot **14** between their upper edges. A silicon seed crystal is then brought into contact with the liquid silicon in the slot. After the mutual effects of melt temperature, withdrawal rate, and crystallization spread across the top of the sheets have been properly adjusted, the silicon crystal can be grown in a ribbon of indefinite length.

The growth of the silicon crystal ribbon is established in thickness and width by the upper horizontal surfaces of the sheets. It is for that reason that the sheets are commonly referred to as a "die." The die accurately controls the crystal ribbon cross section. Growth rate is limited only by the rate at which heat can be removed from the solid-liquid interface.

Orientation of anomalies, such as twinning, can be stabilized immediately at the beginning of the growth process by starting ribbon growth with a seed crystal **15** having the anomalies oriented with respect to one specific plane and direction, namely {110} and <112>. This seed crystal orientation has been proven successful for controlling twinning direction parallel to the ribbon edges and perpendicular to the crystal plane by actual fabrication of ribbon.

The figure illustrates a length of grown crystal **16** withdrawn from the die after the seed crystal. The vertical lines on the {110} plane of the seed crystal and the grown crystal represent the interfaces of crystal planes in mirror image formations, the evidence of the so-called twinning anomaly. This orientation is determinable by conventional x-ray diffraction techniques. The crystallographic direction <112> is in the direction of crystal growth.

As shown, the twinning anomaly present in the preferred orientation in the seed crystal is generally continued in the grown crystal. Other anomalies, such as crystal dislocations, will be similarly controlled in orientation. That reduces or eliminates uncontrolled orientation of the anomalies, and significantly reduces loss of grown crystal. The substantial reduction in any loss of process time reduces wear on the die.

An alternative made of silicon ribbon formation is shown in Figure 17. In this scheme, the dye is beveled to better define the thickness of the grown crystal ribbon. The technique shown in Figure 17 is the so-called inverted Stepanov technique.

The silicon crystal ribbon grown by this process, an inverted Stepanov process, differs from the more conventional Stepanov process only in that the slabs are placed at the bottom of the crucible instead of the top in order that capillary forces that feed the nozzle be aided by the pressure head of the liquid silicon in the crucible.

**Figure 17: Inverted Stepanov Technique for Silicon Ribbon Production**

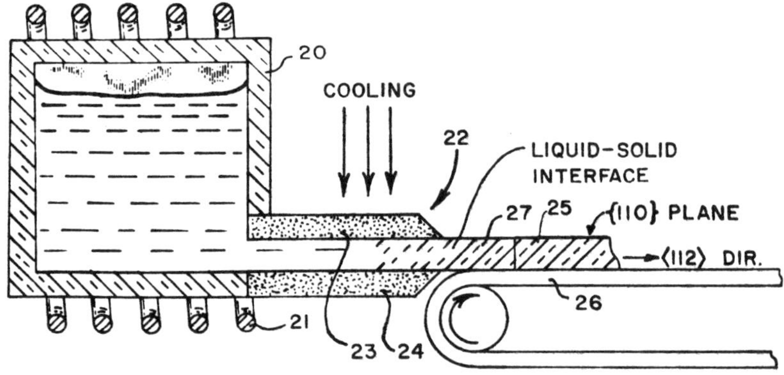


Source: U.S. Patent 4,121,965

In every one of these techniques, the orientation of anomalies is controlled by using a starting seed crystal having a specific {110} crystallographic plane and <112> crystallographic direction. If such a seed crystal is not used, anomalies can occur randomly and in an uncontrolled manner so that there are different orientations of the anomalies along the length of ribbon. As much as 5 feet of ribbon crystal may be grown before the anomalies stabilize in the preferred orientation. That portion of the ribbon cannot be used for manufacture of solar cells and is discarded.

Referring again to Figure 17, a quartz crucible **20** is heated, as by an RF coil **21**, and liquid silicon is fed through a "nozzle" **22** formed by two horizontal and beveled slabs **23** and **24** spaced apart the desired thickness of the crystal ribbon. The fine edges of the beveled slabs define the limits of the liquid silicon. A seed crystal **25** is brought in contact with the melt.

After adjustment of the crucible temperature for the cooling rate of the nozzle, the seed crystal is drawn away from the nozzle, as by a moving belt **26**. The growth of a silicon crystal ribbon **27** is then established. The established rate of growth is maintained by controlling the rate of cooling and the rate of moving the belt to maintain the liquid-solid interface at the tip of the nozzle even as the melt is used, until there is not sufficient melt to keep the nozzle full.

A process developed by *J.-J. L.E. Brissot; U.S. Patent 4,125,425; Nov. 14, 1978; assigned to U.S. Philips Corporation* is characterized in that the melted material is caused to flow over at least one surface of a heated element of a suitable shape, the element, at least at the surface, consisting of a material which is wetted by the melted material, that a seed, preferably a seed crystal of the material to be crystallized, is provided below the lower limit of the surface and at a small distance therefrom, that the upper part of the seed is wetted with the melt and the crystalline tape is formed by drawing away the seed.

Figure 18 illustrates the fundamentals of this process, the upper view showing the initial phases of crystal growth and the lower view showing the process some time thereafter. The figure shows a heated element **1** in the form of a balance knife, the lower part of which is denoted by **2**. Solid silicon, for example, in the form of grains or rods of silicon extending in the direction of the arrows $F_1$ is contacted with the surface **3** opposite to the lower part **2**.

The silicon is heated by means of suitable heating means, high frequency heating means for example, which heating means, denoted by **6**, are shown only diagrammatically. The experiments are preferably carried out in an atmosphere of a noble gas, for example argon or helium. The liquid silicon flows over the element **1**, the surface of which is wetted by the melt and lands at the lower part **2** of the element. A flat seed crystal of silicon is placed at a small distance from the part **2**. A molten zone **5** is formed between the part **2** and the flat seed crystal **4**.

If a gradual displacement (or drawing) in the direction of the arrow $F_2$ is performed, a continuous tape **7** of monocrystalline silicon is gradually grown on the seed crystal **4** of silicon.

An alternative mode of operation is shown in Figure 19 which shows a stage in the growth of a silicon ribbon comparable to the lower view in Figure 18.

Figure 18: Method of Manufacturing Flat Tapes of Crystalline Silicon from a Silicon Melt

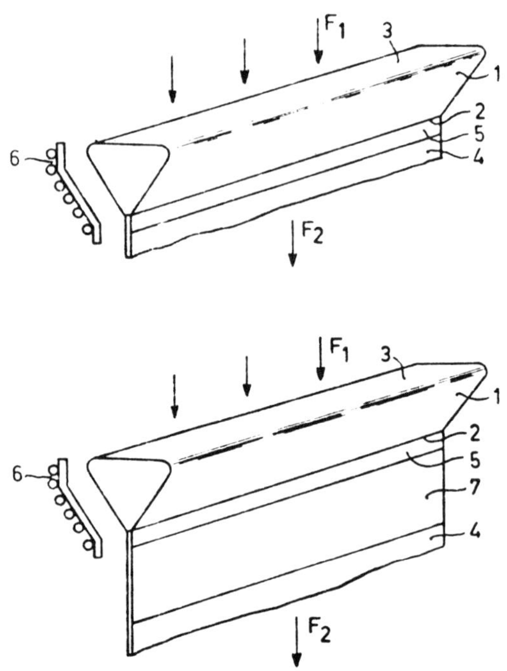


Figure 19: Alternative Mode of Operation of Process Shown in Figure 18

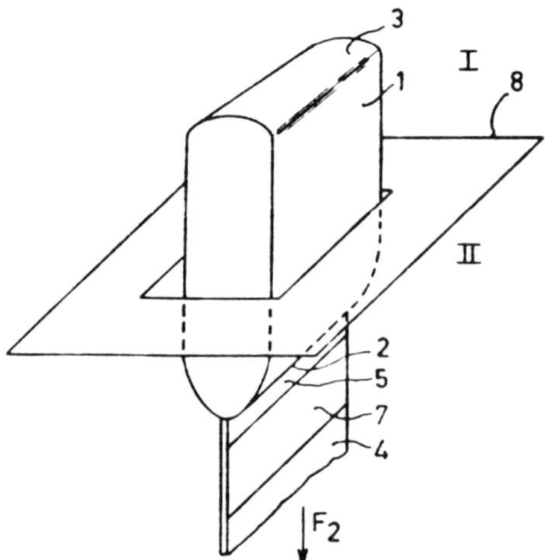


Source: U.S. Patent 4,125,425

According to this embodiment the supply of silicon occurs from gaseous compounds. The spaces referred to by I and II are separated from each other by a gaseous or solid screen which is referred to diagrammatically by **8**. If the screen is solid, it need not be in contact with the heated element; an intermediate space may be present where a suitable gas circulation is maintained. The reaction which makes it possible to obtain silicon takes place in the space referred to by I.

Chlorosilane is supplied, for example, through suitable supply tubes and is reduced by means of hydrogen in the vicinity of the elements. The molten silicon thus formed flows across the element **1**, while the hydrogen chloride formed by the reaction is dissipated. The screen **8** makes it possible to separate the hydrogen chloride from the preferred atmosphere of noble gas in which the crystalline growth of the silicon tape takes place.

One advantage of this process is that the quantity of melt at any instant may be comparatively small, which requires a small amount of electric power.

A second advantage is that the method can generally be discontinued at any instant without any objections, the design of the apparatus used being such that cooling does not result in fracture of those parts of the apparatus which contact the melted material.

Furthermore, crystal drawing takes place by means of a downward movement. Drawing thus is facilitated by gravity in contrast with the methods used so far.

It is to be noted that the lower part of the heated element need not necessarily be rectilinear. If the lower part is rectilinear indeed, the resulting tape is flat. In other cases the tape obtains a shape having a cross section which corresponds to the shape of the lower end of the heated element. In principle the lower end may form a closed curve, the tape-shaped crystal obtaining the shape of a tube.

The atmosphere used may comprise at least one noble gas, for example, argon or helium, to which a given quantity of hydrogen may have been added. It is also possible to operate in a vacuum.

The lower limit of the surface of the heated element along which the melt flows is preferably rectilinear and/or preferably extends horizontally. According to a preferred embodiment, the element along which the melt flows may have the shape of a knife, in particular a balance knife or, according to another preferred embodiment, it may have the form of a plate having parallel or substantially parallel surfaces.

In general the element should be heated at a temperature which is at least equal to the melting temperature of the crystalline material and is preferably not much higher than that melting temperature.

In accordance with the desired thickness of the tape of the crystalline material to be obtained, a suitable thickness of the lower part of the heated element along which the melt flows is preferably adjusted. If desired, thickness may be between a few tens of microns and a few millimeters. A flat seed is preferably used.

The distance between the lower part of the surface along which the melted material flows and the upper part of the seed is chosen in accordance with the desired thickness of the drawn crystalline tape. The rate at which the seed is drawn away is also of influence. If it is drawn away too rapidly, the danger of fracture in the growth of the crystalline tape occurs. Drawing downwards of a silicon single crystal is preferably carried out at a rate in the order of a few millimeters per minute.

It is desirable for the element to be heated in a suitable manner, for example, by high frequency heating or heating by radiation.

The element, at least the surface along which the material flows, consists preferably of a refractory material which in addition to the property of being wetted by the melted material shows the property of being resistant to the action of the melted material.

The refractory material, in particular when melted silicon is used, is preferably selected from the class comprising carbon, for example, pyrolithic carbon, silicon carbide, SiC, and titanium carbide TiC.

It is also possible for the element to consist of a refractory electrically conductive core covered with a thin layer or a refractory nonconductive or high-ohmic material which has a good resistance against the action of the melted material. The material for the thin layer is preferably selected from the class comprising silicon nitride, $Si_3N_4$, and very pure silicon carbide, SiC, for example, of semiconductor quality.

The starting material for the seed crystal can be supplied to the heated element in various manners. According to a preferred embodiment this occurs from solid meltable material which is heated to the melting temperature. According to a further preferred embodiment the meltable material can be prepared by deposition on the heated element from at least one compound of the component(s) of the meltable material.

For the deposition of silicon, suitable silane compounds are to be considered. According to a preferred embodiment decomposition of silane is used. According to another preferred embodiment reduction of chlorosilane by means of hydrogen is used for that purpose.

### Filament Form

An alternative form of silicon which has been proposed for solar cell manufacture is fiber-form silicon.

One technique which combines silicon purification with silicon deposition is the method described by *G.F. Wakefield; U.S. Patent 3,969,163; July 13, 1976; assigned to Texas Instruments, Inc.* This is accomplished by passing conductive fibers such as graphite or the like which are compatible with the later processing steps through an area which is cooled below 700°C and which contains silicon difluoride and a proper N-type dopant. At these temperatures, the silicon difluoride gas will break down into pure silicon which will deposit onto the fiber with the formation of silicon tetrafluoride gas which is then recycled into a further chamber.

In the further chamber, the gaseous silicon tetrafluoride is mixed with the impure metallurgical grade silicon at temperatures above 700°C to form the silicon difluoride gas which is then fed into the former chamber for deposition of pure silicon onto the continuously moving fibers or graphite or the like. A P-type layer can then be formed over the N-type layer in any standard manner, such as by then passing the coated fibers through a further reaction chamber wherein P-type dopant is diffused into the top surface of the N-type layer that has been formed.

The dopants alternatively could be added in the gas stream of $SiF_4$ or the P-type layer formed by ion implantation. In this way, relatively inexpensive P-N junction devices are formed without the requirement of purifying, cutting and polishing a silicon slice in the standard manner. Figure 20 shows a suitable form of apparatus for the conduct of such a process.

**Figure 20: Vapor Deposition Method of Forming Low-Cost Semiconductor Solar Cells**

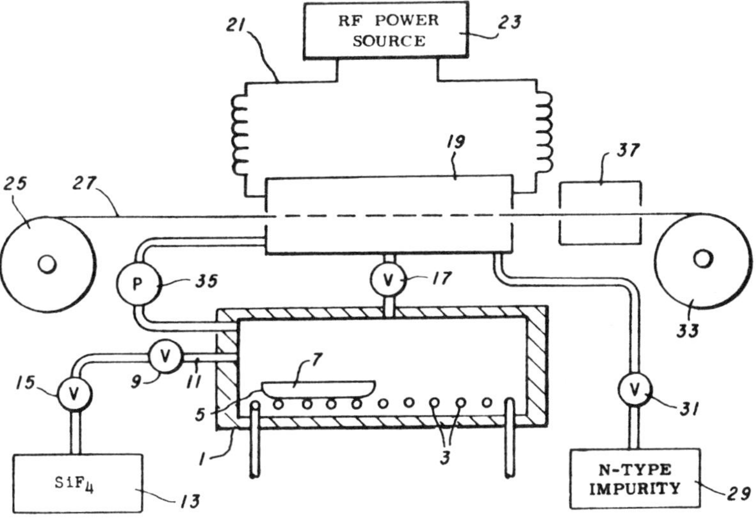


Source: U.S. Patent 3,969,163

There is shown a heater chamber **1** which is heated by heating coil **3** in the range of from about 750° to about 1200°C and preferably 900°C or, alternatively, which can have a heat zone heated to these temperatures. Within this chamber is a vessel **5** containing metallurgical grade silicon **7** and positioned in the heat zone, if such is used.

Also passing into the chamber through a valve **9** is a line **11** through which silicon tetrafluoride is passed from a source of silicon tetrafluoride **13** by a valve **15**. At the elevated temperatures within the vessel **5**, the silicon tetrafluoride, which is in gaseous form, passes over the impure metallurgical grade silicon **7** and forms a silicon difluoride gas which passes through valve **17** into the reaction chamber **19**. The reaction chamber **19**, which can be maintained at atmospheric pressure, though it need not be, is heated by an RF heater **21** which is energized by an RF power source **23** to a temperature of less than 700°C and preferably 600°C.

Also, a spool **25** passes a length of fiber **27**, which can be graphite or any other appropriate fiber, through the chamber **19**. The fiber **27** must be electrically conductive, substantially inert with respect to the silicon, fluorine and silicon fluorides, relatively available and inexpensive and temperature stable to a temperature above the decomposition temperature of the silicon difluoride or the temperature reached in reaction chamber **19**.

At the temperatures within the chamber **19** the silicon difluoride which is passed thereto through the valve **17** decomposes to form pure semiconductor grade silicon and silicon tetrafluoride. A source of N-type impurity **29** sends N-type impurity through valve **31** into the chamber **19** in predetermined amounts so that the silicon along with the N-type impurity is coated onto the fiber **27** to provide a highly doped layer of N-type silicon thereon.

If desired, source **29** could be added to the silicon tetrafluoride input through valve **15** rather than as shown and described. The fiber is continually wound through the chamber **19** from the spool **25** toward the spool **33**, the chamber **19** being at approximately atmospheric pressure but being sufficiently closed to prevent any material leakage of any of the gases therein to the exterior of the chamber **19**. The silicon tetrafluoride, which is formed in the chamber **19**, is passed out of the chamber along the line **35** and recirculated through the valve **9** back to the vessel **5** for further use along with the metallurgical grade silicon to continue the continuous reaction.

It can be seen that a doped layer of silicon is formed on the fibers going directly from metallurgical grade silicon to the coated semiconductor grade silicon layer. The coated fiber is passed through a chamber **37** wherein P-type silicon is then deposited over the N-type silicon or alternatively, P-type impurity gas is provided at a temperature of about 1200°C to form a P-type region on the upper surface of the semiconductor layer to provide a P-N junction therein.

It is apparent that the material finally wound on the spool **33** is a continuous length of conductor over which is a layer of N-type material and another layer of P-type material from which solar cells can be formed. This will merely require the formation of a contact area over the P-type layer to form the cell itself.

Figure 21 shows a solar cell formed from fibrous diodes prepared as described above. There are shown a plurality of diodes **51**, each having a graphite core **53** in contact with the N-type layer **55**. There are also shown contacts **57** connected to the P-type layer **59**. Each of the contacts **57** are coupled together to a line **61** and each of the cores **53** are coupled together to a line **63**.

**Figure 21: Solar Cell Formed from Silicon Deposited on Graphite Fibers**

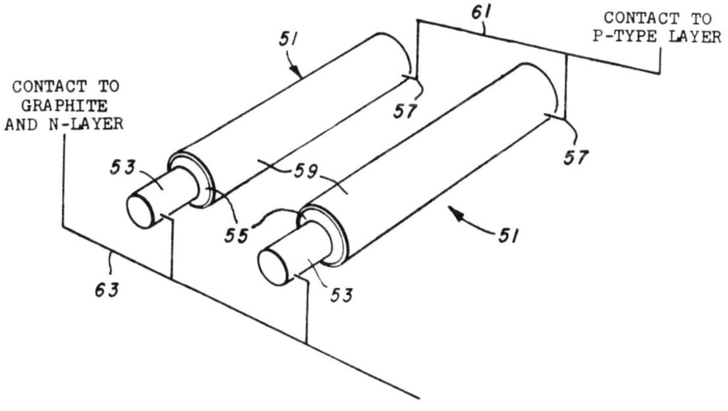


Source: U.S. Patent 3,969,163

A type of construction developed by *J.T. Eliason; U.S. Patent 3,984,256; Oct. 5, 1976; assigned to U.S. National Aeronautics and Space Administration* involves a photovoltaic cell array consisting of parallel collumns of silicon filaments, each being doped to produce an inner region of one polarity type and an outer region of an opposite polarity type to thereby form a continuous radial semiconductor junction. Spaced rows of electrical contacts alternately connect to the inner and outer regions to provide a plurality of electrical outputs which may be combined in parallel or in series.

Such a cell is shown in Figure 22. The unit is made from filaments **10** of silicon having a diameter typically of 0.001 to 0.010 inch. Each filament **10** is initially doped to produce a P-type conductivity internal region, and thereafter an opposite conductivity type doping is applied, as by diffusion, to the outer surface of the filament to produce an N-type conductivity layer. It will be apparent to those skilled in the art that the body of the filament may be doped N-type and the outer portion doped P-type instead of the above manner, chosen for illustration.

Figure 22: Solar Cell Array Made Up of Silicon Filaments

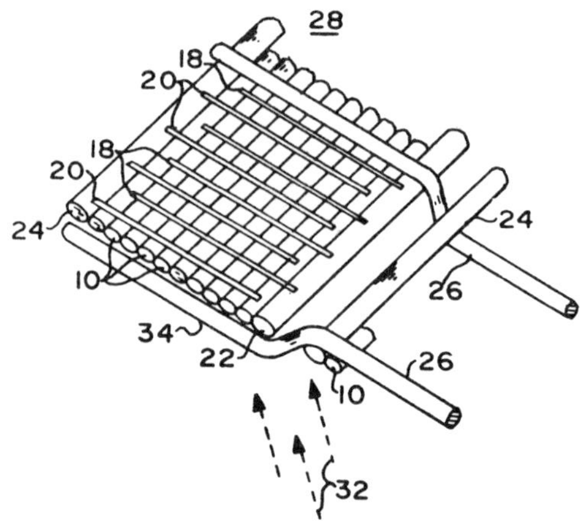


Source: U.S. Patent 3,984,256

The P doping is conveniently achieved by means of boron, and the N doping is achieved by means of phosphorus. A portion of the N-type layer is removed at spaced points, as for example by etching, to expose P material regions. An ohmic contact is then made to the P material region by P contact member **18**, as by diffusing bonding or welding.

N contact member **20** is welded to filament **10** at a spacing typically of 0.020 to 0.200 inch from the nearest contact member **18** to thereby provide one photovoltaic cell unit with an output available between a contact member **18** and contact member **20**. U.S. Patent 3,453,352 generally describes the method of construction of lengths of small cross section semiconductor material.

As shown, the photocell units thus connected are connected in parallel, with contact members **18** being connected to P bus **22** and N contact members **20** connected to N bus **24**. The structure is supported by insulating fibers **26**, which enable the construction of large area arrays consisting of series and parallel electrical interconnected units of the above type, as represented by solar cell blanket **28**. By this technique, photovoltaic cell densities of 100 to 1,000 per square inch may be achieved.

In operation, array **28** would be positioned so that radiation **32** from the sun would strike the underside **34** of the array, and each filament would produce multiple outputs to buses **22** and **24**.

### Tubular Form

In a process developed by *A.I. Mlavsky; U.S. Patent 3,976,508; Aug. 24, 1976;*

*assigned to Mobil Tyco Solar Energy Corp.,* tubular solar cells are provided which can be coupled together in series and parallel arrays to form an integrated structure. Solar energy concentrators are combined with the solar cells to maximize their power output. The solar cells may be cooled by circulating a heat exchange fluid through the interior of the solar cells and the heat captured by such fluid may be utilized, for example, to provide hot water for a heating system. The coolant circulating system of the solar cells also may be integrated with a solar thermal device so as to form a two-stage heating system, whereby the coolant is preheated as it cools the solar cells and then is heated further by the solar thermal device.

The object of developing such structures was to provide solar cells and arrays thereof which have a modular form, can be easily cooled, have structural integrity, can be made by existing techniques, and are capable of withstanding changes in dimensions due to thermal cycling. A further object is to provide a solar cell unit wherein current leakage is minimized by the use of a geometry which minimizes the ratio of exposed active surface area to exposed junction region area.

Still another object is to provide a solar cell module which can be integrated with a solar thermal system. Yet another object is to provide solar cell modules which can be easily and efficiently interconnected physically and electrically. Another important object is to provide solar cells and solar cell arrays of the type described in combination with radiant energy concentrators for maximizing the intensity of radiation received by such cells and also for distributing the concentration of such radiation.

In a typical example of the conduct of this process, a cylindrical substantially monocrystalline P-type silicon tube is grown according to the method described in U.S. Patent 3,591,348. The tube is made with a length of about 6 inches, an outside diameter of 0.50 inch and wall thickness of about 0.01 inch. The interior surface is plated with a 0.001 inch layer of nickel and phosphorus is diffused into the outer surface of the tube to a depth of about 0.5 micron to form an N-type outer region with a distinct P-N junction.

Then aluminum is vacuum-deposited onto the outer surface of the tube in the form of a grid consisting of a plurality of longitudinally and circumferentially extending conductors. The aluminum grid is formed with a thickness of about 4.0 microns. The inner and outer conductors are connected to a measuring circuit and the device irradiated by sunlight. The device exhibits an open circuit voltage of about 0.5 volt and a conversion efficiency of about 10%.

In a process developed by *H. Weinstein and R.H. Lee; U.S. Patent 4,052,782; Oct. 11, 1977; assigned to Sensor Technology, Inc.* several techniques are set forth for promoting oriented crystalline semiconductor growth on glass tubes. In a preferred technique a thin layer or film of aluminum is deposited onto the tube, followed by deposition of some silicon. The structure is heated to the aluminum-silicon eutectic temperature (approximately 477°C) which is below the melting point of glass, then quickly cooled by between about 50° and 100°C. This "super cools" the eutectic, causing the silicon to separate into individual crystalline islands in the aluminum matrix. Subsequently silicon is vapor deposited onto this matrix. The crystal islands serve as growth centers for the newly deposited silicon, promoting oriented crystalline growth thereof. A P-N junction may be formed in the crystalline silicon layer either during or subsequent to its growth.

Other techniques for promoting oriented semiconductor crystal growth include, among others (a) seeding the glass surface with minute silicon particles, (b) depositing silicon oxide or other compound onto the glass at an acute angle to create a wavy surface, and (c) providing an amorphous, liquid-like deposition surface on which the deposited silicon will crystallize normal to the surface since no other orientation is induced by the ultrasmooth surface.

Figure 23 shows one method for growing the semiconductor layer **36** and forming the photovoltaic junction **37**. In this technique, the glass tube **18** first is coated with a layer of aluminum that is sprayed on or applied by wet chemical evaporation. A small amount of semiconductor silicon next is deposited atop the aluminum by a similar technique. The silicon so deposited neeed not be crystalline. The resultant structure is heated together to approximately the aluminum-silicon eutectic temperature of 477°C. This temperature is below the melting point of the glass.

An aluminum-silicon eutectic layer is formed atop the glass. The temperature then is lowered quickly by about between 50° and 100°C (i.e., to between about 427° to 377°C) to produce a "super-cooled" eutectic. The silicon separates from the aluminum and crystallizes to form tiny islands **39** of crystalline silicon within the aluminum film **40**.

**Figure 23: Tubular Silicon Solar Cell**

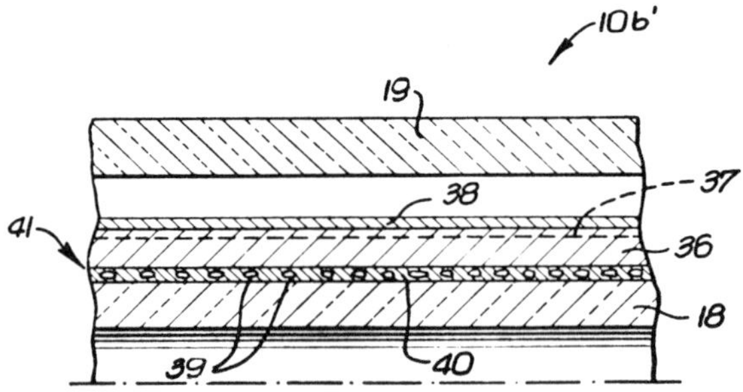


Source: U.S. Patent 4,052,782

The resultant matrix **41** plays two roles. First, the tiny crystalline silicon islands **39** in the matrix **41** promote oriented crystalline growth of the semiconductor layer **36**. Secondly, the aluminum film **40** in the matrix **41** functions as an electrical conductor to the semiconductor junction **37** in the device **10b'**. To grow the semiconductor layer, silane or other gaseous silicon source is vapor deposited onto the matrix. As the silicon deposits it crystallizes in situ. Oriented crystalline growth is promoted by the silicon islands which serve as growth centers. The resultant semiconductor layer, while probably not a single crystal, has a sufficiently oriented crystalline structure to achieve good device performance.

To produce the junction **37**, N- or P-type dopants may be introduced into the silicon layer as it is being grown. For example, during the initial portions of this growth, N-type dopant material may be introduced in vapor form together with the silane or other silicon source. The resultant silicon layer will be of N-type conductivity. Subsequently, a P-type dopant material may be introduced with the silane to produce an overlaying region of P-type conductivity silicon. The junction is at the interface of these two regions. Finally a thin film **38** of tin oxide and indium oxide may be deposited atop the semiconductor layer **36** to complete fabrication of the photovoltaic junction **37**.

Another technique for promoting oriented crystalline semiconductor growth or a glass substrate is shown in Figure 24.

In this "surface alignment" technique a thin layer **42** of material is vacuum deposited onto the unheated glass tube **18** at an acute angle **43** typically on the order of 15° to the surface (i.e., about 75° to a normal from the surface of the tube **18**). The layer **42** that is formed by such acute angle vacuum deposition will not be flat. Rather, its surface **44** will be wave-like with a pitch of microscopic and perhaps molecular order. If a semiconductor material such as silicon now is vapor deposited atop the layer **42**, the wavy surface **44** will promote oriented crystalline growth of the deposited semiconductor.

**Figure 24: Tubular Silicon Solar Cell Illustrating Alternative Preparation Process**

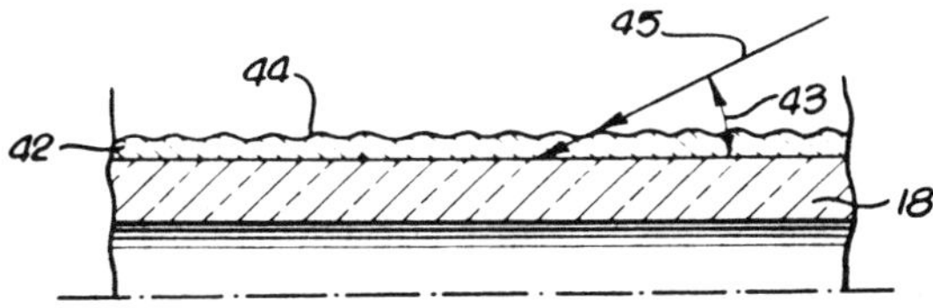


Source: U.S. Patent 4,052,782

In a preferred embodiment, the material of the layer **42** may comprise a combination of silicon oxide and gold. The deposition is accomplished in a vacuum chamber in which the source material is vaporized and directed in a narrow beam (indicated by the arrow **45**) at an acute angle **43** toward the surface of the glass tube **18**.

The tube may be rotated and translated axially during the deposition process to produce an appropriate layer over the entire exterior surface of the tube. By including sufficient gold in the deposited material, the layer will be of relatively low electrical resistivity, and hence can serve as the inner electrode for the P-N junction formed in the subsequently grown silicon semiconductor layer. Alternative source materials for the layer include silicon nitride, silane and metals.

Another technique (not illustrated) for promoting oriented semiconductor crystalline growth on a glass tube is to coat the glass substrate with minute, dust-sized particles of single crystal silicon. This dust may be prepared either by cracking or breaking up a single crystal of silicon into particles which are on the order of one-tenth mil or less. These particles may then be deposited on the cool glass substrate by means of a plasma spray. Or by the use of another procedure, crystalline silicon seeds may be implanted near the surface of the glass tube by means of ion implantation using a silicon gas source such as silane or silicon iodide.

Yet another technique for promoting oriented crystalline growth involves first coating the glass substrate with a thin layer of aluminum. Then, inside a vacuum chamber, the aluminum-coated glass is heated to a temperature somewhat below the aluminum-silicon eutectic temperature of 477°C, but sufficiently high (preferably above about 375°C) so that the aluminum loses its crystallinity and behaves somewhat like a fluid. The effect is that the aluminum surface is smooth, so that when silicon or other semiconductor material subsequently is deposited atop the aluminum layer, there is no preferred direction for crystallization. As a result, the silicon will start to crystallize in a direction normal to the surface.

In other words, the very smooth aluminum surface will promote crystalline growth of the semiconductor film in a symmetrical direction which is normal to the substrate. As before, the deposited silicon may be doped during growth to form the P-N photovoltaic junction. The underlying aluminum layer will serve as the lower electrical contact, and an optically transparent but electrically conductive film may be deposited atop the semiconductor layer to provide the other electrical contact to the junction.

A type of construction developed by *L.C. Garone and K.V. Ravi; U.S. Patent 4,056,404; Nov. 1, 1977; assigned to Mobil Tyco Solar Energy Corp.* basically comprises a substantially monocrystalline semiconductor body in the form of a flat oval tube having a pair of opposed mutually spaced relatively large wall sections connected together by a pair of opposed relatively small side edge sections. The tube has a radiation-receiving outer surface and a photovoltaic junction which is close to the outer surface and is capable of responding to radiant energy passing through the outer surface. Completing the solar cell are a first electrode which is carried on the tube outer surface and a second electrode which is carried on the tube inner surface.

In producing such a device, a substantially monocrystalline tubular body of silicon or other suitable semiconductor material which has the shape of a substantially flat oval in cross section is first formed, e.g., as by the EFG process. A photovoltaic junction is formed near the outer surface of the flat oval tubular body by suitable semiconductor processing techniques. The electrodes are then formed on the outer and inner flat wall surfaces.

### Microspheres

A process developed by *G.F. Wakefield; U.S. Patent 3,998,659; Dec. 21, 1976; assigned to Texas Instruments Inc.,* is one in which generally spherical-shaped semiconductor particles comprising an inner core of one conductivity type and a thin peripheral layer of opposite conductivity type are produced in a fluidized bed reactor. Silicon particles introduced into the reactor are built up to a desired nominal diameter by vapor deposition from a silicon- and dopant-containing atmosphere introduced into the reactor; by changing the dopant constituent, the outer peripheral layer is then deposited.

One use of such particles is in fabrication of a solar cell, wherein an array of the particles is located on an insulating sheet and overlying conductive layers, insulated from each other, make contact respectively with the peripheral layers and with areas of the core regions, exposed by etching.

Figure 25 shows a suitable form of apparatus for the production of such particles. Such an apparatus comprises a cylindrical corrosion-resistant reactor vessel **1** of high purity quartz or fused silica having a quartz distribution plate **2** located towards the bottom thereof. In relation to the embodiment being described, the reactor suitably has a diameter of 75 mm. The reaction vessel includes an inlet **3**, located at the upper end of the vessel, for introducing elemental semiconductor particles into the reactor, and an inlet **4** through which a reducing gas/fluidizing agent and reaction constituents can be admitted into the reactor.

The reaction constituents are derived from sources **6**, **7** and **8** under control of flow valves **9**, **10** and **11** respectively. The sources comprise, respectively, a source of desired semiconductor containing compound, a source of semiconductor dopant compound of one conductivity type (e.g., N-conductivity type) and a source of dopant compound of opposite conductivity type (e.g., P-type). An outlet vent **12** is connected to the top end of the reactor vessel and a discharge outlet **13** for coated semiconductor particles is located at the bottom of the reactor vessel, above the distribution plate.

Operation of the apparatus will be described in relation to the production of N-doped monocrystalline silicon particles having an outer P-doped epitaxial layer. In operation of the apparatus, a desired quantity of high-purity monocrystalline elemental silicon particles **14**, such as would result from the comminution of a monocrystalline silicon rod, of a suitable size, e.g., a diameter within the range 150-350 $\mu$ nominal diameter, are fed through the inlet **3** to form a bed of silicon particles in the reactor, and a stream of pure hydrogen is intro-

Figure 25: Apparatus for Fluidized-Bed Production of Silicon Semiconductor Particles for Solar Cell Use

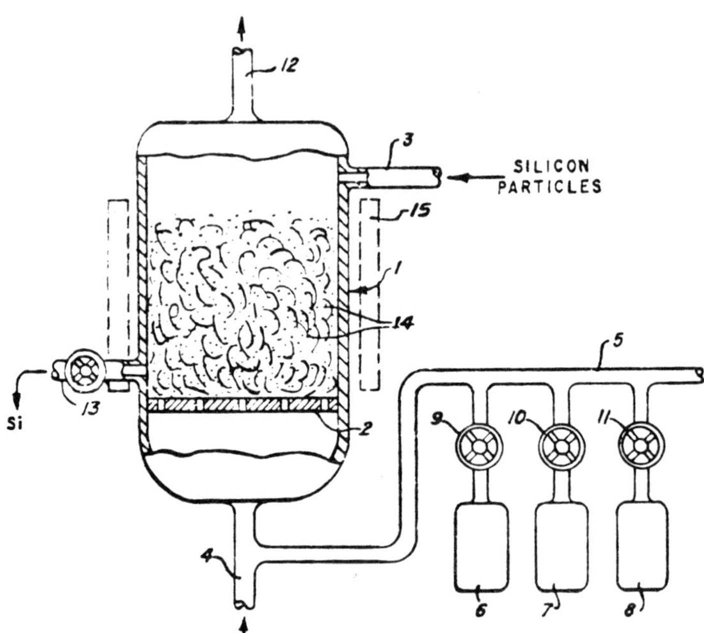


Source: U.S. Patent 3,998,659

duced through the inlet **4** at a flow rate, suitably within the range 25-50 l/min, sufficient to fluidize the silicon particles causing the bed to be significantly expanded in volume and the particles **14** to assume a very agitated motion. The fluidized bed is heated by the heater **15** within the range 950-1250°C sufficient to permit gaseous phase deposition of silicon on the silicon particles.

In order to carry out such deposition, a suitable silicon-containing compound, $SiCl_4$, $SiHCl_3$, $SiH_2Cl_2$, $SiH_4$, and an N-type doping compound e.g., phosphine, arsine, stibine, are fed from the sources **6** and **8** under control of valves **9** and **11** respectively and introduced through inlet **5** into the reactor **1** in a vaporized state. Valves **9** and **11** are set to give concentrations of the silicon compound and of the dopant of about 12-15% and 0.001-0.01%, respectively in the total hydrogen flow. There ensues a chemical reaction between the vaporized compounds entering through inlet **5** and some of the hydrogen introduced through the inlet **4** resulting in deposition of N-doped silicon on the silicon particles in the fluidized bed, the agitated motion of the silicon particles in the fluidized bed facilitating such deposition over the whole surface areas of the particles.

In general, a deposited N-type layer in the thickness range 50-2,500 $\mu$ is suitable.

Typically, the N-doped silicon has a doping level of $10^{15}$-$10^{18}$ atoms/cc. After deposition for a time sufficient to build up the particles to a desired size range, e.g., 650-850 $\mu$ the reaction is changed to effect deposition of a P-doped silicon epitaxial layer onto the particles **14** by closing the flow valve **11** and opening the flow valve **10** to introduce a P-type compound vapor, e.g., borane or boron trichloride into the reactor through inlet **5**, again suitably at a concentration of 0.001-0.01% in the total hydrogen flow. P-type silicon deposition is continued for a time sufficient to build up a desired thickness, e.g., 1-10 $\mu$ P-type surface layer typically having a doping level of $10^{17}$-$10^{20}$ atoms/cc on the silicon particles whereupon such particles may be removed from the reactor vessel **1** through the outlet **13**.

During the abovedescribed reactions, excess hydrogen and reaction by-products are exhausted from the reactor vessel **1** through the vent **12**.

An advantage of this process is that the semiconductor particles are formed and doped with desired conductivity types in a continuous set of process steps without having to be removed from the fluidized zone in which the necessary reactions take place. Consequently, significant manufacturing economics can be realized not only from the point of view of improved throughput time but also to considerable reduction in material wastage since the doped particles obtained as a result of the process are directly suitable for processing into semiconductor devices and steps such as crystal growing, slicing and polishing of slices and breakage of slices into chips are eliminated.

In June 1979 a breakthrough in solar cells was claimed by Texas Instruments, Inc. based on the application of such microspheres to a substrate, much like sand is applied to form sandpaper. Defects in a few spheres have a minimal effect on overall system performance. The thus-coated substrate is then bathed in a liquid electrolyte, producing hydrogen which can be stored and used as a power source.

### Polycrystalline Silicon

Work is also under way to develop low-cost techniques for producing sheets of polycrystalline silicon with crystal grains large enough in size and oriented properly to allow the production of cells with acceptable efficiencies. Some processes are yielding individual crystal grains 1 to 10 mm on a side, blurring the distinction between single and polycrystalline devices. Perfection of such processes could lead to more efficient use of the silicon feedstock and replaces the expensive task of growing a perfect crystal with a casting, forming, or depositing process which may be much less expensive.

Some of the methods proposed would create a thin layer of polycrystalline material on a solid backing, or substrate, thus eliminating the cost and material waste associated with slicing the silicon material into wafers (3). Concepts include:

1. Dipping a substrate into molten silicon, covering the substrate with a thin coating of polycrystalline silicon. (Honeywell).

2. Continuously pulling a thin sheet of silicon from the surface of a pool of molten metal where silicon vapor is being deposited. (General Electric).
3. Depositing silicon from a silicon-halide gas directly onto a hot substrate. It may be possible to combine the final stage of silicon purification (distillation) with this deposition process. (Rockwell International and Southern Methodist University).
4. Forming heated silicon material under pressure in rollers. (University of Pennsylvania).
5. Casting blocks of polycrystalline material and slicing the blocks. (Crystal Systems, Salem, Mass.).

Devices developed by *E. Sirtl and C.G. Currin; U.S. Patent 3,900,943; Aug. 26, 1975; assigned to Dow Corning Corp.* are solar cell devices formed from bulk silicon deposited in the form of columnar crystallites bounded by substantially vertical grain boundaries. Junctions are formed across crystallites and along grain boundaries. The grain boundaries are made substantially nonconductive by diffusion from one side of the sheet. $P^+$ and $N^+$ layers are provided as contact areas for electrodes. Deposition of silicon takes place directly from decomposition of silicon-containing vapors onto a nonsilicon substrate sheet.

Very similar ground is covered by *E. Sirtl and C.G.Currin; U.S. Patent 3,953,876; April 27, 1976; assigned to Dow Corning Corp.*

As pointed out by Sirtl and Currin, for some time thought has been given to developing cheaper methods of producing solar cells. One approach which has been tried but found to result in an extremely inefficient unit is the use of bulk polycrystalline material. Grain boundaries between crystallites in such materials prevent proper transfer of charges in the material. Dopants tend to follow grain boundaries when diffused into the material.

Accordingly unoriented P-N junctions appear around grains effectively isolating charges. Further, heavy metal impurities tend to concentrate in the grain boundaries and along with discontinuities in the grain boundaries contribute to recombining of electron-hole pairs resulting in no electrical output from the system.

Their improved process, as described in the above-cited patents, however, is based on the discovery that when polycrystalline silicon is deposited on a smooth, flat substrate in relatively fine grain form, after several microns of disoriented growth a natural selection process takes place resulting in subsequent growth in the form of columnar crystallites in the $\langle 110 \rangle$ growth direction separated by $\{111\}$ grain or twin boundaries developed generally perpendicular to the surface. The crystallites are each monocrystalline in nature and continue growing in height as deposition continues.

Each crystallite therefore has the potential to be made into an active semiconductor device by doping with impurities to form one or more P-N junctions therein. Inasmuch as dopant diffusion tends to follow grain boundaries faster than diffusion in the bulk material, electrical isolation between crystallites can be achieved as desired. For solar cell arrays heavy doping from both sides to create a $P^+$ layer on one side and an $N^+$ layer on the other effectively acts as contacts for the N-type and P-type areas adjacent the junctions.

The combination of planar junctions across each crystallite combined with vertical junctions extending substantially vertically part way down the grain boundaries provides the opportunity for any electron-hole pair formed by a striking photon to find a nearby P-N junction. Electrodes and reflective and/or protective coatings can be applied by prior art techniques.

The array thus formed is much more economically produced than prior devices requiring large areas of silicon and can be made without the problems inherent in bulk polycrystalline silicon solar cells which have heretofore been manufactured.

Figure 26 shows a suitable form of apparatus for producing such improved polycrystalline silicon on a laboratory basis. It shows a reaction chamber **11** having inlet means **12** and outlet means **13** for introduction and exhaust of gases to and from the chamber. The chamber may be made of quartz or molybdenum, for example, but may alternatively be any other material capable of retaining its integrity in contact with gaseous silicon, hydrogen, silicon hydrides and chlorides at temperatures at least as high as 1200°C.

**Figure 26: Dow Corning Reactor Design for Polycrystalline Silicon Deposition on a Laboratory Scale**

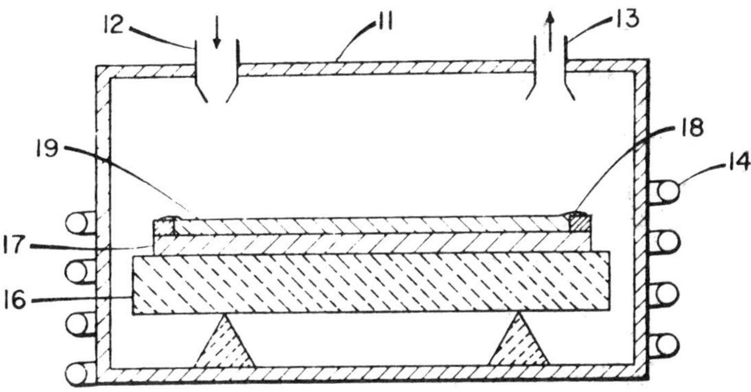


Source: U.S.. Patent 3,953,876

Heating means such as a high frequency electrical coil **14** are disposed in or around the reaction chamber **11**. Mounted within the reaction chamber is a susceptor body **16** which may be, for example, graphite. The susceptor element **16** acts to couple with the high frequency heating coils **14** to supply heat to a substrate element **17** positioned on top of the susceptor. Alternatively, electrical resistance heating or radiant heating can be used.

The substrate sheet **17** should also be a material which is capable of withstanding temperatures above 1200°C in gaseous atmospheres of hydrogen, silicon, silicon hydrides and chlorides. A preferred substrate is highly polished tantalum in a thickness of approximately 1 mm. However, the thickness is not critical except to the extent that thicker substrates tend to modify heat flow and the adherence of silicon to the tantalum varies according to substrate thickness.

Preferably, a mask **18** which may, for example, be made of molybdenum is used to mask the edges of the tantalum substrate **17** to prevent silicon deposition from extending over the edge of the tantalum thereby causing adherence. However, particularly in the case where the tantalum is to remain on the device, made from the silicon being deposited, the mask may be omitted.

According to this process, silicon is deposited on the substrate by introduction in a hydrogen stream through the inlet **12** of silicon-containing gases decomposable at temperatures between about 1025° and 1200°C. Preferred gases for this purpose include trichlorosilane, dichlorosilane, silicon tetrachloride, hexachlorodisilane, and silane, or monosilane as it is sometimes called. While a number of these gases decompose at temperatures below 1025°C the higher temperatures are preferred for crystallographic reasons. The ratio of silanes to hydrogen in the feed stream should be between 0.5 and 10 mol percent for best results.

As is fairly common in silicon deposition processes the feed gases are directed by the inlet **12** into contact with the substrate while it is heated to the proper temperature range resulting in a deposition layer **19** of silicon on the substrate. Spent gases are removed from the chamber through the exhaust or chamber outlet **13**.

While the deposition of silicon from the abovementioned raw materials is not unusual, it has been found that by proper choice of substrate surface, and conditions within the reactor along with long-term deposition that deposition initially occurs in a relatively unoriented form and then by a natural selection process begins to take the form of columnar or dendritic crystallites growing in the <110> growth direction separated by {111} grain or twin boundaries developed generally perpendicular to the substrate surface. These crystallites are each monocrystalline in nature and continue growing in height as deposition continues.

The following gives one specific example of the conduct of such a batch process. A 1 mm thick tantalum substrate material approximately 3 cm by 3 cm square was masked with molybdenum to provide an opening 2 cm by 2 cm. This was placed on a Speer SX-4 graphite susceptor plate and heated to a temperature of 900°C. Trichlorosilane in a hydrogen stream in a mol ratio of 2% trichlorosilane was flowed into the reaction chamber in which the susceptor was placed at a rate of 2½ liters per minute for a period of 5 minutes. During this 5-minute period arsine in argon was introduced to the hydrogen-trichlorosilane feed stream in a ratio of 14 ppm by volume of arsine to argon and 40,000 ppb by volume of arsine to the hydrogen-trichlorosilane mixture. After 5 minutes the temperature of the susceptor was raised to 1050°C and the arsine level was reduced while the same flow rate and mol ratio of trichlorosilane in hydrogen was continued for 95 minutes.

The result was a layer of silicon 160 microns thick which adhered firmly to the tantalum substrate. The arsine level during the latter deposition in three different runs which were otherwise identical was varied with concentrations of 870 ppb by volume, 3,750 ppb by volume, and 16,000 ppb by volume. Microscopic examination of the cross section of the silicon sheet showed approximately 150 microns of substantially vertical columnar silicon crystallites averaging approximately 10 microns in diameter with substantially no horizontal grain boundaries.

Boron was then diffused into the silicon sheet having the arsine level of 870 ppb and diffusion was made from the top thereof to a nominal depth of 0.4 micron. Electrodes were applied by normal solar cell techniques and the solar cell thus produced showed a short circuit current output of 5-10 mA with an open circuit voltage of 100 mV. This is equivalent to a solar cell efficiency of 0.2%.

The efficiency of this early device produced in a laboratory apparatus obviously must be improved to give an economically competitive cell.

A process developed by *T.L. Chu; U.S. Patent 3,961,997; June 8, 1976; assigned to the U.S. National Aeronautics and Space Administration* is one whereby low-cost polycrystalline silicon cells supported on substrates are prepared by depositing successive layers of polycrystalline silicon containing appropriate dopants over supporting substrates of a member selected from the group consisting of metallurgical-grade polycrystalline silicon, graphite and steel coated with a diffusion barrier of silica, borosilicate, phosphosilicate, or mixtures thereof such that P-N junction devices are formed which effectively convert solar energy to electrical energy.

In single crystal silicon the diffusion lengths of the electrons and holes are sufficient to allow the majority of photon-generated carriers to diffuse to the junction and be "collected" as electrical current. On the other hand, in polycrystalline material the diffusion lengths are determined primarily by the crystallite size, and therefore, most of the electron-hole pairs diffuse to grain boundaries where they recombine and are thus lost to the electrical current.

A type of construction developed by *E. Crisman and W.F. Armitage, Jr.; U.S. Patent 3,978,333; August 31, 1976* overcomes this problem by providing for electrical contact with each, or at least substantially each, individual crystallite or crystal grain, and therefore permits current collection from those areas which, in the single crystal base, do not have direct electrical contacts.

Figure 27 shows a prior art type of electrical contact (a) and the improved type of contact suitable for use with polycrystalline silicon (b).

Referring to Figure 27a, there is shown a conventional single crystal solar cell comprising a single crystal base wafer **10**, preferably of silicon, the top surface of same, or the surface to which illumination is to be applied, having diffused thereon a suitable impurity layer **12** to produce a P-N junction **14** near the surface to be illuminated and in a plane parallel to the surface. A bottom surface **16** of electrically conductive metallic material is deposited on the underside of the wafer and functions as one of the ohmic contacts for the cell. The other ohmic contact **18**, also a thin metal deposition, is applied to the top surface of the layer **12**.

**Figure 27: Electrical Contacts Suitable for Single Crystal Silicon (a) and Polycrystalline Silicon Solar Cells (b)**

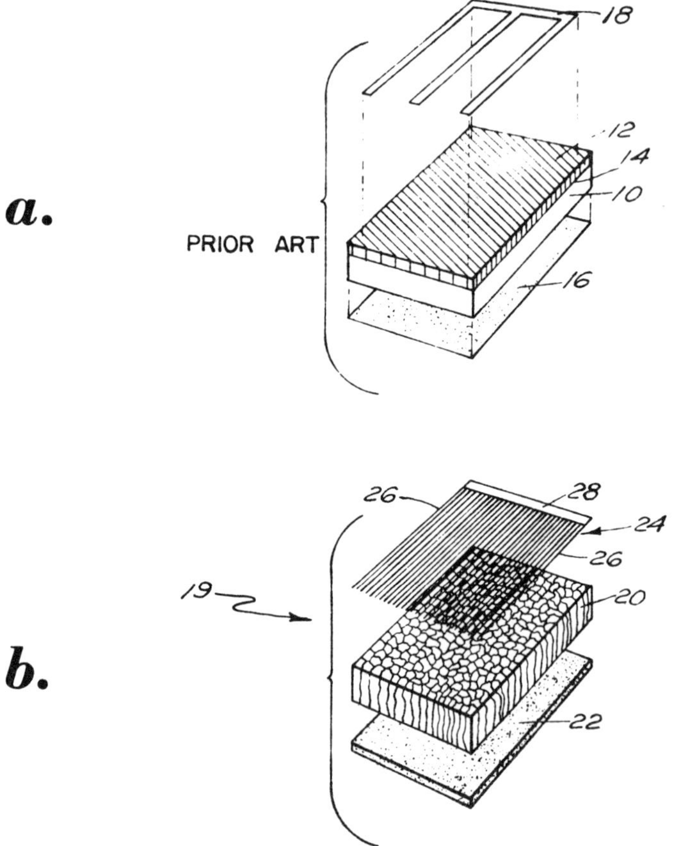


Source: U.S. Patent 3,978,333

In Figure 27b, there is shown a solar cell **19** comprising a polycrystalline base wafer **20** preferably of silicon, which preferably is deposited or grown on substrate **22**, which substrate may function as the back or bottom surface electrical contact for the cell **19**, assuming that the substrate is of electrically conductive material. The deposition of the polycrystalline silicon wafer on the substrate can be performed at moderate temperatures (less than the melting temperature of silicon) by decomposition of silicon-bearing gases, such as $SiCl_4$, $SiH_4$, etc., or at high temperatures by sheet casting of molten silicon.

It is desirable to control the parameters affecting crystallite growth in the wafer so that the crystallites grow to a predictable average size, which size is on the order of a diffusion length of the minority carriers in the base material.

The next step in the manufacture of the cell **19** is the deposition of a fine line grid pattern **24** on the top or front surface of the polycrystalline wafer **20**, it being understood that this is the surface to which illumination is to be applied. It is important to note that the grid effects a rectifying junction on the polycrystalline wafer and at the same time functions to collect and carry current from each crystallite. Thus, the affixation of the grid to the polycrystalline wafer achieves two functions, i.e., effecting of the desired P-N junction, and, at the same time, the grid functions as the ohmic contact on the front or top surface of the wafer.

A critical feature of this process is the fact that the grid is arranged so that the fingers **26** thereof make contact with substantially each and every crystallite or grain of wafer **20**, this being why it is desirable that the wafer be formed with grains or crystallites of generally the same size, where possible. It will be understood, however, that even where the polycrystalline base **20** comprises a few relatively large, central crystallites with smaller surrounding crystallites, the cell **19** will still work efficiently so long as the smaller crystallites are each contacted at least once by the fingers **26**, to the extent that this is reasonably possible.

Since the grid is functioning as an ohmic contact, obviously the fingers must be interconnected by a conducting bar, such as shown at **28**. It will be understood that the grid may take other forms such as the waffle pattern or a concentric pattern. The important feature in any of these arrangements is that the grid contact substantially all of the individual crystallites of the wafer, and it is also important that the fingers of the grid be kept as narrow as reasonably possible so as to insure maximum exposure of the top surface of the base **20**.

It will be understood that the grids, of course, can be produced by any conventional techniques, such as the commercial photo-lithographic technique used in the microelectronic industry, or by direct vapor deposition through a suitable evaporation mask, or by laser evaporation of a thin film of metal deposited over the entire surface.

Once the metal grid has been deposited on the polycrystalline surface, a suitable heat treatment can be applied thereto to form an alloyed P-N junction or to form a Schottky barrier. For example, if the polycrystalline wafer is N-type silicon, then the grid may be aluminum or an aluminum-bearing alloy. By heating the grid and wafer to a temperature of 576°C or over, which temperature is the aluminum-silicon eutectic temperature, and subsequently cooling, a P-type layer will form on the N-type substrate, creating an abrupt P-N junction. The residual, unalloyed aluminum left on the P-type layer serves as the ohmic contact for the surface of the cell to be illuminated.

If, however, the combination is heated below the aluminum-silicon eutectic temperature, but high enough to allow the aluminum to make intimate contact with the silicon—usually in the range of 500°C—then an aluminum-silicon metal-semiconductor, or Schottky barrier junction, will be formed.

Where an alloyed P-N junction is being formed, it will be understood that if the polycrystalline base is N-type silicon, then the grid must be aluminum or an aluminum-bearing alloy or other metal which is a P-type dopant, such as, for example, boron, indium or gallium. If the polycrystalline base is P-type silicon, then an N-type dopant must be used for the grid, such as an antimony and gold alloy. If a Schottky barrier is being formed, it has been found that whether the polycrystalline base is N-type silicon or P-type silicon, the grid may comprise aluminum, silver, gold, platinum, platinum silicide, tungsten or molybdenum.

It will be understood that the ohmic contact **22** provided on the unilluminated or underside of wafer **20** may be applied by any standard commercially known process, or, as previously indicated, it may be the substrate on which the polycrystal **20** is grown. It is thought that an aluminum-silicon alloyed P-N junction appears to offer the most attractive commercial possibilities for cell **19**, primarily for economic reasons. It is thought, however, that formation of a Schottky barrier may possibly result in higher efficiency cells due to the possibility of using different metals in combination with silicon.

An apparatus developed by *W. Keller and K. Reuschel; U.S. Patent 4,108,714; Aug. 22, 1978; assigned to Siemens AG, Germany* is one in which polycrystalline silicon is melted within an inductively heated crucible having a slit-like aperture at the bottom thereof and a stream of molten silicon is controllably fed through such aperture into a nip between a pair of spaced-apart heat-controlled rollers which convert the molten stream into a ribbon of a desired thickness, which then solidifies and may be severed into desired planar bodies.

Figure 28 shows a suitable form of apparatus for the conduct of such a process. As shown there, a crucible **2**, for example, composed of quartz or other suitable material, is provided with a polycrystalline silicon melt **4**. The crucible is annularly encompassed by a heating coil **3** which is operationally coupled with an energy source (high frequency current) for inductively heating the melt. The crucible includes a slit-like aperture **9** at the bottom thereof and molten silicon exits through the aperture **9** as a stream.

**Figure 28: Siemens Apparatus for Producing Polycrystalline Silicon Ribbons**

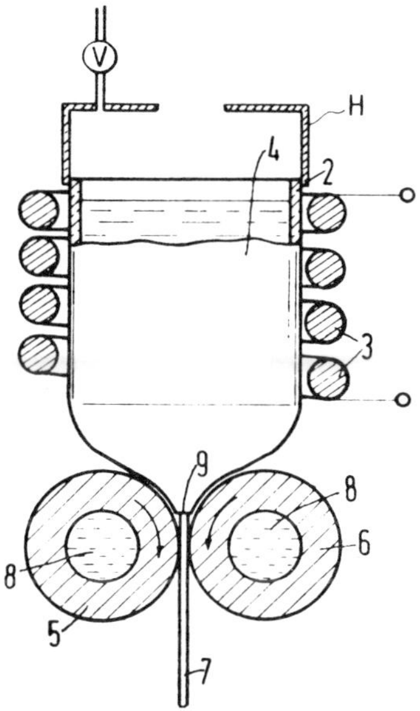


Source: U.S. Patent 4,108,714

A pair of horizontal and parallelly disposed spaced-apart rollers **5** and **6** are mounted for rotation in the direction indicated by the curved arrows and are positioned below the aperture so as to receive the stream of molten silicon into the space or nip between the rollers.

During rotation, the rollers convert the silicon stream into a wide flat silicon ribbon or band **7** of a desired thickness in accordance with a desired solar cell thickness, generally in the range of about 0.1 to 0.3 mm.The rate of rollers is adjusted so that the peripheral surfaces thereof travel at a linear rate of about 30 to 200 cm/min. As shown, the rollers may be hollow cylinders which are composed of a material selected from the group consisting of silicon and silicon carbide or at least the outer surfaces thereof may be composed of a silicon-compatible material.

A heat-exchange fluid **8** may be circulated within the rollers so as to maintain the temperature of the rollers at about 900° to 1300°C. As the molten silicon stream travels through the nip between the rollers, it solidifies as a macropolycrystal and forms a solid ribbon or band, which may be immediately severed into the desired plate-shaped silicon bodies or may be wound onto a storage core or the like for future use.

The metering or controlled feeding of the molten mass or stream fed to the nip between the rollers may be accomplished by providing a pressure differential between the interior of the crucible and the exterior thereof. For example, a gastight cover member **H** may be secured onto the upper end of crucible **2** and a pressurized gas controllably fed into the crucible so as to force the molten silicon out through aperture **9** at a controlled rate. As shown, the cover member may comprise a hollow housing which is gas-impermeable and includes a valve controlled gas passageway so that an operator may readily adjust the pressure difference between the gas in contact with the interior of the crucible and the ambient atmosphere surrounding the crucible. In this manner, the amount of molten material required for the production of a certain thickness ribbon may be precisely apportioned.

After solidification, the flat silicon ribbon or band may be severed, for example, by cutting or grinding through only the thickness dimension of such ribbon. No further work along the flat or planar surfaces thereof is required. Alternatively, the solidified silicon ribbon **7** may be wound onto a storage reel or the like for later use.

A process developed by *P.S. Kotval and H.B. Strock; U.S. Patent 4,124,410; Nov. 7, 1978; assigned to Union Carbide Corp.* yields silicon substrates, both multigrained and single crystal, prepared by the slag oxidation of partially purified silicon platelets and further treated to obtain low-cost refined metallurgical silicon having suitable properties for desirable solar cell applications.

The partially purified platelets are obtained by the precipitation thereof from a solution of metallurgical grade silicon in a molten liquid solvent, such as aluminum, with treatment to remove adherent impurities therefrom. Multigrained refined metallurgical silicon (RMS) boules useful for the substrates, can be pulled by the Czochralski method from a melt of the refined silicon following the slag oxidation. By unidirectional solidification of the refined silicon-slag melt following slag oxidation, a multigrained, directionally solidified refined metallurgical silicon (DS/RMS) is obtained and is also suitable for such solar cell applications.

Multigrained, DS/RMS boules pulled from a melt of the DS/RMS constitute another useful refined silicon material suitable for solar cell substrate purposes. By remelting and redirectionally solidifying the DS/RMS, a further, single crystal product useful as the planar diode and solar cell substrate can be obtained by pulling DS/RMS boules from a melt of the twice directionally solidified material. The substrates from all of such materials, and the overall planar diode and solar cell products, are low-cost products having desirable solar cell properties.

While the multigrained RMS and DS/RMS, and the single crystal DS/RMS, materials thus utilized as solar cell substrates have a substantially higher impurity content level than in conventional high-purity semiconductor grade material, the materials are nevertheless of sufficient purity to permit their utilization as useful solar cell substrates. The epitaxial type solar cell products have an N-on-P-on-P substrate configuration, while the diffusion-type products have pentavalent impurities diffused in the low-cost substrates to form a P-N junction therein.

A process developed by *B. Authier, R. Griesshammer, F. Köppl, W. Lang, E. Sirtl and H.-J. Rath; U.S. Patent 4,131,659; Dec. 26, 1978; assigned to Wacker-Chemitronic Gesellschaft für Elektronik-Grundstoffe mbH, Germany* is a process for producing large-size, self-supporting plates of silicon deposited from the gaseous phase on a substrate body. It comprises heating a graphite substrate to deposition temperature of silicon, which is deposited on the substrate from a gaseous compound to which a dopant has been added until a layer of about 200 to 650 μm has formed, subsequently melting 40-100% of this layer from the free surface downward, resolidifying the molten silicon by adjustment of a temperature gradient from the substrate body upward, and finally separating the silicon therefrom. The plates so formed are used primarily for making solar cells.

A suitable form of apparatus for the conduct of this process is shown in Figure 29.

A suitable drive means, e.g., a set of rollers **1** advances a steel plate **2** extending through the entire apparatus; on the plate, a plurality of individual, abutting graphite plates **3**, preferably of pyrographite nature, are continuously travelling through the device at a rate of 1-15 cm per minute, preferably 3-7 cm/min. Steel plate **2**, or a similarly dimensioned ceramic or quartz plate which is sectionally and selectively heatable from below by a resistance heater comprising coils or tubes, has the same width as the graphite plates **3** and is equipped with a butt edge, so that the plates **3** are propelled through the apparatus with the formation of a continuous band-shaped deposition surface. The width of plates **3** depends on the dimensioning of the apparatus and corresponds substantially to the width of the deposition chambers minus the wall thickness.

**Figure 29: Process for Producing Large-Size, Self-Supporting Plates of Silicon**

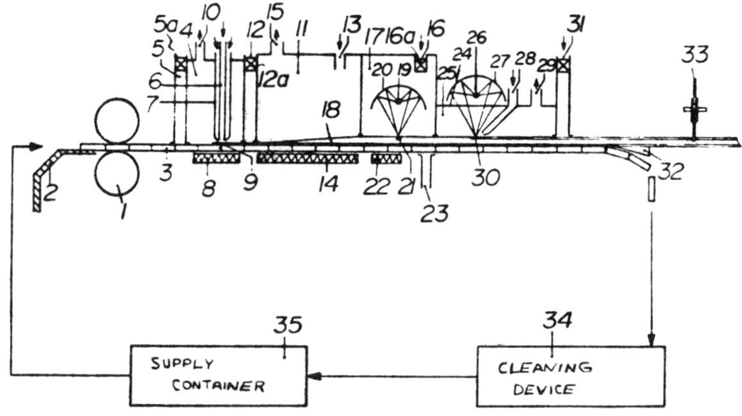


Source: U.S. Patent 4,131,659

Advantageous sizes are about 5-20 cm for the width, as well as 5-20 cm for the length, and 0.1-1 cm for the height, having a parallelogram-shaped, preferably square-shaped, cross section in the direction of movement.

In addition to the described mode of operation, the graphite plates may be placed on a conveyor belt which is sectionally heatable to different temperatures and moved through the apparatus thereon. Yet, another possibility is to transport the abutting plates while they glide over two guide rails, the heating being brought about by direct passage of current, or contact-free by heating elements of ceramic or quartz arranged below the plates.

Before entering the first deposition chamber **4**, the graphite plates **3** pass through a gas trap **5** provided with a pressure-maintaining valve **5a** where they are circulated by an inert gas, e.g. argon, under a pressure of 0.1-0.3 bar above the pressure of the first chamber **4**. This arrangement avoids the possibility that reactive gases from the chamber could escape to the outside or that air from the outside enters the chamber.

In that chamber, the plates are met by a gas mixture blown through a narrow double tube, the inner tube **6** admitting, e.g., monosilane, advantageously in an amount of about 300-400 parts by volume of hydrogen, and the outer tube admitting ammonia, either undiluted or likewise, preferably in 400 parts by volume of hydrogen. The molar ratio calculated on the components monosilane: ammonia is preferably 1:2-1:4.

The gaseous mixture is blown out of the narrow slit-shaped double tube whose width corresponds substantially to the clear width of the chamber and which extends with its gas outlet opening to about 0.5-3 cm, preferably 1-2 cm, above the surface of the graphite substrate at a pressure of 1-2 bar, preferably 1.1-1.4 bar; the rate at which the gas passes is adjusted in such a manner that an intermediate layer of silicon nitride **9** is deposited on the plates in a thickness of 1-5, preferably 1.5-3 $\mu$m; while the plates pass through the chamber, the temperature of the plates is raised to about 750° to 1200°C, preferably 950° to 1050°C, by heating device **8**. Reaction gases escape from the chamber through exhaust tube **10**.

After the first deposition chamber which is only required when the formation of an intermediate layer between the substrate body and the silicon layer is intended, the band of graphite plates passes a gas trap **12** with a pressure-maintaining valve **12a**; there, an inert gas, especially argon, is blown at a pressure of 0.1-0.3 bar above the pressure of the two adjacent chambers **4** and **11**, which are preferably maintained at an almost equal pressure, onto the surface of the intermediately coated band of graphite plates, in order to prevent the reaction and decomposition gases from escaping from the two chambers.

Chamber **11** is the large deposition chamber where the deposition of silicon on the substrate is taking place. The gas for the deposition is admitted through tube **13**. It consists of a mixture of, e.g., about 80-95 parts by volume of hydrogen and about 5-20 parts by volume of, for instance, trichlorosilane, the total, of course, adding up to 100% by volume. To this is added an amount of about 50-100 ppm diborane in hydrogen. Altogether, the amount of dopant added should be so calculated that in the deposition gas, an atomic ratio of boron: silicon will be 1:$10^5$ to 1:$10^7$. The size of the gas admitting tube **13** is not critical, provided only its sectional area permits the introduction of 1-2 bar, preferably 1.1-1.4 bar, in such an amount that during the passage of plates **3** through the large deposition chamber **11**, doped silicon will be deposited in a thickness of 200-650 $\mu$m, preferably 300-500 $\mu$m, the deposition occurring on the plates while they are heated up to 1000°-1250°C, preferably 1050°-1150°C by heating means **14**. The reaction gases and unreacted deposition gases escape through a tube **15** from chamber **11**.

After leaving the deposition chamber, the coated graphite plates enter a recrystallization chamber **17** which is filled with inert gas, e.g., argon, by means of a tube **16** having a pressure-maintaining valve **16a**, the pressure being higher by 0.1-0.3 bar as compared to the one obtained in chamber **11**. In chamber **17**, a halogen rod-containing lamp **19** extending over the entire width of the band of plates **3** is mounted, wherein a reflector **20** having an elliptical curvature focuses radiation, preferably on a small area over the width of the band, whereby 40-100% of the thickness of the silicon layer **18** is melted.

The basic load of the energy supply is furnished through the substrate body, namely, by heating the band by heating device **22**, raising the temperature to about 1150° to 1250°C. By a gas admission tube **23** provided below the band of graphite plates, an inert gas, e.g., argon of room temperature, thus, about 25°C, is blown as coolant onto the band, thereby dissipating the heat of crystallization and causing an oriented growth of the resolidifying silicon from the substrate body toward the free surface.

Next, the band of graphite plates leaving recrystallization chamber **17** with the resolidified silicon layer thereon, or as the case may be, also carrying the intermediate layer **9**, enters a doping chamber **25** covered by a quartz plate **24**. Above the quartz plate, a lamp **26** having a halogen rod is arranged, and above the lamp an elliptically curved reflector **27**, which superficially melts the resolidified silicon layer **18** to a depth of 0.5-3 $\mu$m, and a dopant is admitted having opposite polarity to the one already present in the silicon layer.

The dopant is introduced by a gas inlet tube **28** extending to about 0.5-3 cm, preferably 1-2 cm, toward the remelted area of the silicon layer. A suitable dopant is, e.g., phosphine, which is admitted in the amount of 50-100 ppm in hydrogen. This dopant has to be introduced in a quantity which will not only compensate the previously present dopant, but must provide one phosphorus atom for $10^3$ to $10^7$ silicon atoms. The reaction and residual gases are withdrawn through tube **29**. The silicon layer **30** remelted to a width of 1-2 mm absorbs phosphorus from the decomposition of the phosphine with formation of a P-N transition to the adjacent silicon layer.

After having passed a gas trap **31** which is similarly built to traps **5** and **12**, the deposited silicon layer is lifted off the graphite plates **3** with a sharp spatula **32** and is cut into individual plates with a cutting device **33**, preferably along the seams of the substrate plates. The graphite plates are then superficially cleaned with steel brushes or by sandblasting in a cleaning device **34**, added to with new graphite plates from the supply container **35** and returned for new deposition to the apparatus.

## SILICON-HYDROGEN "ALLOYS"

An intriguing recent development is the discovery that an amorphous silicon-hydrogen "alloy" can be used to construct photovoltaic cells with useful efficiencies. Efficiencies of 5.5% have been measured and 15% efficiency may be possible (3). The hydrogen apparently attaches to "dangling" silicon bonds, minimizing the losses that would otherwise result at these sites. Acceptable performance is possible, despite the large number of remaining defects, because the amorphous material is an extremely good absorber of light; test cells are typically 1 micron or less in thickness. The properties of this complex material are not well understood.

The materials developed by Energy Conversion Devices, Inc. and their president, S.R. Ovshinsky, are presumably amorphous semiconductor alloys of silicon, fluorine and hydrogen (11) but explicit details of their preparation are not yet available.

Amorphous silicon materials may be fabricated by the process described by *D.E. Carlson; U.S. Patent 4,064,521; Dec. 20, 1977; assigned to RCA Corp.*

Figure 30 shows a glow discharge apparatus which is suitable for fabrication of such an amorphous silicon photovoltaic device. The glow discharge apparatus **30** includes a vacuum chamber **32** defined by a vacuum bell **34**, typically of a glass material. In the vacuum chamber is an electrode **36**, and a heating plate **38** spaced from and opposite the electrode. The electrode is of a metallic material having good electrical conductivity such as platinum and is in the form of a screen or coil. The heating plate is a ceramic frame which encloses heating coils which are energized from a current source **40**, external to the vacuum chamber.

A first outlet **44** into the vacuum chamber is connected to a diffusion pump, a second outlet **46** is connected to a mechanical pump, and a third outlet **48** is connected to a gas bleed-in system which is the source of the various gases utilized in the glow discharge process. While the first outlet is described as being connected to a diffusion pump, it is anticipated that a diffusion pump may not be necessary since the mechanical pump connected to the second outlet may evacuate the system to a sufficient pressure.

In the fabrication of the photovoltaic device, the substrate **12**, e.g, antimony, is placed on the heating plate, and the substrate is connected to one terminal

**Figure 30: Glow Discharge Apparatus Suitable for Fabrication of Amorphous Silicon Photovoltaic Devices**

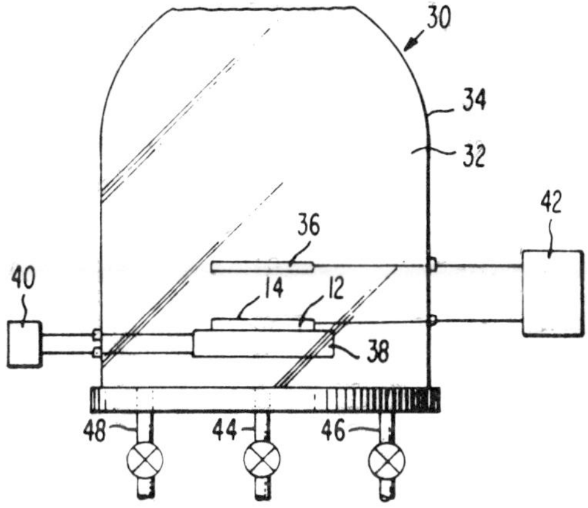

Source: U.S. Patent 4,064,521

of a power source **42** and the electrode is connected to an opposite terminal of power source. A voltage potential therefore exists between the electrode and substrate. The power source can be either dc or it can be ac, i.e., in the low frequency range, for example, 60 hertz, or it can be RF, i.e., in the high frequency range, for example, on the order of megahertz.

Typically, when the power source is dc, the electrode is connected to the positive terminal of the power source, and the substrate is connected to the negative terminal of the power source. Thus, the electrode functions as an anode and the substrate functions as a cathode when the power source is energized. This is referred to as a cathodic dc operation. However, in a dc operation the substrate and electrode can be of the opposite polarities described, i.e., the substrate is the anode and the electrode is the cathode, which is referred to as anodic dc operation.

It has been discovered that the deposition rates are somewhat higher in the cathodic mode than in the anodic mode. Furthermore, RF glow discharge operation can be accomplished in electrodeless glow discharge apparatus of a type well known to those in the art, e.g., a capacitive RF glow discharge system and an inductive RF glow discharge system. However, more uniform deposition over a large area, i.e. greater than 10 $cm^2$, is attained in dc or ac glow discharge than in electroless RF glow discharge.

After the substrate and electrode are connected to the power source the vacuum chamber is evacuated to a pressure of about $10^{-3}$ to $10^{-6}$ torr, and the substrate is heated to a temperature in the range of 150° to 450°C by energizing the heating coil to the heating plate.

Next, silane, $SiH_4$, is bled into the vacuum chamber to a pressure of 0.1 to 5.0 torrs and as a result, the substrate temperature is raised to a value in the range of 200° to 500°C.

To initiate the glow discharge between the electrode and the substrate, resulting in the deposition of the amorphous silicon of body **14** onto a surface of the substrate, the power source is energized. For deposition of the body **14** the current density between the electrode and substrate should be in the range of 0.3 to 3.0 mA/cm$^2$ at the surface of the substrate. The deposition rate of the amorphous silicon increases with the vapor pressure of the silane and the current density. For a silane pressure of 1.5 torrs, a current density of 1 mA/cm$^2$ at a cathodic substrate **12** and the substrate **12** at 350°C, the deposition of 1 micron of amorphous silicon occurs in less than one minute.

Once the glow discharge is initiated for the dc cathodic mode, electrons are emitted from the substrate and strike silane molecules, $SiH_4$, both ionizing and disassociating the molecules. The positive silicon ions and the positive silicon hydride ions, such as $SiH^+$, are thus attracted to the substrate **12**, which is the cathode and silicon containing some hydrogen is thereby deposited on the substrate. It is believed that the presence of hydrogen in the body of amorphous silicon is beneficial to good electronic properties. It has been found that a body **14** of amorphous silicon fabricated by a glow discharge in pure silane is of a slightly N-type conductivity, when deposited on a substrate **12** which is at a temperature above 100°C.

The temperature of the substrate in the glow discharge process may influence the composition and structure of the material deposited thereon due to the effects of auto doping, eutectic formation and induced crystallization, e.g., deposition on a single crystalline silicon substrate at temperatures above about 500°C results in the deposition of a polycrystalline silicon, and deposition on a gold substrate at a temperature above 186°C results in induced crystallization of the deposited silicon.

After deposition of the amorphous silicon, the wafer of substrate **12** and body **14** is placed in a state-of-the-art evaporation system and then a metallic region is evaporated onto the body **14**. Likewise, an electrode of grid form and an antireflection layer are deposited on the metallic region by state-of-the-art evaporation and masking techniques. The entire processing may be accomplished in a single system accommodating both glow discharge and evaporation.

Fabrication of the photovoltaic device is completed by the connecting of wire electrodes (not shown) to the substrate **12** and grid electrode for connection to external circuitry.

## GERMANIUM

Germanium has a band gap of approximately 0.7 electron volt. This corresponds to photons of electromagnetic energy having a wavelength of approximately 1.8 microns. This is well into the infrared range. A germanium photovoltaic cell has a relatively poor sensitivity to electromagnetic energy having a shorter wavelength. Silicon by comparison has a band gap of approximately 1.1 electron volts and responds more favorably to electromagnetic energy having a shorter wavelength such as in the visible and ultraviolet regions of the spectrum.

However, one type of cell design using germanium has been described by *A.H. Smith; U.S. Patent 3,615,855; Oct. 26, 1971; assigned to General Motors Corp.* It is prepared by epitaxially depositing on one surface of a monocrystalline slice of P-type silicon a P-type silicon-to-germanium transitional region; epitaxially depositing an N-type germanium layer on the transitional region to form a P-N junction; lapping and polishing to an optically smooth finish the opposite surface of the silicon slice and bonding a light permeable current collecting grid to the optically polished surface of the silicon slice. Also, an ohmic contact is bonded to the epitaxial layer of N-type germanium to complete the device.

A germanium photovoltaic device which was developed by *R.W. Beck; U.S. Patent 3,620,829; Nov. 16, 1971; assigned to General Motors Corp.* is one based on the discovery that the maximum efficiency is realized when water is adsorbed on the surface of the device, and that as the amount of adsorbed water varies, so does efficiency of the cell. Hence, if a germanium photovoltaic device is used in an ambient having relatively low humidity conditions, such as a vacuum, the adsorbed humidity is lost and the device commensurately degrades. Analogously, if the device is operated in an ambient in which the relative humidity varies, the efficiency of the device varies with the change in humidity.

It was further found that the adsorbed water produces its beneficial effect by reducing surface recombination velocity to a minimal value. Efficiency of a photovoltaic device is highly dependent upon surface recombination velocity at the light-gathering surface. Hence, when a voltaic device is operated under decreasing humidity conditions, the adsorbed water evaporates permitting surface recombination velocity to increase. As the surface recombination velocity increases, efficiency of the device decreases. Moreover, under vacuum conditions virtually all of the adsorbed water can be lost, so that the surface recombination velocity is not suppressed at all.

Also, it appears that the adsorbed water maintains surface recombination velocity at a low value due to its polar or ionic character. In any event, if one replaces the water on the germanium surface with another compound of a similar degree of polar bonding one can maintain surface recombination velocity at a minimal value even under extremely low humidity conditions. In this manner one can stabilize surface recombination velocity at a low value under varying humidity conditions and even preclude degradation of germanium photovoltaic devices under vacuum conditions.

The improved operation described by Beck is accomplished by applying an insulating coating to the active surface of a germanium device, such as the light gathering surface of a photovoltaic cell, of a material which is about as ionic in character as water and which is nonvolatile and inert under the conditions to which the device will be subjected.

Antimony trioxide was found to be the preferred material for use in a layer from 1,000 to 10,000 Angstroms thick.

## GROUP II COMPOUNDS (Cu, Cd, Pb SULFIDES AND TELLURIDES)

$CdS/Cu_2S$ are much better absorbers of light than crystalline silicon, so cells made from these materials can be thinner and tolerate smaller crystal grains than was possible with the crystalline silicon. Commercial $CdS/Cu_2S$ cells will probably be 6 to 30 microns thick, and the crystal structure produced with a relatively simple spray or vapor deposit process is large enough to prevent grain structure from significantly affecting cell performance.

The primary drawback of most of the "thin film" cells is their low efficiencies. Research is proceeding rapidly in a number of areas, however, and a number of thin film cells may be able to achieve efficiencies greater than 10%.

Older designs of cadmium cells degraded and failed relatively rapidly, but accelerated lifetime tests on modern designs appear to indicate that cells hermetically sealed in glass with proper electrical loading could have a useful life of decades.

Research is also under way on a number of other materials which may be used to manufacture inexpensive thin film cells. Experimental cells have been constructed which substitute indium phosphide or copper indium selenide for the copper sulfide in the $CdS/Cu_2S$ heterojunction, and efficiencies greater than 10% have been demonstrated in single-crystal laboratory cells made with these materials (3).

As reviewed by Hovel (2), the main advantage of CdS solar cells is low processing and materials costs. Their main disadvantages are low efficiency and the steps necessary to ensure stability. The most important work associated with these cells is aimed toward increasing efficiency from the present 8 to 10% or more. The nature of the disadvantages of these cells and possible remedies are described. An excellent bibliography is included in the review.

A photovoltaic cell design developed by *N. Nakayama, K. Yamaguchi and E. Hirota; U.S. Patent 3,492,167; Jan. 27, 1970; assigned to Matsushita Electric Industrial Co., Ltd., Japan* consists of an N-type cadmium sulfide sintered plate and an electrochemically deposited P-type thin layer on one surface of the sintered plate, the layer consisting of cadmium, copper and sulfur, and being a CdS-like structure with an accompanying copper sulfide structure. Electrodes are applied respectively to the P-type thin layer and to another surface of the sintered body, yielding means to generate a photovoltaic effect when light is irradiated on the P-type thin layer.

Materials developed by *R.A. Ruehrwein; U.S. Patent 3,496,024; Feb. 17, 1970; assigned to Monsanto Co.* are epitaxial films of large single crystals of inorganic compounds. Epitaxial films which may be prepared in accordance with the process described are prepared from volatile compounds and elements of beryllium, zinc, cadmium and mercury with volatile compounds and elements of sulfur, selenium and tellurium. Typical compounds within this group include the binary compounds beryllium sulfide, zinc selenide, cadmium telluride, mercury selenide and cadmium sulfide.

As examples of ternary compositions within the defined group are those having the formula $ZnS_xSe_{1-x}$ and $CdS_xSe_{1-x}$, x having a numerical value greater than zero and less than 1.

The process epitaxial films are characterized as graded energy gap crystals. A graded energy gap crystal is characterized by a nonuniformity of composition which results in a corresponding nonuniformity in the forbidden energy gap of the material. The nonuniformity of the forbidden energy gap may be one of gradual increase or decrease in a given direction in a linear or nonlinear manner or any other type of profile. The range over which the forbidden energy gap can vary is governed by the elemental components that make up the crystal. For the process materials the energy gap of the crystal may vary from about 0.025 to about 4 eV.

The following example illustrates the formation and deposition of an epitaxial graded gap film on N-type CdTe as the substrate. A polished crystal of N-type CdTe one millimeter thick and containing $1 \times 10^{17}$ carriers/cc is placed in a fused silica reaction tube located in a furnace. The CdTe crystal is placed on a silica support inside the tube. The reaction tube is heated to 600°C and a stream of hydrogen is directed through the tube for 15 minutes to remove oxygen from the surface of the CdTe.

A stream of hydrogen is then directed through a reservoir of $CdBr_2$ maintained at about 600°C thus vaporizing the $CdBr_2$ which is then carried by the hydrogen through a heated tube from the reservoir to the reaction tube containing the CdTe substrate crystal.

Meanwhile, separate streams of hydrogen are conducted through separate tubes containing in one of them a reservoir of $TeCl_4$ heated to about 330°C and in the other a reservoir of $InCl_3$ (as a doping component) heated to about 170°C. From the heated tubes the $TeCl_4$ and $InCl_3$ are carried by the hydrogen on through the tubes to the reaction tube. In the system the mol fractions of the $CdBr_2$, $TeCl_4$ and $H_2$ are 0.05, 0.15 and 0.80, respectively.

The separate streams of vaporized $CdBr_2$, $TeCl_4$ and $InCl_3$ conjoin in the fused silica reaction tube where a reaction occurs between the cadmium and tellurium components in which a single crystal film of N-type cadmium telluride begins to form on the substrate crystal of CdTe.

After the film begins to form, the hydrogen flow rate through the $CdBr_2$ reservoir is gradually reduced while at the same time the flow rate of a hydrogen stream through a separate tube leading to the reaction tube and containing a reservoir of mercury heated to about 260°C is started and gradually increased so that the mol fraction of the sum of the two Group II components is maintained about constant in the reaction tube.

During the interval a single crystal deposit of $Cd_xHg_{1-x}Te$ is formed on the substrate with x decreasing from one to zero as the deposition proceeds. When the hydrogen flow rate through the $CdBr_2$ reservoir has been reduced to zero, the hydrogen flow through all of the tubes is terminated. The epitaxially grown crystal removed from the reaction tube is composed of cadmium telluride on one (bottom) face and mercury telluride, N-type on the opposite (top) face and contains about $10^{17}$ carriers/cc. The side edges of the crystal after lapping them are of graded composition from CdTe to HgTe.

Metallic leads are attached to two aforesaid graded composition opposite edges of the crystal and the crystal is placed in a magnetic field of about 50,000 gauss. Upon irradiating the CdTe face of the crystal with radiation from a hot body heated to about 2000°C, electrical potential is generated in the crystal. It is found that the efficiency of conversion of the radiant energy falling on the crystal to electrical energy withdrawn from the crystal is about 20%.

A type of cell construction developed by *N. Nakayama, E. Hirota, T. Shiraishi and T. Yamanaka; U.S. Patent 3,520,732; July 14, 1970; assigned to Matsushita Electric Industrial Co., Ltd., Japan* is a copper sulfide-cadmium sulfide device whose preparation is illustrated by the following example.

Cupric sulfide (CuS) in a quartz tube is melted at 1180°C for 2 hours in nitrogen gas atmosphere and thereafter slowly cooled to room temperature. The obtained material is a polycrystalline melted body of cuprous sulfide ($Cu_2S$), in chalcocite structure. A further purification is carried out by a zone-refining method by using an evacuated quartz of $10^{-2}$ mm Hg. The obtained purified body is a P-type semiconductor and exhibits a resistivity of $10^{-1}$ ohm-cm at room temperature and a mobility of charge carried of 3 $cm^2$/volt-sec. A plate of disk type having dimensions of 3 mm thickness and 2.0 $cm^2$ area is prepared.

A commercially available cadmium sulfide (CdS) powder is pressed into pellets and sintered at 800°C in nitrogen gas atmosphere for 2 hours. Thus-obtained sintered material is an N-type semiconductor and exhibits a resistivity of $10^{-1}$ ohm-cm. A mobility of charge carrier of this N-type semiconductor is of 30 $cm^2$/volt-sec. A plate of disk type having dimensions of 0.35 mm thickness and 0.75 $cm^2$ area is prepared from this sintered body of N-type CdS.

The surface of the disk of both P- and N-type materials are polished into optically flat condition and then slightly etched with dilute HCl solution. The combined disks of both N-type and P-type semiconductor are put in an alloy die, enveloped with graphite powder, and hot-pressed into a single laminated body at a temperature of 400°C and at a pressure of 200 kg/cm$^2$ for 10 seconds and thereafter quenched. The hot pressing apparatus is of known construction and includes a system of high frequency induction heating and is operated by hydrostatic pressure.

An indium electrode is applied on the N-material, and a gold electrode is applied on the P-material. The photovoltaic effects of the thus-prepared cell is measured by exposing the cell in normally incident sunlight. The cell generates an open-circuit photovoltage of 0.5 volt and a short-circuit photocurrent of 16 mA per unit area of light incident surface. The conversion efficiency of the cell is 3%.

A process developed by *N.E. Heyerdahl and D.J. Harvey; U.S. Patent 3,531,335; Sept. 29, 1970; assigned to Kewanee Oil Co.* involves preparing a compound semiconductor film having controlled resistivity, involving the steps of supporting a substrate, such as glass, molybdenum, etc., in an atmosphere-controlled chamber, simultaneously vaporizing in the chamber a Periodic Group VI material such as sulfur, selenium or tellurium, and a Periodic Group II material such as zinc, cadmium or mercury, while the substrate is heated to a temperature between 150° and 500°C, allowing the vaporized materials to form a compound film on an exposed surface of the substrate, and, after formation of the compound film, discontinuing vaporization of the Group VI material, allowing the film to cool while being bombarded by vapor of the Group II material.

The resultant films have low resistivity and serve as an effiient photovoltaic cell having a low ratio weight to external generated electrical energy, so that they have become widely proposed for uses where weight is a factor.

Figure 31 shows a suitable form of apparatus for the conduct of this process. This apparatus may take a variety of structural forms; however, in accordance with the illustrated embodiment, apparatus **10** includes a somewhat conventional vacuum chamber in the form of a bell jar **12** resting upon a lower base **14**. The upper surface **16** of the base is provided with an appropriate sealing material, and as it contacts the lower edge **18** of the bell jar, the jar is hermetically sealed in accordance with conventional practices.

Within the vacuum chamber there is provided material-receiving vessels **20, 22, 24** formed from a heat-resistant material and surrounded by heating elements **21, 23,** and **25,** respectively. These vessels are somewhat tubular in shape and have upwardly facing mouths **26, 27,** and **28,** respectively. Various materials placed within these vessels or boats are heated by the heating elements to form a vapor which is directed upwardly from the boats toward a substrate holder **32** positioned directly above the boats and adapted to support a substrate **34.** This substrate is directly aligned with the vapor emitting from the boats.

Since the boats are tubular with upper mouths, the vapors impinge on the substrate where they crystallize in a manner to be hereinafter described in detail. Of course, the holder may or may not be a solid crylindrical structure. In prac-

**Figure 31: Apparatus for Preparing Cadmium Telluride Films of Controlled Resistivity**

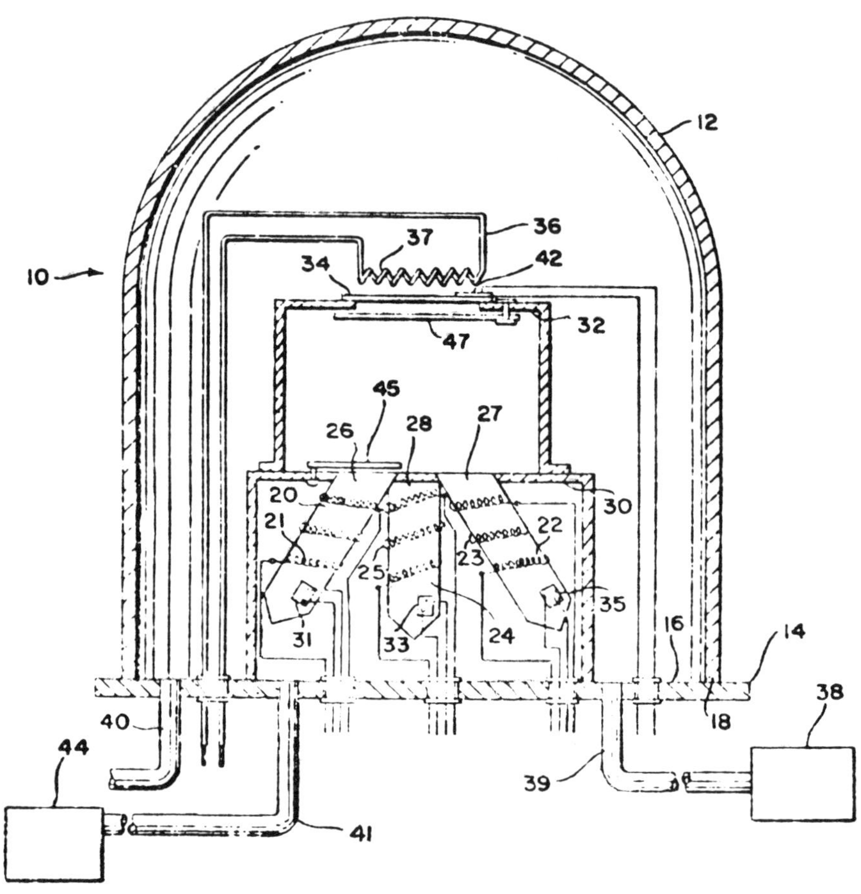


Source: U.S. Patent 3,531,335

tice, the support includes an upper plate with lower support arms fastened within the bell jar.

The temperature of materials being vaporized within the boats must be accurately controlled; therefore, thermocouples **31, 33** and **35** are secured in heat-conducting relationship with the boats. These thermocouples are connected onto appropriate control devices or thermometers (not shown) so that the temperature of each boat is known and can be accurately controlled. The particular control arrangement for maintaining a given temperature within each boat is well known in the art; therefore, it has been omitted for the purpose of simplicity. An electric resistance heater **36** is positioned above the substrate **34** and includes a heating element **37**. A thermocouple **42** is positioned on the substrate so that the temperature of the substrate is known at all times and can be varied by controlling the heater.

The vacuum chamber or bell jar is evacuated by a somewhat standard evacuating system **38** which includes a mechanical vacuum pump and a diffusion pump. The system is communicated with the interior of the vacuum chamber by a conduit **39** so that the pumps forming the system can lower the pressure within the vacuum chamber to a value approaching $1.0 \times 10^4$ mm of mercury. The interior of the vacuum chamber is also communicated with an atmosphere port **40** and a conduit **41** connected with a supply **44** of inert gas. A valve (not shown) selectively closes the vacuum chamber or communicates the vacuum chamber with either atmosphere or the supply **44**.

Boat **20** is provided with a shutter **45**. This shutter may be swung back and forth to expose the mouth of the boat. It is also possible to provide the other boats with shutters; however, this is not necessary to the successful operation of this process. A shutter **47** is provided beneath the substrate **34** to selectively cover or uncover the exposed lower surface of the substrate for controlling the actual bombardment of the substrate with the vapors from the boats. Shutter **47** is generally over the substrate except during disposition of crystalline film thereon and during cooling of the substrate to a preselected temperature.

In accordance with a more specific aspect of this process, a Group II material or a Group II-VI compound, such as zinc, cadmium, mercury, zinc sulfide, zinc selenide, zinc telluride, cadmium sulfide, CdSe, CdTe, HgS, HgSe or HgTe, is deposited in a boat **20**. A Group VI material, or a Group II-VI compound, such as sulfur, selenium, tellurium, ZnS, ZnSe, ZnTe, CdS, CdSe, CdTe, HgS, HgSe, or HgTe, is deposited in boat **22**. If desired, the boat **24** receives a Group III material, such as indium, gallium or aluminum. Each of these materials is semiconductor grade, which is preferably 99.999% pure.

The substrate is heated by the heater to a temperature between 150° or 200°C and 500°C. Below the lower or minimum temperature or temperatures, the cadmium or tellurium, or the other materials used, will tend to condense on the surface of the substrate without forming a compound polycrystalline film. This, of course, forms an unwanted barrier on the film. Above the higher or maximum temperature, the polycrystalline film tends to evaporate from the exposed surface of the substrate.

Each boat is heated to a temperature which will evaporate the material within the boats at a predetermined rate. When cadmium telluride is to be crystallized onto the substrate, boat **20**, which includes the cadmium, is heated to a temperature between 350° and 500°C. Tellurium within boat **22** is heated to a temperature in the range of 400° to 500°C. This causes the cadmium and tellurium to evaporate together and travel to a position at least adjacent the exposed surface of the substrate.

The substrate, which may be lime glass, molybdenum or a similar material, is held at the proper temperature to assure compounding and crystallization of the cadmium and tellurium as cadmium telluride on the substrate. In practice a thin sheet of molybdenum is used because of its flexibility, light weight and affinity for the compound film.

If a dopant, or impurity, is to be formed with the polycrystalline structure, the dopant material is placed in boat **24**. Preferably, indium is used as the dopant, which coacts with the compound film to control the conductivity of the film. By doping the film with indium, a donor material, the film is N-type. Of course, the film could be doped with other materials, or the temperature of the substrate and the evaporation of the elements could be controlled to vary from stoichiometry, so that the material may be either P-type or N-type. When indium is used to control the conductivity, boat **24** is heated to a temperature of 850° to 1000°C.

The materials within the heated boats are compounded and deposited as a polycrystalline film on the substrate. By using the time-rate method of control, the thickness of the film may be maintained within preselected limits. The temperature of each boat determines to a great extent the evaporation rate and the impingement rate of the material within the boats. These rates are maintained for calculated periods to obtain the desired film thickness.

Heating elements **23**, **25** of boats **22**, **24** are deenergized after the film has reached the desired thickness. Thereafter the vapor from boat **20** fills the chamber and the heater is deenergized. This allows the substrate to cool in an atmosphere of the Group II material. After the substrate has cooled to a temperature of about 200°C, emission of the cadmium vapor from the boat **20** is stopped by interrupting current flow through the heating element **21** and/or covering boat **20** with shutter **45**. This process reduces the resistivity of the film, especially when the film is formed from cadmium telluride. This allows the film to be used as an efficient photovoltaic cell or other semiconducting unit.

A technique developed by *K. Yamashita; U.S. Patent 3,568,306; March 9, 1971; assigned to Matsushita Electric Industrial Co., Ltd., Japan* produces photovoltaic cells by use of powdered polycrystalline cadmium sulfide, cadmium selenide or their solid solution so as to easily make, in any desired shape and at low cost, photovoltaic cells which have a high sensitivity to visible rays, are resistive to relatively high temperatures, and have a stable service life.

A process developed by *T.N. Duy and W. Palz; U.S. Patent 3,884,779; May 20, 1975; assigned to Societe Anonyme de Telecommunications, France* provides for the controlled formation of the layer of copper sulfide of a cadmium sulfide photocell by immersion in a solution of cuprous ions. The photocell being produced is maintained throughout the duration of its immersion in the solution at a constant potential at least equal to that of a pure copper electrode immersed in the same solution.

A somewhat similar process is described by *T.N. Duy, W. Palz and J. Vedel; U.S. Patent 3,885,058; May 20, 1975; assigned to U.S. Secretary of the Navy*.

A device developed by *K. Bogus and S. Mattes; U.S. Patent 3,888,697; June 10, 1975; assigned to Licentia-Patent-Verwaltungs GmbH, Germany* is a thin layer photocell which comprises a semiconductor body of cadmium sulfide with a layer of cuprous sulfide thereon and then an additional layer containing metal on the cuprous sulfide layer.

The specific improvements claimed for this particular cell design are increased efficiency and increased stability at high temperatures. Such a cell design is shown in Figure 32 in sectional and perspective views.

The support body **1** is comprised, for example, of plastic and is coated on the upper surface side provided for the accommodation of the semiconductor body with a layer **2** of silver or another metal of good conductivity. On to the plastic support body is evaporated a cadmium sulfide layer **3** which is about 20 to 80 $\mu$m thick. This cadmium sulfide layer is of the N-type conductivity and has for example a specific resistance in the region between 1 and 100 ohm-cm.

**Figure 32: Cadmium Sulfide-Copper Sulfide-Copper Solar Cell Design**

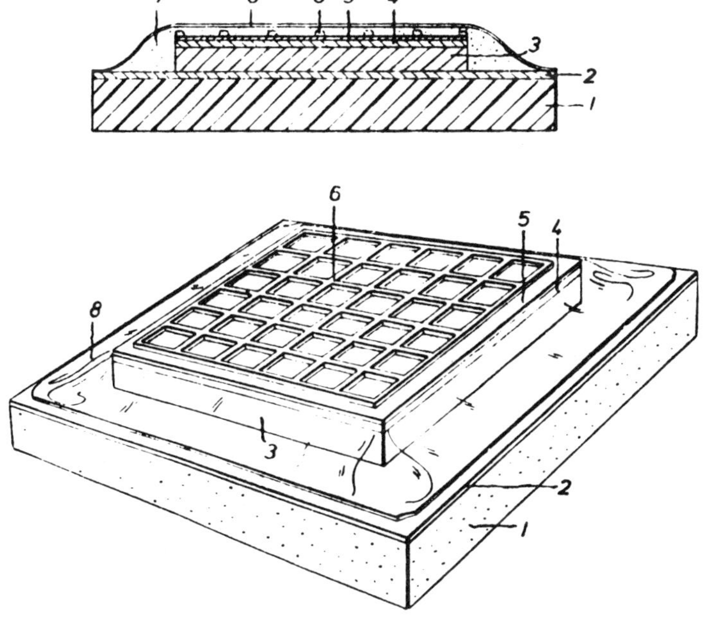


Source: U.S. Patent 3,888,697

A copper sulfide layer is to be produced on the free upper surface of the CdS layer and this layer is for example 500 to 2000 A thick. This is effected in an advantageous manner in that the cadmium sulfide is dipped after a prior surface cleaning, into a solution containing positive copper ions. Such a solution comprises for example copper chloride, ammonium chloride and a reducing agent.

In this solution, which, for example, is warmed to a temperature of 90°C, the CdS layer is transformed on the surface into copper sulfide. The stoichiometric composition of this copper sulfide layer is decisive for the photoelectric efficiency of the cell. In this case efforts are made that the copper sulfide layer comprise, as far as possible, large regions of $Cu_2S$ and the other possible copper-sulfur compounds be extensively suppressed. The coppe sulfide layer is given the reference numeral **4**.

Now in the further course of the manufacturing process, the photocell can be subjected first to a tempering by which a photoelectric efficiency corresponding to the typical prior art in the order of magnitude of 5 to 6% is achieved. The temperature necessary in this treatment and the duration of its effect on the photocell is dependent on the stoichiometry and the structure of the copper sulfide layer. The photocells are for example tempered at a temperature of approximately 200°C for 10 minutes.

However, this intermediate tempering can also be given up when carrying out the manufacturing method. A further upper surface layer **5** containing copper is applied to the copper sulfide layer. This layer consists preferably of metallic copper with a thickness of approximately 20 to 30 A. The thicker this layer is, the longer the semiconductor arrangement must be tempered in a subsequent process step so that an optimal efficiency of the cell is set up. The necessary tempering time can be easily determined by experiment.

It has been shown that, with a copper layer thickness of approximately 20 to 30 A, photocells with good electrical properties can be produced, wherein the optimal photoelectric efficiency is set up after tempering of about 15 minutes' duration at approximately 200°C. The tempering is effected for example in air or in vacuo. The efficiency could be increased for example from 5.5 to 7.3%. The reason for this should be sought in the improvement of the stoichiometric composition of the copper sulfide layer.

The additional layer **5** containing copper can be evaporated on to the copper sulfide upper surface in vacuo or deposited chemically or electrolytically. Furthermore a layer enriched with copper can be produced on the copper sulfide layer in that the arrangement is tempered in a reducing atmosphere. This is effected for example by treating the copper sulfide layer in a discharge field in an atmosphere of hydrogen.

To finish the photocell a contact grating or grid-like electrode **6**, which comprises, for example, gold, is applied to the upper surface layer **5** containing copper and is glued thereto with the help of a suitable adhesive. After that, a transparent foil **8** coated with a transparent adhesive **7** is laid on the semiconductor arrangement and pressed together with this. This pressing process preferably takes place at a temperature at which the adhesive becomes plastic so that on cooling and hardening of the adhesive, the semiconductor body, metal electrode and foil stick together firmly and intimately. The transparent and glued-on foil **8** is preferably stuck with its edge to the upper surface of the support body **1** so that the thin-layer photocell is protected on all sides against outer influences.

Whereas known cells are stable only up to about 70°C, cells designed by this process remain stable to over 100°C and show only a slight reversible decrease in the short-circuit current. From this it follows that the photocells can be used also in space conditions to convert solar energy into electrical energy, since the useful life is considerably increased and the time degradation of the cell is greatly reduced.

A cell developed by *J.F. Jordan and C. Lampkin; U.S. Patent 3,902,920; Sept. 2, 1975; assigned to D.H. Baldwin Company* is a large-area photovoltaic cell comprising a layer of multicrystalline cadmium sulfide, about 1 to 2 microns thick, formed by simultaneously spraying two suitably selected compounds on a uniformly heated plate of Nesa glass. The process then consists of forming a coating of $Cu_2S$ by spraying two suitable compounds over the cadmium sulfide layer while the latter is heated, to form a photovoltaic heterojunction, applying thereover a layer of $CuSO_4$, and applying electrodes of Cu and Zn, respectively, to separated areas of the layer of $CuSO_4$, and heating the cell to form a cuprous oxide rectifying junction under the copper electrode by reaction of the Cu electrode with the $CuSO_4$, while diffusing the zinc through the body of the cell.

The diffusion of the zinc provides a negative electrode coplanar with the positive copper electrode, eliminating any need for introducing mechanically complex provision for making a connection to the Nesa glass, while the use of a rectifying positive electrode enables use of a layer of CdS only 1 to 2 microns thick, rather than the usual 20 microns, despite the fact that such thin layers tend to have pinholes, which in the prior art render the cells inoperative but in this case do not.

Figure 33 shows the process for forming a cadmium sulfide layer on glass. The glass plate **10** is sprayed while the plate is floating in a bath **20** of molten metal, specifically tin. The glass plate is not wet by the tin, so that when the glass plate is removed from the molten tin bath, after it is sprayed, the underside of the plate is clean, or easily cleaned. The spray is provided via an oscillating nozzle **21**, which repeatedly retraces a planar path designed uniformly to cover the plate with spray. The spray is a true water solution of cadmium chloride and thiourea. As the fine droplets of the spray contact the hot surface of the glass plate, the water is heated to vaporization and the dissolved material is deposited on the plate, forming CdS, plus volatile materials, and the CdS, if it has nucleating areas available, grows as small crystals.

**Figure 33: Apparatus for Producing Cadmium Sulfide Layer on Glass Substrate**

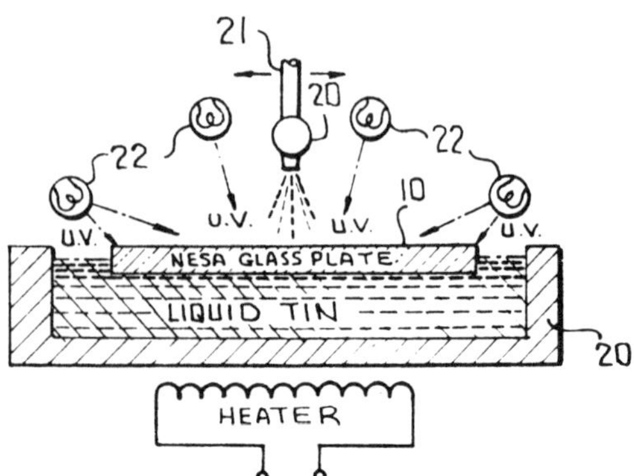


Source: U.S. Patent 3,902,920

The nucleating areas are provided by the tin oxide, and if the spray is sufficiently uniform and sufficiently slow, and if the temperature of the glass surface is adequately high and uniform, crystal growth is uniform and all the crystals have nearly the same spatial inclinations, so that a uniform layer of nearly identical microcrystals exists. It has been found that irradiating the crystals, as they grow, with high-intensity UV light, from source **22**, assists in the crystal growing process and produces a higher yield of near-perfect layers than is otherwise the case.

A process developed by *F.A. Shirland; U.S. Patent 3,975,211; Aug. 17, 1976; assigned to Westinghouse Electric Corporation* is one in which a $Cu_2S$ thin film is epitaxially formed on a CdS film by vacuum deposition in a heterojunction forming relationship. By a first method a $Cu_2S$ layer on the order of 0.01 micron in thickness is formed on a CdS polycrystalline thin film by dipping in a solution of cuprous ions. The CdS film itself is less than 5 microns thick and rests on a conductive substrate.

After the dipping step the $Cu_2S$ film is increased to a thickness on the order of 0.1 micron by vapor evaporation of an additional amount of $Cu_2S$. By a second method both the CdS and $Cu_2S$ are entirely vapor-deposited on a substrate to achieve approximately the same final structure as the first method.

A device developed by *J.A. Duisman; U.S. Patent 4,143,235; March 6, 1979; assigned to Chevron Research Company* is one consisting of a cadmium sulfide photovoltaic cell of improved efficiency comprising a barrier layer and cadmium sulfide-containing bilayer, the bilayer being formed by depositing at a first temperature an initial layer of cadmium sulfide in interfacial contact with the substrate and then depositing a subsequent layer of cadmium sulfide at a second temperature which is at least 20°C below the first temperature. Such a device is shown in sectional elevation in Figure 34.

**Figure 34: Chevron Research Design of Cadmium Sulfide Solar Cell**

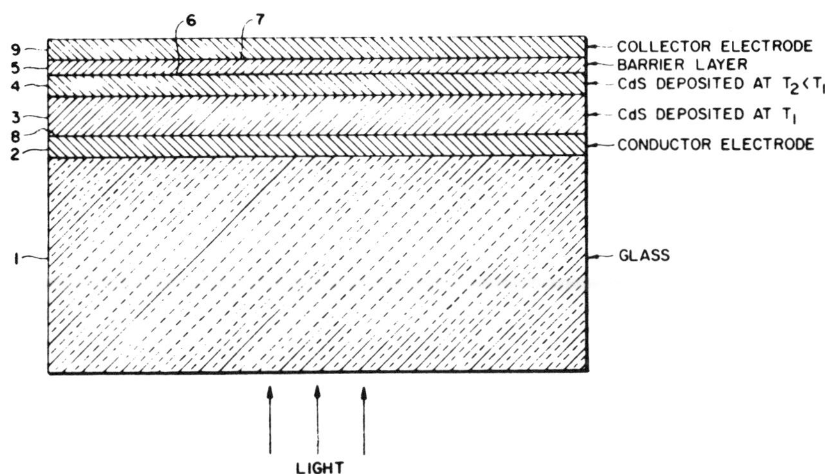


Source: U.S. Patent 4,143,235

## GROUP III-V COMPOUNDS (Ga, In ARSENIDES AND PHOSPHIDES)

Gallium arsenide has a higher theoretical PV efficiency than silicon because its excitation threshold is better matched to the energy in the sun's spectrum. GaAs cells can be used to achieve high efficiencies in intense radiation, particularly if they are covered with a layer of $Ga_xAl_{1-x}As$ which has the effect of reducing surface and contact losses. Efficiencies as high as 24.5% have been measured for such devices operating in sunlight concentrated 180 times (3). In addition, both theory and experiment show that the efficiency of GaAs cells is reduced less by high temperature than is the efficiency of silicon cells.

Although it has been recognized as desirable to utilize the semiconductor material GaAs for a converter of solar energy to electrical power, the theoretical efficiency predicted for this device has not been obtainable because of difficulty of providing a P-N junction sufficiently close to a surface receptive of the solar radiation. It has been determined theoretically that a GaAs solar cell can have approximately 25% conversion efficiency. Although this theoretical efficiency is greater than the theoretical efficiency for a silicon solar cell, e.g., approximately 15%, the latter has been developed considerably for practical uses because the problems involved have generally been satisfactorily addressed.

In a technique developed by *J.M. Woodall; U.S. Patent 3,675,026; July 4, 1972; assigned to IBM,* however, this problem is said to have been overcome, permitting attainment of high efficiency in a GaAs cell.

The basis of this development is the finding that for maximum power output with GaAs semiconductor material, it is necessary that the P-N junction be as close as possible to the radiation receptive optical surface because the lifetime for diffusion of electron-hole pairs which are created by absorption of photons is relatively small compared to the comparable lifetime for diffusion of electron-hole pairs for Si. The window, e.g., $Ga_{1-x}Al_xAs$, should be effectively transparent to permit the desired photons to reach the P-N junction; it should have sufficient electrical conductivity to obtain the required current; and it should also provide a barrier for the cosmic rays in the solar radiation which cause radiation damage to the solar cell.

Figure 35 shows a suitable form of apparatus for the liquid phase epitaxial growth of a $Ga_{1-x}Al_xAs$ window layer on a GaAs substrate for a solar cell.

Quartz chamber **110** is provided within which the preparation of the compound is obtained. Orifice **112** is the inlet for a high-purity inert gas used during the steps of the procedure. After having served its intended purpose, the inert gas introduced via the orifice exits from the chamber via orifice **114**. A crucible **116** of $Al_2O_3$ is established within the chamber; the components of the desired window, e.g., for $Ga_{1-x}Al_xAs$, the components of the ternary compound Ga, Al, and As are established as a liquid in equilibrium at a given temperature in the crucible. The heat source whereby the liquid **118** is raised in temperature and the heat sink whereby the temperature of the liquid is lowered are not shown.

For convenience, a vertical tubular electric furnace with temperature control can be used for both the heat source and heat sink, with the ambient environment providing sufficient temperature for cooling. Quartz tube **120** is introduced into the chamber via orifice **122**. Removable cap **124** is placed on top of its tube. The quartz tube is connected by coupling **125** to a graphite piece **126** which has a tube portion **128** therein connecting to the tube portion of tube **120**. Orifice **130** of tube portion **128** exits just above the surface of liquid **118**. The graphite portion is machined to have a low extending portion **132** upon which a solid substrate, e.g., single crystalline GaAs layer **134**, is affixed by the thrust of screw **136**.

A crucible is selected which does not react with the components of the liquid at the temperature of growth of the crystalline compound according to the practice of the process. A suitable pressure of the inert gas **111** introduced at orifice **112** is maintained in the chamber to inhibit vapor formation of highly volatile components in the liquid and further to preclude any undesirable reactions in the liquid with contaminants that might otherwise be introduced into the chamber. Illustrative inert gases suitable for the gas are argon and helium. Another gas which is inert for the liquid **118** consisting of the components Ga, Al and As, is high-purity forming gas, e.g., 10% $H_2$ + 90% $N_2$.

In an illustrative operation for growing a layer of $Ga_{1-x}Al_xAs$, the crucible is loaded with the components Ga, Al and As for a suitable liquid in equilibrium at a given temperature, e.g., 20 grams Ga, 0.150 gram Al, 3.0 grams pure GaAs, and 0.040 gram of determining P-type dopant Zn.

Figure 35: Apparatus for Growth of Window Layer on GaAs Substrate by Liquid Phase Epitaxy in Solar Cell Manufacture

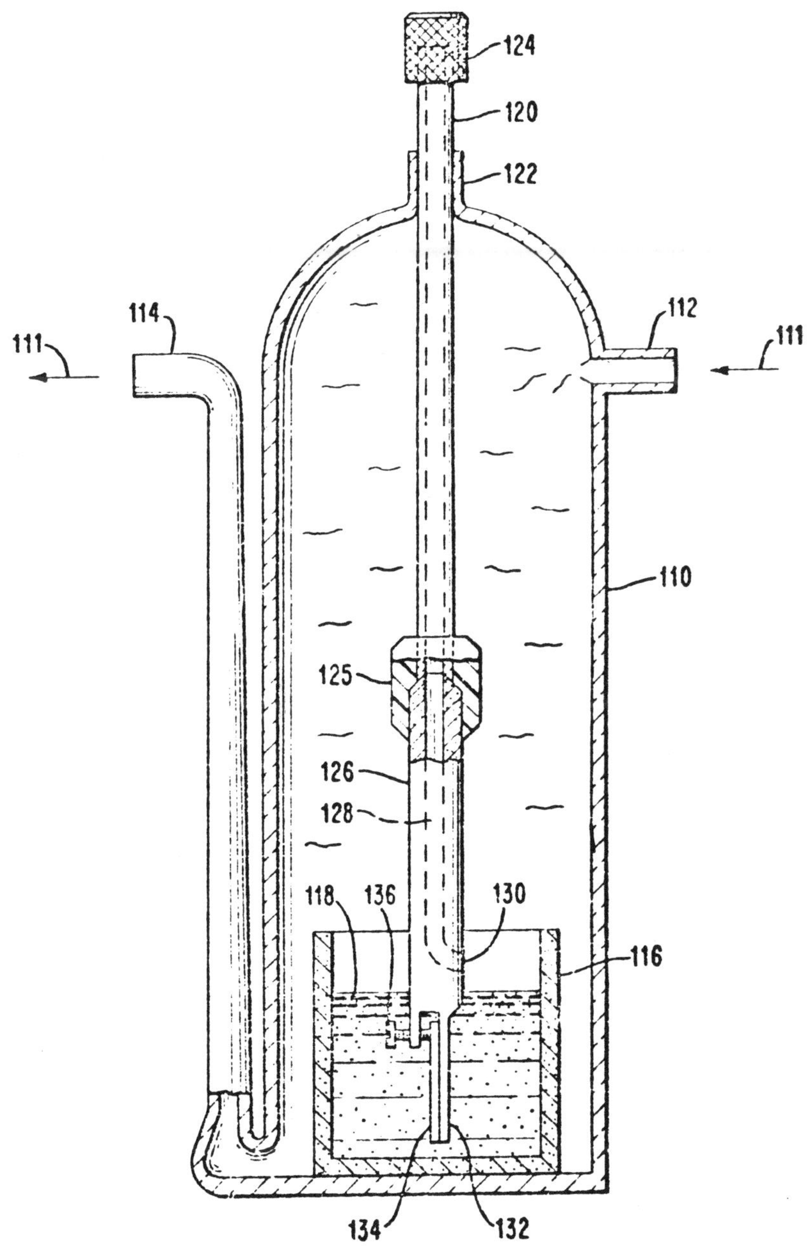


Source: U.S. Patent 3,675,026

The crucible is introduced in the chamber through a port, not shown. A substrate **134** of N-type GaAs doped with Si, with the surface main face perpendicular to the <100> crystalline direction is affixed to the extending portion **132** and the composite structure of tube **120**, graphite portion **126** and GaAs substrate **134** is established in the chamber above the liquid. The chamber is flushed with inert gas and a suitable pressure maintained. The entire chamber is placed into an isothermal furnace maintained at a given temperature for equilibrium of the liquid, e.g., 950°C. A suitable time is permitted to elapse so that the liquid achieves equilibrium at the given isothermal temperature, e.g., 30 minutes. Substrate **134** is then immersed in the liquid and a period of time is allowed to elapse so that the substrate achieves equilibrium with the liquid at the operational temperature.

Conveniently, temperature of the liquid can be lowered slightly before introducing the substrate, e.g., lowering by 20°C, and after the substrate has been introduced into the liquid the temperature is raised somewhat, e.g., by 10°C, so that the temperature at which the initiation of the growth is to occur is at a preselected temperature, e.g., 950°C. The raising of the temperature by 10°C also results in good wetting of the melt to the GaAs substrate **134**. For a uniform composition of a grown layer of $Ga_{1-x}Al_xAs$ on the substrate, a particular cooling rate is selected, e.g., from 0.5°C per minute, and the cooling at this rate is continued to 930°C until a required layer of thickness of the crystalline compound is obtained. As a result of this growth schedule, the zinc diffuses into the GaAs in such a manner to form a P-N junction, 1 to 2 microns from the GaAs optical surface. The depth of the P-N junction can be further increased by subsequent heat treatment if required for certain operational circumstances.

This structure can now be formed into devices by forming square or rectangular pieces by conventional cleaving or sawing procedures. An individual piece is then electrically contacted by alloying ohmic contacts at the device surfaces. The ohmic contact on the $Ga_{1-x}Al_xAs$ window surface should be made as small as possible to allow the maximum amount of incident radiation to penetrate the device. Suitable ohmic contact materials are Sn-Au alloys for the N-type GaAs and Au or Au-In alloys for the P-type $Ga_{1-x}Al_xAs$ window.

This process provides a reliable method for forming a P-N junction in a wafer of GaAs which is close to the wafer surface and also forms a protective window which is substantially transparent to solar radiation. In the practice of this process, a P-type layer of $Ga_{1-x}Al_xAs$ with a band gap of approximately 2.1 eV is grown onto an N-type GaAs substrate with a band gap of approximately 1.4 eV.

During fabrication of the device, the P-type dopant in the $Ga_{1-x}Al_xAs$ layer diffuses into the GaAs substrate thus forming a P-N junction in the GaAs near the growth of interface. Illustratively, such structures produce 30% more power from solar radiation than a standard Si cell of the same dimensions. Since the window layer of $Ga_{1-x}Al_xAs$ of a solar cell according to this process acts as a shield for undesired radiation and an electrical contact path to the GaAs P-N junction, the weight is less than for a comparable Si device which requires separate components to serve these functions.

A device developed by *W.D. Johnston, Jr.; U.S. Patent 3,982,265; Sept. 21, 1976; assigned to Bell Telephone Laboratories, Inc.* is a high-efficiency solar cell having N-type aluminum arsenide grown on a P-type gallium arsenide substrate and protected by a layer of anodic oxide. The aluminum arsenide is deposited by vapor phase epitaxy by reacting high-purity arsine, hydrogen chloride and aluminum at approximately 1000°C in an all-alumina reactor tube system. The aluminum arsenide layer is protected from deterioration by first anodizing it in pure water and phosphoric acid at pH 2.0 with a current density of 2-8 $mA/cm^2$ at room temperature.

Second, the anodic oxide so formed is annealed at about 450°C for at least 20 minutes in dry nitrogen. The oxide layer also acts as an antireflective coating. A portion of the oxide layer is etched away to expose a region of the aluminum arsenide to which an electrical contact is applied. The other contact is made to the substrate. Such a solar cell design is shown in Figure 36.

Gallium arsenide P-type substrate layer **2** underlies N-type aluminum arsenide layer **4** deposited adjacent thereto. A heterojunction **3** is formed at the metallurgical interface between the semiconductor layers. The N-type carrier concentration in layer **4** suitably exceeds about half the carrier concentration in the P-type substrate and has a value of $N > 5 \times 10^{17}$ carriers per cubic centimeter for a quantum efficiency approaching 100%. An anodic oxide coating layer **5** is produced on the aluminum arsenide layer according to a method such as that described below so that the layer is protected from deterioration for long periods of time. Since the anodic oxide of aluminum arsenide has a refractive index (1.8) which is approximately the square root of the refractive index (3.3) of aluminum arsenide itself, the oxide is advantageously utilized as an antireflective coating as well.

**Figure 36: Bell Lab Solar Cell Device Having Aluminum Arsenide Grown on Gallium Arsenide Substrate**

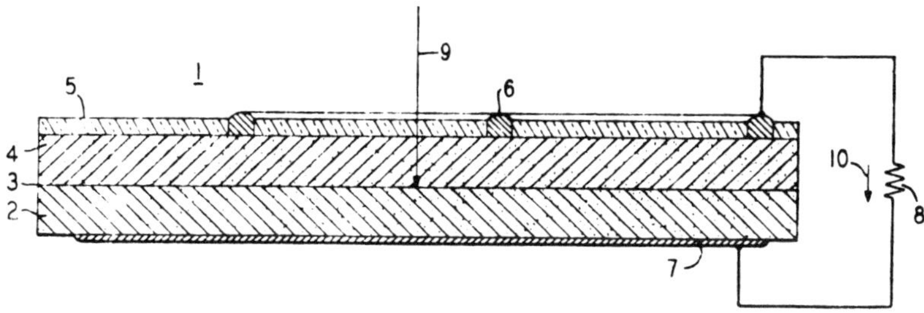


Source: U.S. Patent 3,982,265

Electrical contacts **6** and **7** attached to layers **4** and **2**, respectively, are connected to a load resistor **8** or other suitable power-consuming device. Thus, when incident light indicated by ray **9** impinges upon solar cell **1**, it passes through the transparent anodic oxide layer and high band gap aluminum arsenide layer **4** to be absorbed near the heterojunction **3**, producing an electric current **10** in the load **8**.

The N-type aluminum arsenide (AlAs) layer is vapor phase epitaxially grown on a 400-micron thick, zinc-doped P-type gallium arsenide substrate wafer **2** by a hydrogen chloride (HCl) transport process similar to that described in the paper "Vapor Growth and Properties of AlAs" by M. Ettenberg, et al., *Journal of the Electrochemical Society,* Volume 118, No. 8, August 1971, pp 1355-1358.

High-purity (99.999%) aluminum is held in alumina boats in a horizontal alumina tube 4 feet long, $^{11}/_{16}$ inch inside diameter and 1 inch outside diameter, through which 0.5 to 2% anhydrous HCl by volume in hydrogen carrier gas passes at 1000° to 1100°C at about 450 $cm^3$/min. The aluminum is taken up and carried as a chloride through perforations in a closed end of the tube having 18 holes of $^1/_{16}$ inch diameter. The stream formed thereby reacts with 1 to 2% arsine ($AsH_3$) in hydrogen carrier gas which passes thereto at about 500 cubic centimeters per minute along the outside of the alumina tube inside a larger 6-foot long alumina tube of $2^7/_{16}$ inch inside diameter. Gas-tight 99.8% pure alumina components are used throughout.

The aluminum arsenide which is formed is vapor phase epitaxially grown at constant temperature to a thickness of 3 to 40 microns to form the layer **4** on a cleaned crystallographic (100)-oriented gallium arsenide substrate wafer **2** illustratively having a carrier concentration P = $1.5 \times 10^{18}$ and having its surface normal oriented at 0°-10° from the tube axis. Layers of approximately 20 microns thickness can be grown in one hour at a substrate temperature of 1030°C. The aluminum arsenide layers are found to be N-type with carrier concentrations varying from about $2 \times 10^{17}$ at a substrate temperature of 980°C to about $5 \times 10^{18}$ at 1100°C, a suitable value of $8 \times 10^{17}$ occurring at 1030°C.

It is important that contaminants be eliminated from the system. High-purity reactants and use of high-purity nonreacting structural components in the reaction zone are essential. A structure essentially devoid of fused silica and incorporating the all-alumina tubes described above may be advantageously employed in this regard. The alumina tubes are supported in a manner allowing for thermal expansion so that mechanical stresses and consequent breakage are avoided.

After vapor growth, the grown device having an exposed aluminum arsenide or other aluminum-Group V component layer is removed from the reaction tube and placed in an anodization bath so a protective anodic oxide coating may be applied as illustrated by step **21** of the method depicted in Figure 37. The bath of water is adjusted in pH with phosphoric acid ($H_3PO_4$). A pH in the range of 1 to 4 may be used; a pH of 1.5 to 2.5 is even better for AlAs; and a pH of about 2 is excellent for AlAs. The best value of pH in the 1 to 4 range depends on the composition of the aluminum-Group V component compound to be anodized, however. Uniformity of oxide growth on aluminum arsenide is found to be degraded if the bath is impure, particularly with respect to halide and nitrate ion concentrations, so that the use of a multifiltered deionized type of water is preferable.

The anodization apparatus may be of a type familiar to the art wherein a source of electricity including a compliance voltage source and a current-limiting resistance is connected with its anode connected to the device to be anodized and with the cathode placed nearby in the electrolyte bath. See, for instance, B. Schwartz, et al; U.S. Patent 3,882,000; May 6, 1975. Because aluminum arsenide is far more reactive in water than the gallium compounds discussed in the patent, the electric current should be caused to flow preferably within seconds after the aluminum arsenide device is placed in the electrolyte bath. The bath temperature may be any temperature up to about 70°C for which the bath is liquid. Room temperature may conveniently be used.

It is found that the anodic oxide thickness obtained is about 10 A/V of compliance voltage available at the source of electric current. Thus, if a quarter-wavelength antireflective coating of about 900 A is to be grown, a compliance voltage of approximately 90 volts is suitable. Layer quality is excellent when the current density employed is 2 to 8 $mA/cm^2$ of surface to be anodized. Growth of oxide layers up to 2500 A occurs within minutes. The oxide so formed is believed to be a hydrated glass involving a mixture of oxides of aluminum and arsenic ($Al_2O_3$ and $As_2O_3$).

After the anodically oxidized device is removed from the electrolyte bath, the anodic oxide coating layer is annealed as illustrated by step **22** of Figure 37, during which step the water of hydration is believed to be driven off from the glass. The annealing step is effective at temperatures higher than about 350°C up to a limit which is approximately 520°C. At the upper temperature limit devitrification by loss of arsenic or other Group V component and consequent deterioration of the oxide coating on aluminum arsenide and other aluminum-Group V compounds may become evident.

An annealing period of at least 20 minutes is suitable at 400°-450°C and two hours is ample at all temperatures in the effective range, with a greater length of time doing little harm. An atmosphere of dry nitrogen may be provided, although any other atmosphere is suitable which is inert relative to the anodic oxide and essentially free of hydrogen, water vapor, or other hydrogen-containing compounds.

The anodic oxide resists deterioration in contact with photoresists, waxes, glycol phthalate and shellac-based cements, such as are used in device fabrication and which deteriorate unprotected aluminum arsenide in varying degrees.

Figure 37: Process for Anodic Oxide Coating of Device of Figure 36

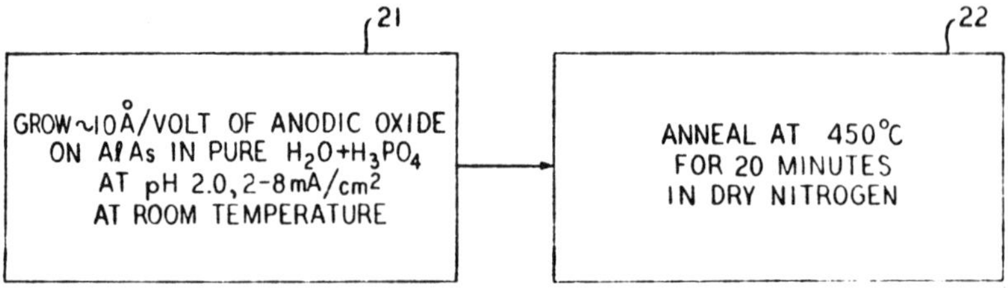

Source: U.S. Patent 3,982,265

Standard photolithographic techniques are employed in defining an area on the anodic oxide layer **5** of Figure 36 and etching the oxide to expose the aluminum arsenide layer so that a finger contact or other electrical contact may be formed thereon. An etching dip in 2 to 10% hydrofluoric acid (HF) in water solution for approximately 2 seconds is suitable for removing the oxide layer in the contact area without removing the underlying portion of the aluminum arsenide layer as well.

Ohmic contact **6** is gold, nickel-tin, gold corresponding to the N-type conductivity of layer **4**; and contact **7** is gold, zinc, gold corresponding to the P-type conductivity of layer **2**. Contacts **6** and **7** are deposited by electroplating, then sintered at 450°C in dry nitrogen for a few minutes, and connected with indium solder to gold wire leads.

A cell design developed by *R.J. Stirn; U.S. Patent 4,053,918; Oct. 11, 1977; assigned to U.S. National Aeronautics and Space Administration* is a Schottky barrier solar cell consisting of a layer of wide band gap semiconductor material such as AlGaAs on which a very thin film of semitransparent metal is deposited to form a Schottky barrier. The layer of the wide band gap semiconductor material is on top of a layer of narrower band gap semiconductor material, to which one of the cell's contacts may be attached directly or through a substrate. The cell's other contact is a grid structure which is deposited on the thin metal film.

A cell design developed by *W.P. Rahilly: U.S. Patent 4,070,205; Jan. 24, 1978; assigned to the U.S. Secretary of the Air Force* is an improved gallium arsenide solar cell provided by forming a $P^+$ layer on top of a wafer of plural vertical P-N junction eutectic gallium arsenide crystal by liquid phase epitaxial growth of P-doped GaAs followed by liquid phase epitaxial growth at $Al_xAsGa_{1-x}$ on the surface of the vertical P-N junction substrate. The deposited GaAs layer with P dopant and the $Al_xAsGa_{1-x}$ layer forms horizontal P-N junctions with the N-type vertical regions. An $N^+$ region is formed on the solar cell backside by ion implantation of an N dopant followed by a pulse electron beam current of the implanted region.

Another cell design proposed by *W.P. Rahilly; U.S. Patent 4,116,717; Sept. 26, 1978; assigned to U.S. Secretary of the Air Force* is an improved gallium arsenide solar cell provided by ion implanting both the top and bottom of a plural vertical P-N junction eutectic gallium arsenide cell body to obtain an electrical drift field, with multiple ion implants progressively larger in dose and progressively lower in implant energies to provide a P-type ion implanted top layer having a common connection to all P regions of the cell body and an N-type ion implanted bottom layer having a common connection to all N regions of the cell body. The implanted regions of the cell are pulsed electron beam annealed at room temperature.

A technique developed by *R.L. Moon; U.S. Patent 4,126,930; Nov. 28, 1978; assigned to Varian Associates, Inc.* is one in which aluminum gallium arsenide is used as a transparent, conducting contact layer on the exposed surface of a gallium arsenide photovoltaic cell. Increased conductivity for the high current generated when concentrated solar radiation strikes the cell, is provided by doping the AlGaAs layer with magnesium. During the formation of the layer, Mg diffuses into the gallium arsenide to form a P-type layer and a P-N junction. Figure 38 shows a suitable form of apparatus for producing the AlGaAs device.

**Figure 38: Epitaxial Growth Apparatus for AlGaAs Manufacture**

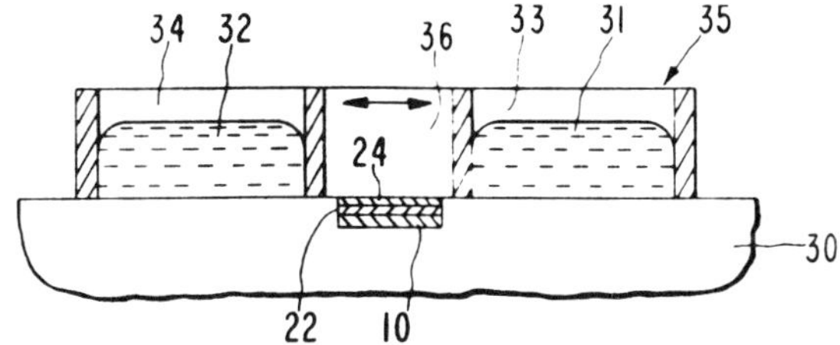


Source: U.S. Patent 4,126,930

Substrate **10**, in a recess in inert base **30**, is positioned in an empty bin **36**. Materials **31** and **32** for growing the epitaxial layers are in open-bottomed bins **33** and **34** in inert holder **35** on base **30**. The apparatus is heated in an inert atmosphere in a surrounding furnace to melt **31**, the bin **33** slid over the substrate and the temperature lowered to grow liquid layer **22** until desired thickness is achieved. Second layer **24** is grown similarly with material **32** in bin **34**.

*L.M. Fraas, K.R. Zanio and R.C. Knechtli; U.S. Patent 4,128,733; Dec. 5, 1978; assigned to Hughes Aircraft Co.* produce a gallium aluminum arsenide-gallium arsenide-germanium solar cell where deposition of a gallium aluminum arsenide layer establishes a first P-N junction in the GaAs of one band gap energy on one side of a gallium arsenide substrate, and deposition of germanium establishes a second P-N junction in Ge of a different band gap energy on the other side. The P-N junctions are responsive respectively to different wavelength ranges of solar energy to thus enhance the power output capability of a single wafer (substrate) solar cell. Utilization of the Group IV element germanium, as contrasted to compound semiconductors, simplifies the process control requirements relative to known prior art compound semiconductor processes, and germanium also provides a good crystal lattice match with gallium arsenide and thereby maximizes process yields. This latter feature also minimizes losses caused by the crystal defects associated with the interface between two semiconductors.

Figure 39 is a schematic process diagram of the chemical vapor deposition (CVD) system used to deposit the epitaxial layers in the manufacture of such a device and Figure 40 is an enlarged view of one of the pyrolytic chambers used in the CVD system.

As will be seen, dopant gas flow lines of the CVD system are connected for the pyrolytic deposition of P- and N-type epitaxial layers of germanium on chosen substrates, but this CVD system can, if desired, be modified by those skilled in the art for the pyrolytic deposition of either gallium arsenide or gallium aluminum arsenide as well.

The system includes a main vacuum deposition chamber **90** which is connected to a turbomolecular pump (TMP) **92** by means of a suitable gate valve **94**. The pump is capable of pulling vacuums on the chamber of the order of $10^{-10}$ torr. A gas flow line **96** also extends as shown into the main vacuum chamber and is connected through three gas flow controllers **98**, **100**, and **102**, respectively, to N- and P-type dopant sources **104** and **106** and to a supply **108** of germane, $GeH_4$.

The gaseous compound germane, $GeH_4$, when combined with the appropriate P- or N-type dopant and heated in the chamber to a suitable elevated temperature such as 500°C or greater, decomposes in the pyrolysis decomposition chamber **118** in which the substrate wafers are located to produce epitaxial layers of germanium. Thus, P- and N-type epitaxial layers of germanium may be deposited on selected substrates with controlled levels of impurity doping, and this epitaxial deposition of Ge by the pyrolytic decomposition of germane is defined by the following equation:

$$GeH_4\,(gas) \xrightarrow[\Delta]{} Ge(solid) + 2H_2\,(gas)$$

A vertical coupling **110** extends as shown between a rotary feedthrough member **112** and a rotary support plate **114**. The rotary support plate rests on three bearings **109**. Individual CVD pyrolysis chambers **116** and **118** are suitably located and secured at the periphery of the rotary support plate, and each CVD chamber contains multiple horizontal compartments **120** therein for receiving a plurality of single-wafer solar cell structures which are ready for epitaxial deposition. The two CVD pyrolysis chambers are positioned adjacent to controlled heater plates **122** and **124**, respectively, and these heater plates are used to bring the CVD pyrolytic deposition temperature up to a desired elevated level, typically

Figure 39: Chemical Vapor Deposition Apparatus for Epitaxial Layer Manufacture

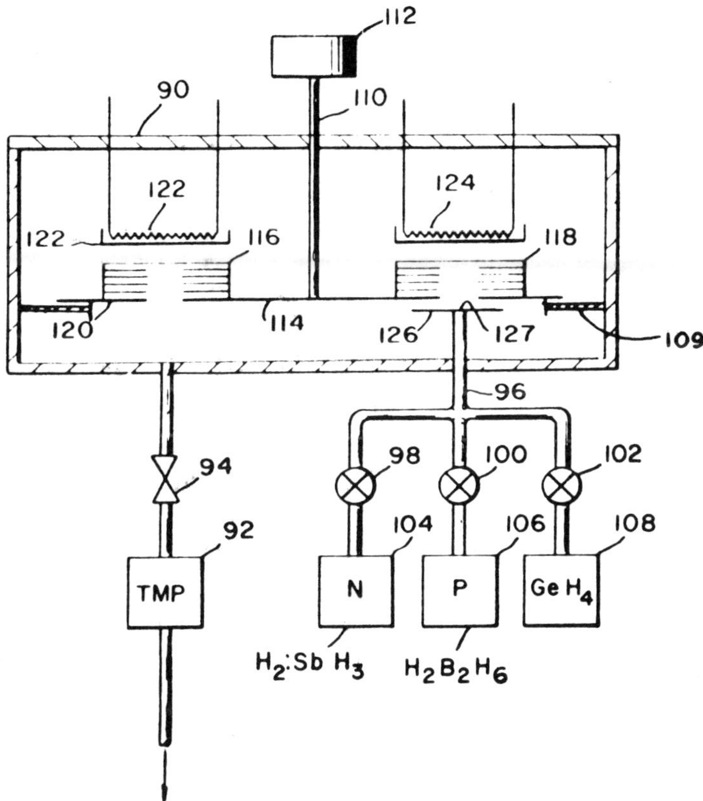


Figure 40: Enlarged View of Pyrolytic Reaction Chamber of Figure 39

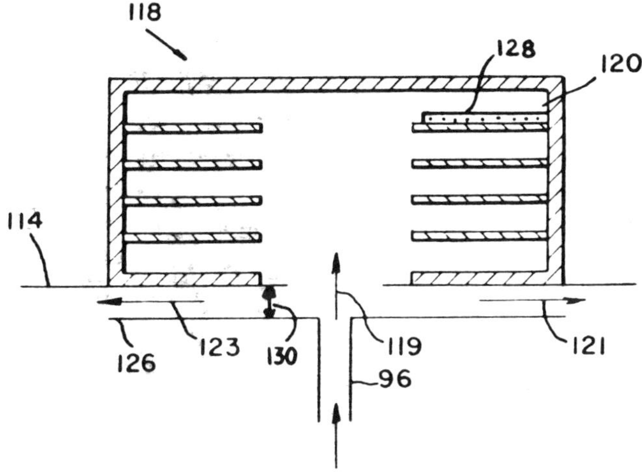


Source: U.S. Patent 4,128,733

on the order of about 525°C for a corresponding pyrolysis chamber **118** total pressure of about $5 \times 10^{-3}$ torr.

This total reactant pressure of $5 \times 10^{-3}$ torr in the chamber **118** can be compared with the residual (impurity) gas pressure in the chamber of about $10^{-8}$ torr and this in turn will allow the epitaxial deposition of high-purity Ge films on the semiconductor substrate. The use of the pyrolytic decomposition chamber **118** within the larger vacuum chamber **90**, with its controlled vacuum condition, i.e., $10^{-8}$ torr residual gas pressure, enables the creation of sufficiently high reactant gas pressures within the chamber **118** ($5 \times 10^{-3}$ torr) to produce acceptably high Ge film growth rates and sufficiently low to produce highly uniform Ge films on the closely stacked semiconductor substrates.

For these above parameters of temperature and pressure, the epitaxial growth rate of germanium on gallium arsenide substrates will be about 2.0 micrometers per hour. However, higher pressures of $GeH_4$ can be used to obtain faster growth rates for Ge.

Referring to Figure 40, as well as Figure 39, the appropriately doped P- or N-type germane gas passes through the main gas flow line **96** and into the CVD pyrolysis chamber **118**. The $GeH_4$ gas flow will be in the direction of arrows **119**, **121** and **123**, and the gas traveling the path indicated by the arrow **119** will enter all of the horizontal wafer mount compartments **120** of the pyrolytic decomposition chamber.

The appropriately doped germane gas will undergo thermal decomposition in these compartments to thereby epitaxially deposit monocrystalline layers of P- or N-type germanium on the stacked GaAs semiconductor wafers, and the gases within the decomposition chamber **118** which do not thermally decompose therein will eventually exit this chamber along the path of the horizontal arrows **121** and **123** as indicated. For a more detailed discussion of the decomposition of germane, $GeH_4$ per se, reference may be made to an article by S.A. Papasian et al, *Journal of the Electrochemical Society,* 115, 965 (1968) and to an article by M. Davis et al, *Journal of Applied Physics,* 27, 835 (1956).

For each molecular surface collision within the decomposition chamber **118**, there is a small probability that a $GeH_4$ molecule will react to form a Ge-adsorbed atom. Thus, to enhance this probability, exceedingly low pyrolysis chamber pressures are desirable. Relative to other prior vacuum pyrolysis systems such as the one disclosed, for example, by B.A. Joyce and R.R. Bradley, *Phil. Mag.* 14, 289 (1966), this is made possible herein by mounting the chamber **118** as shown on top of the rotary plate **114**.

The rotary plate is in turn closely spaced from the top plate **126** of the gas inlet line **96** by a small adjustable gap **130**. This small gap is typically about 1/16 inch and allows a high pressure ratio to be established between the decomposition chamber and the outer vacuum chamber **90** which surrounds it. Using this mounting arrangement, residual gas pressures in the chamber in the $10^{-8}$ torr range should be possible.

The $GeH_4$ vacuum process described herein offers several advantages over other prior techniques. Specifically, it has the advantage of permitting the use of a relatively low semiconductor substrate **128** temperature on the order of about 525°C, and this avoids the potential for the degradation of the GaAs/GaAlAs substrate and the probable occurrence of interdiffusion and plastic deformation between adjacent semiconductor layers. The possibility of vapor back-etching, which is present in prior methods which utilize halogens, is eliminated.

Additionally, since a carrier gas, e.g., $H_2$, is present only at low pressures ($\leqslant$100 millitorr), gas impurities in the system can be routinely monitored using a residual gas analyzer. This low carrier gas pressure also allows high wafer stacking densities and uniform coatings, as described for example by R.S. Rosles in *Solid State Technology,* 63, (April 1977). These are notable advantages over standard hydrogen transport processes which operate typically at 1 atmosphere. Furthermore, when this process is compared to conventional vacuum-evaporation methods, it has the further advantage that contaminants originating from a hot Ge evaporation source are eliminated.

For the positions of the two CVD chambers shown in Figure 39, the CVD chamber **116** is being preheated and will, upon completion of the epitaxial deposition of GaAs wafers in the CVD chamber **118,** be rotated counterclockwise as indicated to assume the rightside position in the chamber **90** between the plates **124** and **126**. This position will be maintained for a predetermined time and at a CVD deposition temperature and appropriate vacuum level within the chamber.

The N-type dopant in the impurity source **104** may advantageously be introduced into the main gas flow line **96** by the conventional hydrogen transport of one of the Group V hydrides, such as arsine, $AsH_3$, phosphine, $PH_3$ or stibine, $SbH_3$, into the vacuum chamber **90**. The P-type dopant in the impurity source **106** may be introduced into the main gas flow line **96** by the hydrogen transport of one of the Group III organometallics, such as $Ga(CH_3)_3$, $In(CH_3)_3$ or $Al(CH_3)_3$, or by the hydrogen transport of diborane. For each dopant source **104** and **106**, hydrogen gas may be used as a carrier gas to combine the required amount of P- or N-type impurity with the germane from the source **108**.

Finally, it should be noted that this pyrolysis process can be applied to the deposition of GaAs semiconductor layers simply by using $Ga(CH_3)_3$ and $AsH_3$ gases as the main source gases rather than $GeH_4$. Similarly, using appropriate metal organic vapors and hydride gases, a large variety of other semiconducting layers can be fabricated.

Various other modifications may be made in the above-described process. For example, the P- and N-type layers of the solar cells may be reversed, or the device geometry and ohmic contact locations may be changed. Additionally, the process described herein is not apparatus-limited by the particular film deposition equipment disclosed, and may instead be modified, for example, to utilize a continuous treatment wafer in-wafer out type of epitaxial deposition system.

Such a system might include, for example, a conveyor belt which is rigged to enter the left-hand wall of the vacuum chamber and carry a plurality of separate pyrolysis chambers thereon. These individual pyrolysis chambers would then be brought into a particular CVD reaction zone, heated to a desired elevated temperature, and then exposed to the reactant gas streams in much the same manner as described above. Thereafter, the conveyor belt or other equivalent substrate transport mechanism would carry the treated wafers out of the vacuum deposition chamber and into an auxiliary cooling chamber.

## MIXED SEMICONDUCTOR COMPOUNDS

According to *K.J. Bachmann, E. Buehler, J.L. Shay and S. Wagner; U.S. Patent 3,988,172; Oct. 26, 1976; assigned to Bell Telephone Laboratories, Inc.*, solar cells showing improved efficiency, amounting to about 14% for overall solar power conversion, are obtained by annealing InP/CdS solar cells in a slightly reducing atmosphere for about 15 minutes in a temperature range preferably from about 550° to about 600°C. In an annealing temperature range from 400° to 625°C, an inversely dependent adjustment of annealing time is found desirable. The atmosphere preferably comprises mainly a substantially inert component and typically comprises an $H_2$ + $N_2$ mixture, such as forming gas (15% $H_2$ + 85% $N_2$).

A solar cell of the type described is shown in Figure 41. It includes a P-type InP substrate **11**, an N-type CdS layer **13** on sustrate **11**, and an Au/Zn/Au contact **12** to the InP. Figure 41 also shows the cell in place in an annealing oven **21**.

P-type InP substrates for single crystal solar cells are grown by the liquid encapsulant technique. Cd or Zn is added as an acceptor impurity to achieve controlled P-type conductivity. The single crystalline ingot is oriented for cutting (111) wafers, which are then polished with a final Syton polish on one face. The P-type InP portion of the heterodiodes had ~2 Ω-cm resistivity. The back contact **12** to the InP side of the final diode can be applied before or after growing the N-type CdS layer **13**. When the contact is applied before, the Syton polished face is embedded in wax, and three films, i.e., Au, Zn, and Au are successively electrodeposited on the InP. After removal of the wax, and boiling first in acetone, then in methanol, the metallic contact is annealed for ~5 minutes at 475°C in forming gas.

When no contact is applied to the InP wafer before the CdS layer growth, the wafer is cleaned after the Syton polishing by boiling in acetone and then in methanol. The InP wafers, with or without contact, are now ready for CdS growth. They are immediately mounted, in a laminar flow hood (not shown), on a substrate holder which is then transferred to the CdS growth station. Syton-polished InP wafers which had been slightly etched in cold HCl before a final methanol rinse may also be used.

**Figure 41: Apparatus for Annealing Solar Cells Made of InP/CdS**

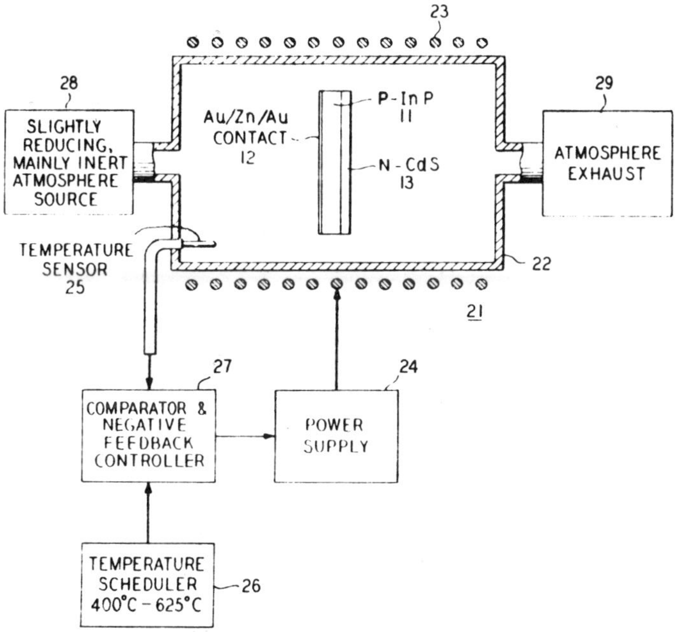

Source: U.S. Patent 3,988,172

The apparatus for CdS growth is a modification of earlier apparatus developed by D. Beecham, *Review of Scientific Instruments,* 41, 1654 (1970) and S. Wagner, *Applied Physics Letters,* 22, 351 (1973); *Journal of Applied Physics,* 45, 246 (1974). After mounting the InP substrate holder therein, the vacuum station used for the growth of CdS films is closed and evacuated with oil-free pumps to a pressure $<1 \times 10^{-5}$ torr. The coaxial isothermal CdS source (not shown), which contains elemental cadmium and elemental sulfur, is then heated to 350°C. Simultaneously the InP substrate holder is heated to 200°-250°C. During the heat-up, a shutter separates the source compartment from the substrate.

When source and substrate have reached the set temperature, the shutter is opened and a 5 to 10 μm thick N-CdS film is grown on the P-InP wafer. The rate of film growth is monitored with a quartz crystal oscillator. It is about 0.15 μm/min. When the desired film thickness has been reached the shutter is closed and both the source and the substrate heaters are turned off. The substrate holder is allowed to cool to room temperature, the vacuum station is filled with nitrogen, opened, and the samples are removed.

Samples **11, 13** without Au/Zn/Au contact **12** are waxed with the CdS layer **13** down. Then the contact **12** is made by electroplating, and is sintered as described earlier. On the CdS surface, no contact is illustrated because, in initial experiments, it was not deposited until the annealing step was completed.

Now as to the process for increasing the efficiency of heterojunction InP/CdS solar cells to 14%, this process in particular increases the efficiency of inadvertently inferior cells also to 14%. This latter unexpected advantage obviously increases the yield in the production of solar cells and thereby reduces costs. The effectiveness of this process may be of greatest value in some future production of thin-film InP/CdS solar cells by increasing the efficiency of continuous sections to a uniformly high value.

Briefly, in one successful experimental example, the process involved annealing an InP/CdS solar cell in "forming gas," (15% $H_2$ + 85% $N_2$) for 15 minutes at 600°C. This temperature is near optimum since lower temperatures are somewhat less effective, and temperatures of 650°C or more catastrophically destroy the cells. Forming gas or a variant thereof ($H_2$ + $N_2$) is essential since annealing in air at these temperatures severely degrades the cells's performance. Slightly lower temperatures and correspondingly longer annealing times should be as effective in increasing the efficiency, although less convenient. Nevertheless, for temperatures from 550° to 600°C, a 15-minute anneal is approximately sufficient to gain most of the improvement obtainable.

The efficiency of a given InP/CdS solar cell is directly related to its particular current-voltage characteristic as measured in the dark. The previously measured departures of the current density versus voltage curves for previously disclosed heterodiodes from an ideal shape can be attributed to impurities and defects at the interface between the CdS and the InP. Such defects and associated excess current in the heterojunction region could probably be eliminated by more sophisticated growth procedures in which either the InP substrate was vapor etched or an InP layer (P-type) was deposited, in situ just prior to the CdS growth; but such procedures may be disadvantageous for large-scale commercial use. Clearly, there is a variability to the previously measured departures from the ideal, but all such departures can be greatly reduced by the following processing.

The heterojunction device **11-13** is placed in the annealing oven **21**, which includes the air-tight chamber **22**, heating coil **23**, power supply **24**, temperature sensor **25**, temperature scheduler **26**, and comparator and negative feedback controller **27** acting on power supply **24**, all of conventional type. To an input port of the oven is connected the source **28**, which supplies a controlled stream of slightly reducing, mainly inert, gas atmosphere, which flows over and around the heterojunction device and out through the output port, which is connected to the atmosphere exhaust **29**.

Although the oven had a door (not shown), the same process is applicable to a continuous flow assembly line for the heterojunction devices if the assembly line flows into and out of the plane of the paper. The entire assembly line could be made in one air-tight chamber to preserve the integrity of the annealing process.

There is evidence that a slightly reducing i.e., nonoxidizing atmosphere is desirable to the annealing process, because preliminary tests of annealing in air (an oxidizing atmosphere) showed marked increases in CdS resistivity. Such CdS resistivities would lead to excessive internal power dissipation in the solar cells.

## OXIDE CERAMICS

A device developed by *P.S. Brody; U.S. Patent 3,855,004; Dec. 17, 1974; assigned to U.S. Secretary of the Army* is a ferroelectric ceramic photoelectric device capable of producing high voltages of at least 500 volts per linear inch of active material. The device is produced by first subjecting ferroelectric ceramic to a poling process whereby a high voltage is applied across electrodes attached to the edge of a wafer of the ceramic for a specific period of time. The amount of voltage produced by a particular specimen of the ceramic is controlled by the amount of voltage that is applied during the poling process and the length of time that it is applied. The ceramic utilized for producing this high voltage photoelectric device is selected from the group consisting of barium titanate and lead zirconate, and a solid solution of lead titanate and lead zirconate doped with lanthanum.

The device functions at high voltage because it is polycrystalline. A condition for operation is that the crystals therein be ferroelectric in the operating temperature range. Constant high voltage output of the device is obtained independent of illumination intensity. This is accomplished by connection of a high resistance element across the electrodes of the device. The photocurrent may also be made to vary linearly with the magnitude of the intensity of illumination. This is obtained by connecting a low resistance element across the electrodes of the device.

A device described by *A.M. Glass and D. von der Linde; U.S. Patent 3,975,632; Aug. 17, 1976; assigned to Bell Telephone Laboratories, Inc.* is one in which photovoltaic generation of voltages well above the band gap results upon absorption of radiation by a dipolar dopant within a transparent polarized pyroelectric body. Generation is by a charge transfer mechanism in accordance with which electrons are transferred from excited absorbing species. A photovoltage of greater than a thousand volts has been observed in $Fe^{++}$-doped $LiNbO_3$.

## ORGANIC MEDIA

Organic semiconductors constitute a class of materials that has been extensively investigated as a possible substitute for the conventional crystalline semiconductor materials. In particular, since the photosensitivity of organic semiconductors is well known, many attempts have been made to produce therefrom photoelectric devices that could be used to replace the expensive single crystal devices now being used. The single crystal semiconductor solar cells now in use are quite expensive, but their use persists because the best organic semiconductor solar cells have efficiencies many orders of magnitude too low.

Even if the organic semiconductor solar cell were to have an inferior electrical efficiency, its lower cost would make it competitive, provided that the efficiency differential is not too great. In the solar cell application, reduced efficiency results in a collection area penalty. Where collection area is a primary factor, such as in spacecraft applications, a more expensive cell will be tolerated. In other words, a basic device cost penalty will be accepted under certain conditions. However, when space is a lesser factor, such as in ground-based systems, a moderate area penalty is acceptable because the cost penalty is no longer justified.

Organic dyes in general have proven to be semiconductors, and they are photoelectric in varying degrees. They display a photovoltaic response when operated in a suitable cell structure. Unfortunately, these organic dyes are essentially insulators and, therefore, produce cells that have very high impedence values. In an effort to develop more useful cells, many materials and methods of processing have been investigated, along with processes for making suitable cells.

The most widely used known fabrication method is to dissolve the dye in a suitable solvent and then cast a thin film by solvent evaporation. Two such films cast upon transparent conducting surfaces can be pressed together to produce a photovoltaic cell. If the layers are thin enough, the high volume resistivity effect is reduced to a lower level. However, if the films are too thin, insufficient optical absorption occurs. Accordingly there is an optimum film thickness for any particular material. The two conducting surfaces provide the electrical connections, and light can be applied to the dye through either surface. If the light is to be applied through only one surface, the other one can be made opaque. A metal support plate can then be used.

Such materials as eosin, rose bengal, fluoroescein, erythrosin, crystal violet, malachite green, tetracene, pentacene, aceanthraquinoxaline, poly-n-vinylcarbazole, metal polyphthalocyanines and others have been used. They have been fabricated into suitable structures by casting from a solvent, vacuum evaporation, pyrolysis, and hot and cold powder compact pressing. While successful cells have been fabricated, none have produced efficiencies sufficiently high to compete with conventional cells. The area penalty in such cells is too great.

One device, that developed by *A. Golubovic; U.S. Patent 3,530,007; Sept. 22, 1970; assigned to U.S. Secretary of the Air Force* is a photoelectric device comprising a photoconductive organic layer disposed between and interconnected to two metal electrodes. Upon exposure to illumination, the photoconductive organic material generates a voltage between the electrodes, thus providing a system for use as a solar cell or a photosensitive circuit element. The cell is responsive to distinct wavelengths of incident radiation in the ultraviolet, visible and infrared regions.

Figure 42 shows such a solar cell in perspective and cross-sectional views. There is shown a solar cell comprising a transparent substrate **10** made of glass, quartz, mica, plastic or other suitable light transparent substance having electrical insulating properties. The cell is prepared under high vacuum conditions of

**Figure 42: Solar Cell Including Aceanthraquinoxaline Photosensitive Material**

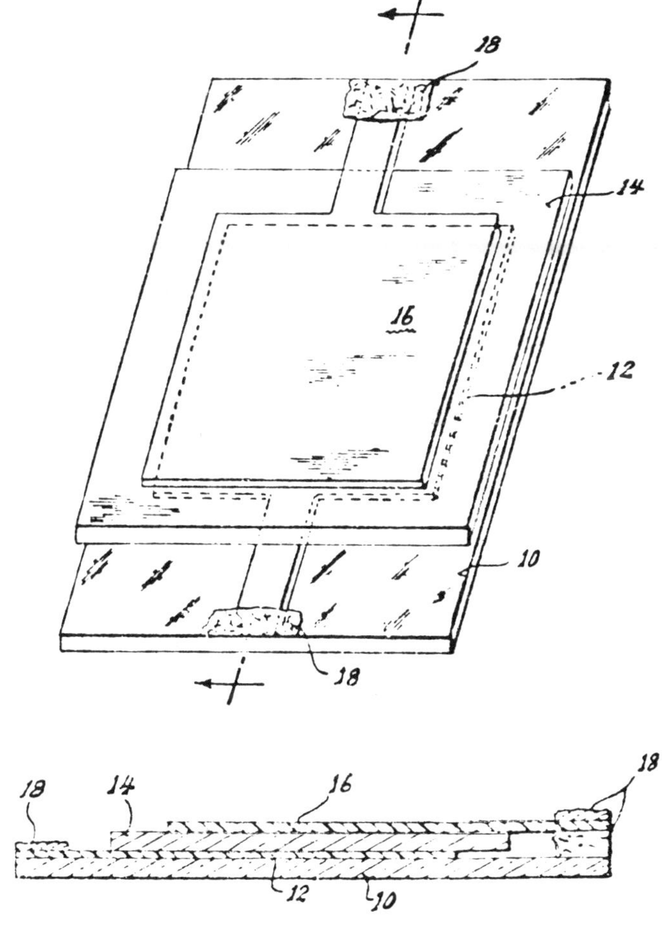


Source: U.S. Patent 3,530,007

about $10^{-6}$ mm of mercury by successive fast depositions of a first or base metal electrode **12**, a light sensitive layer **14** consisting of one or more depositions of a similar or different photoconductive organic material and a second or front metal electrode **16**. A suitable conducting material **18** such as a silver paste is applied to the electrodes to serve as convenient electrical connections for electrical leads not shown. A transparent conductive glass or plastic could also be used as a combination base electrode and substrate rather than utilizing a separate electrode **12** and substrate **10**.

The electrodes **12** and **16** in any single cell are of different metals or other conductive materials which differ in their electronegative potential. A very useful combination consists of a transparent aluminum base electrode and a silver or gold front or top electrode. Deposition of the photoconductive organic thin layer **14** is best achieved by fast evaporation from a stainless steel cup in high vacuum ($<10^{-5}$ mm Hg) applying a temperature close to the melting point of the organic material. Under such conditions a very thin (1-50 $\mu$) pinhole-free layer is obtained, the surface of which is glassy smooth. Such a surface facilitates the direct deposition of the top electrode **16**. Each layer is deposited through an appropriate movable mask divided into three or four parts.

The organic materials used for the preparation of the photoconductive cells are essentially intrinsic photoconductive semiconductors and of different basic chemical structures such as fused aromatic systems, azoaromatic systems and photoconductive dyes. In cells with double organic layers, the combination of an organic photoconductor with charge transfer complexes, metal complexes and free radicals is used.

The following examples of various organic solar cells prepared in accordance with the foregoing principles may better serve to illustrate the development.

*Example 1:*

Substrate: Glass
Base electrode: Aluminum 20% transparency
Organic material: Aceanthraquinoxaline (chromatographically pure)
Front electrode: Gold

The aluminum electrode was negative and this cell produced an open voltage of 1.06 volts and a current of $0.345 \times 10^{-6}$ A/cm$^2$.

*Example 2:*

Substrate: Glass
Base electrode: Aluminum 23.8% transparency
Organic material: Tetracene (crystallized from xylene)
Front electrode: Gold

The aluminum electrode was negative and this all produced an open voltage of 1.0 and a current of $0.775 \times 10^{-6}$ A/cm$^2$.

*Example 3:*

Substrate: Glass
Base electrode: Aluminum 14.3% transparency
Organic material: Pentacene
Front electrode: Gold

The aluminum electrode was negative and this cell produced an open voltage of 0.75 and a current $0.537 \times 10^{-6}$ A/cm$^2$.

In the following Examples 4 and 5 TEA + (TCNQ–)(TCNQ) is a triethyltetracyanoquinodimethane complex salt.

*Example 4:*

Substrate: Glass
Base electrode: Aluminum 23.10% transparency
Organic material: 1st layer aceanthraquinoxaline; 2nd layer obtained by the sublimation of TEA + (TCNQ–)(TCNQ) ion radical complex salt.
Front electrode: Gold

The aluminum electrode is negative and this cell produced an open voltage of 0.79 volt and a current of $0.750 \times 10^{-6}$ A/cm$^2$.

*Example 5:*

Substrate: Glass
Base electrode: Aluminum 10.85% transparency
Organic material: 1st layer tetracene; 2nd layer sublimed TEA + (TCNQ–)(TCNQ)
Front electrode: Gold

The aluminum electrode is negative and this cell produced an open voltage of 0.34 volt and a current of $0.22 \times 10^{-5}$ A/cm$^2$.

*Example 6:*

Substrate: Glass
Base electrode: Aluminum 27.65% transparency
Organic material: 1st layer tetracene; 2nd layer-2,7-dinitrofluoren-$\Delta^{9a}$-malonitrile
Front electrode: Gold

The aluminum electrode is negative and this cell produced an open voltage of 0.85 volt and a current of $0.13 \times 10^{-5}$ A/cm$^2$.

A process developed by *R.E. Kay and E.R. Walwick; U.S. Patent 3,844,843; Oct. 29, 1974; assigned to Philco-Ford Corp.* is one in which a photovoltaic cell is fabricated from an active medium comprising an organic semiconductor in a gel. When a film of such material is sandwiched between transparent conducting electrodes a solar cell is obtained. The electrical output is greatly in excess of that obtained from prior art organic semiconductor solar cells of the same area.

Such a device is shown in Figure 43. In the device shown, glass plate **1** is coated with a transparent, conductive film **2** of tin oxide by the well-known chemical vapor pyrolysis process. The film **2** may be doped with antimony oxide to lower its electrical resistance. Such tin-oxide-coated glass plates are available commercially. One well-known version is known as Nesa Glass. A mixture of organic semiconductor, gel agent, and solvent is cast upon the conductive film **2** and the solvent allowed to partially evaporate. The resulting layer **3** of gel contains the organic semiconductor in a form that is highly responsive to optical energy. Glass backing plate **4** carries on its upper surface a conductive layer **5**. It is pressed against the surface gel layer **3** after the solvent is sufficiently evaporated to produce the desired gel.

**Figure 43: Solar Cell with Organic Semiconductor Contained in a Gel**

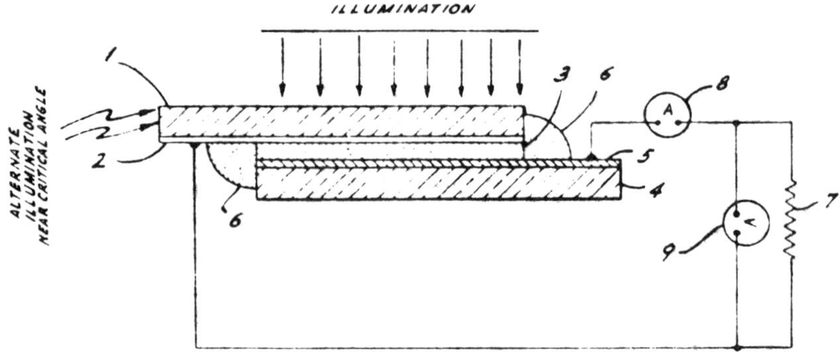


Source: U.S. Patent 3,844,843

Alternatively, gel layers can be cast on both the front plate film **2** and back plate film **5** and the gel surfaces pressed together. In still another alternative a bead **6** of suitable cement, such as an epoxy resin, can be cast around the completed cell to seal the device and hold glass plates together. Film **5** is desirably a metal that acts as a nonrectifying gel contact. Platinum has proven to be suitable, but any metal that is nonreactive and does not produce insulating surface layers will be useful. If the cell is to be illuminated from the back side, film **5** must be thin enough to be partly transparent. If no back side illumination is desired, film **5** can be made as thick as desired, or the entire back plate could be made of metal and used in place of glass plate **4** and film **5**.

Excessive solvent evaporation can produce degraded cell performance and since additional evaporation can occur over extended periods of time, a solvent retention mechanism may be desired. The epoxy bead form of construction shown above is effective if the casting operation is done properly. Alternatively, the addition to the gel of glycerol, a well-known humectant, will limit solvent evaporation and prevent excessive drying even over long periods of time in exposed cells.

Typically the cell is illuminated through glass plate **1**, as shown, to provide area illumination. This is the method contemplated for most solar cell operation. However, the gel-dye material is not strongly optically absorptive, and radiant energy normal to the cell surface may not be completely absorbed at the critical region of the dye-tin oxide interface.

Alternatively, the cell may be illuminated near the critical angle from the end, via the tin-oxide-coated glass **1**, as shown. Such end illumination results in better excitation of the gel because of multiple surface reflections at the tin oxide film. For ordinary construction multiple attenuated reflections from film **2** are obtained if the angle of light incidence is near the critical angle. The multiple reflections cause the illumination to traverse a greater length of sensitive material and thereby produce better optical absorption.

The other circuit elements illustrated may be used to determine the power that the solar cell is capable of providing. Resistor **7** acts as the load for the cell. Ammeter **8** and voltmeter **9** monitor the cell output current and voltage. Output power is calculated by multiplying the voltage by the current. The resistance of load resistor **7** can be varied while maintaining constant illumination to evaluate cell performance. For example, the internal resistance of the cell can be calculated by observing the voltage-current characteristics of two load values. Also, the resistance of load **7** can be varied to determine the value that produces maximum power output.

In a typical example, a cell was constructed as shown in Figure 43. The front electrode was tin-oxide-coated glass having a resistance of about 250 ohms per square. The rear electrode was an opaque film of bright platinum on glass. The active material was cast from a water solution of 6% by weight crystal violet and 5% by weight agar. This solution was prepared by dissolving the agar in boiling water and then adding the crystal violet. The hot solution was poured upon the rear electrode, whereupon the excess solvent quickly volatilized. The front electrode plate was then pressed against the exposed surface of the gel while it was still warm in such a manner as to avoid entrapment of air bubbles.

This could be observed through the transparent electrode during its application. Excess gel material will be expelled from between the plates and can easily be trimmed off after the gel-dye solution cools. The resulting structure is sufficiently coherent to withstand handling. The film after the above treatment was typically about 0.1 mm thick. The dark resistance of a 10-$cm^2$ cell measured about $1.4 \times 10^4$ ohms. The maximum open circuit photovoltage was about 0.425 volt. Direct sunlight produced about 0.5 mW output or about 0.005 $mW/cm^2$. Considering incident sunlight at 100 $mW/cm^2$ this represents an efficiency of about 0.05%. The better prior art devices produced less than about a millimicrowatt ($10^{-9}$) per square cm.

A cell design developed by *C.W. Tang, A.P. Marchetti and R.H. Young; U.S. Patent 4,125,414; Nov. 14, 1978; assigned to Eastman Kodak Company* consists of an electrically insulating binder, an organic dye, and an organic photoconductor. The photovoltaic element has superior conversion efficiencies compared to other organic photovoltaic elements.

In a typical example of the preparation of such a cell, an organic solar cell was fabricated as follows: A coating solution 1.4% by weight of solids in a mixed solvent of 1,1-dichloromethane and 1,2-dichloroethane (in 1:1 weight ratio) was prepared by sequentially dissolving in 40 g of the mixed solvent 112 mg of a thiapyrylium dye salt 2,6-diphenyl-4-(4-dimethylaminophenyl)thiapyrylium hexafluorophosphate, 294 mg of a high-molecular-weight bisphenol A polycarbonate having an inherent viscosity of about 2.3 and 294 mg of an organic photoconductor, 4-di-p-tolylamino-4'-[4-(di-p-tolylamino)styryl] stilbene.

A clean piece of Nesatron glass about 2.5 cm x 2.5 cm was spun on a turntable at about 2,000 rpm. While spinning, a small quantity (about 0.5 ml) of the coating solution was poured on the conducting surface of the Nesatron glass. A thin and apparently uniform film was obtained upon evaporation of the solvent. The thickness of the dry film was about 2000 A.

The spin-coated film was then subjected to vapor treatment by a toluene vapor to induce aggregation of the thiapyrylium dye and the bisphenol A polycarbonate polymer. Vapor treatment for 2 min was sufficient to induce essentially complete aggregation.

To complete the fabrication, an evaporated layer of indium having an area of 1 $cm^2$ was applied on top of the aggregated coating.

The photovoltaic cell was illuminated through the semitransparent Nesatron electrode with a broad-band illumination of 100 $mW/cm^2$ intensity provided by an unmodified Kodak 600H slide projector containing a tungsten light source and filtered through a Corning 3384 filter, which cuts off light of wavelength shorter than 500 nm. The open-circuit voltage developed between the Nesatron and the indium electrodes was 800 mV and the short-circuit current was 1.0 $mA/cm^2$. The maximum power delivered to a resistive load of 1000 ohms was about 0.2 mV, representing a power conversion efficiency of 0.2%.

A solar cell developed by *A.K. Ghosh and T. Feng; U.S. Patent 4,127,738; Nov. 28, 1978; assigned to Exxon Research & Engineering Company* consists of at least two electrodes (one of which must be substantially transparent to the light), each electrode being made of different materials and in which one electrode comprises an element that has a work function (generally expressed in electron volts) greater than that of aluminum (e.g., gold or silver) and the other electrode comprises an element that has a work function equal to or less than that of aluminum (e.g., aluminum or magnesium).

Sandwiched between and in contact with the electrodes is a photoresponsive organic layer comprising at least one organic compound which, in general, has the capacity to sensitize or desensitize silver halides, titanium dioxide, zinc oxide, cadmium sulfide, selenium and polyvinyl carbazole. (Examples of the organic compounds are the cyanine dyes, especially the merocyanine dyes.) The electrode comprising an element having a work function equal to or less than aluminum forms a Schottky barrier with the organic layer. Optionally, an insulating film is interposed between the Schottky barrier electrode and the organic layer.

A cell design developed by *S. Chen; U.S. Patent 4,138,532; Feb. 6, 1979; assigned to Optel Corp. and Grumman Aerospace Corp.* is a photogalvanic cell having a light transparent electrode and a spaced counterelectrode separated by an electrolyte. The electrolyte includes an n-methylphenazine dye system which not only contributes to the conversion of light to electrical energy but also is capable of storing electrical charge after light is removed.

Figure 44 shows such a prior art device of a related type using a dye system. In the left-hand compartment, a dye **D** (which should be a good donor in the ground state) will be excited by light so that the excited state $D^*$ is able to reduce water to hydrogen leaving the oxidized dye $\mathbf{D^+}$. In the right-hand compartment, a dye **A** (which should be a good acceptor in the ground state) will be excited by light so that $A^*$ will be able to oxidize water to oxygen leaving the reduced dye $\mathbf{A^-}$. The two compartments are then coupled together electrochemically so that $\mathbf{A^-}$ can spontaneously reduce $\mathbf{D^+}$ thus restoring the dyes to their original states.

The dark reactions following the photochemical electron transfers may be rather complex and specific catalysts capable of storing electrochemical equivalents will be required. J.R. Bolten in a published paper in *Proceedings VIII International Conference on Photochemistry,* Edmonton, Canada, August, 1975 discovered a reaction which partially meets the requirements for the dye **A**. The dye is the N-methylphenazinium cation ($NMP^+$). At wavelengths less than 500 nm $NMP^+$ is able to photooxidize water yielding the reduced dye $NMPH^+$ and $OH^-$ radicals.

The prior art device, however, was a multicompartment cell for water photolysis. The process device is shown in Figure 45. A dye system is used in the electrolyte which achieves spectral sensitivity in the region ranging from 500 nm down to ultraviolet. This is in contrast to an iron-Thionine photogalvanic cell which is not sensitive below 500 nm. The present photogalvanic cell should exhibit a greater power conversion efficiency. A second distinct advantage of the device over the iron-Thionine system relies upon the fact that the dye utilized in the present electrolyte, namely, N-methylphenazine, is capable of achieving charge storage in addition to energy conversion from light to electrical forms of energy.

**Figure 44: Prior Art Cell Design Using Dye System for Water Photolysis**

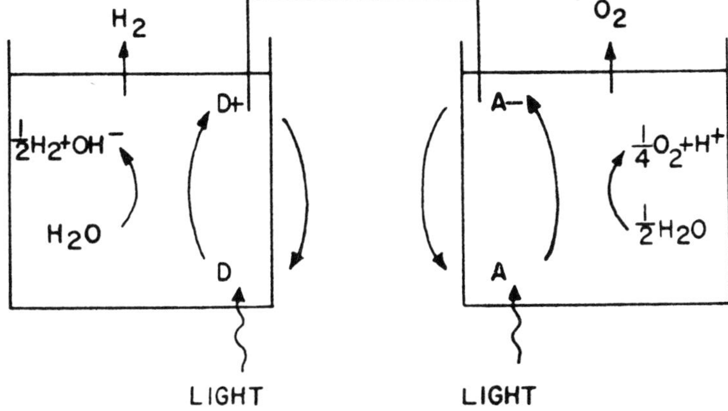


Source: U.S. Patent 4,138,532

**Figure 45: N-Methylphenazine Photogalvanic Cell**

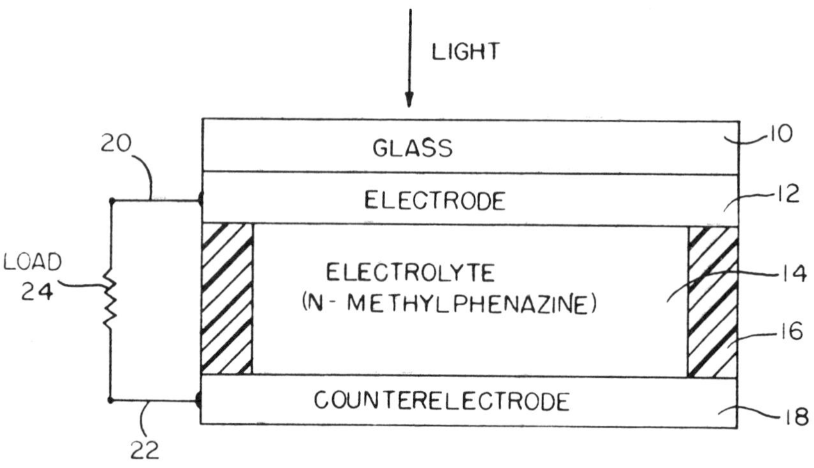


Source: U.S. Patent 4,138,532

As shown in Figure 45, the cell design includes a light transparent substrate material such as glass **10**. A thin film or layer of conducting material is deposited onto the glass to function as electrode **12**. The electrode is fabricated from a transparent thin film conducting material such as $SnO_2$. A prefabricated material including the glass and the electrode deposited thereon is commercially available and is known as Nesa glass.

A counterelectrode **18** is positioned in spaced registry with the electrode and may be fabricated from a disc of carbon or a platinized carbon member. In the event of the latter, the platinum may be deposited on the carbon material by conventional metalization techniques.

The aqueous material between the electrode and the counterelectrode is the electrolyte which includes the N-methylphenazine dye system. The electrolyte is an aqueous medium which includes a conventional acid, such as sulfuric acid. The aqueous medium has N-methylphenazine methosulfate added thereto, the latter being commercially available from a number of sources such as Eastman Kodak Co., and is available in powdered form.

An annular wall **16** is transversely disposed between and connected to the electrode and counterelectrode and serves the purpose of sealing the electrolyte **14** within the device as well as supporting the internal cell components. The wall may be fabricated from most any suitable inert and electrically insulative material such as epoxy or polyethylene.

Wires **20** and **22** respectively are connected to the electrode and counterelectrode. A load **24** is connected between these wires and draws current from the device when the latter is exposed to light.

The transparent electrode is a thin film semiconductor and upon irradiation by light shorter than 500 nm, the transparent electrode becomes negatively charged while the counterelectrode becomes positively charged. As a result, electric current will be drawn by load **24**.

It has been found that the particular dye system including the N-methylphenazinium ions permits the cell to also store electrical energy. Accordingly, the cell could be charged without an attached load and after irradiating energy ceases, a load may be connected across the cell for power.

The stability of the present N-methylphenazine photogalvanic cell may be improved by adding to the electrolyte a suitable redox couple, such as Fe(II)-Fe(III).

The suspected mechanism for the theoretical operation of the device is that the light excites the molecules of the dye to reach excited states. The excited molecules or their photoproducts eject electrons and these are collected by the electrodes or stored in the electrolyte for later use by a connected load.

# Materials Availability

The statement was made in the review by the U.S. Office of Technology Assessment that: "Photovoltaic energy production is unlimited by scarce materials, since cells can be manufactured from silicon, one of the Earth's most abundant elements." While silicon is in plentiful supply (about one atom in five in the Earth's crust is a silicon atom), current production rates of purified silicon are not adequate to support a large solar industry however.

Enthusiasm about cells based on materials other than silicon must be tempered to some extent by uncertainties about the limits imposed on their production by U.S. and world supplies of component materials.

Domestic supplies of both cadmium and gallium will be sufficient to supply annual production rates in excess of several thousand megawatts a year but production beyond this level could tax domestic supplies and, in the case of cadmium, known world reserves limit the ultimate potential of $CdS/Cu_2S$ devices to about the level of current U.S. energy consumption (3).

U.S. production of cadmium in 1970 could support the annual manufacture of cell arrays with a peak output of 5,000 to 20,000 MWe (assuming cells are 10% efficient). World production is about five times greater than U.S. production. The higher figure applies to the spray process for manufacturing cells, which results in cells about 6 microns thick. Significant increases in cadmium production may be achievable, however, if demand increases. Identified U.S. reserves of cadmium are sufficient to produce cells with an annual output equal to the current U.S. consumption of electricity. Known world reserves are about five times greater than U.S. reserves.

Gallium supplies should not place a significant constraint on the use of GaAs devices. U.S. consumption of gallium in 1973 was 8.5 metric tons per year, most of which is imported. If this amount of GaAs were used to produce 50-micron-thick cells used in tracking collectors producing concentration ratios of 1,000, the arrays would have a peak output of about 10 GWe at an efficiency of 20%. These arrays would have an average output of about 1% of the average output of all electric-generating facilities in the U.S. in 1977. Domestic production of gallium could be increased substantially if an attempt is made to extract the gallium associated with coal, aluminum, and zinc ores. A recent study indicated that the United States could produce about 600 metric tons per year of gallium from these sources. So the gallium resource is ample. World resources of gallium are about $1.1 \times 10^5$ metric tons (3).

# Toxic Hazards of Solar Cell Materials

Silicon is nontoxic but the manufacture of silicon devices with present techniques involves the use of a number of hazardous chemicals ($PH_3$, $BCl_3$, $H_2S_2$, HCl, HCN). Existing state and federal laws should be sufficient to ensure that releases of these materials into the air and water are kept to acceptable levels, although vigilance will be needed to ensure compliance with these regulations. Meeting the standards may add to the cost of the devices (3).

Both cadmium and arsenic are toxic and while it may be possible to reduce the hazards they present to manageable proportions and while both materials are already used extensively in commercial products and manufacturing, it clearly will be necessary to examine these hazards with some care before recommending an energy system which would significantly increase the use of these materials near populated areas.

Cadmium is a cumulative heavy-metal poison with a half-life in the human body of 10 to 25 years. Cadmium poisoning is believed to lead to accelerated aging, increased risk of cancer, heart disease, lung damage, birth defects, and other problems. Chronic exposure to airborne cadmium can lead to emphysema and other respiratory problems. Cadmium, however, is used in many commercial industrial products; it is used for corrosion protection on bolts and screws, in paints, plastics, rubber tires, motor oil, fungicides, and some types of fertilizers. Increased use of cadmium resulting from large-scale manufacture of $CdS/Cu_2S$ cells could result in increased release of cadmium from mining, refining, and manufacturing of the cell, and it would increase the amount of cadmium present in products located near populated areas, thereby increasing the risk of exposure in the event of fires, accidents, or the demolition of buildings.

More stringent stardards may be needed both in manufacturing processes involving the material and in permitting usage of the material in commercial products. At present, cadmium is not subject to any environmental controls, but under both the Toxic Substance Control Act (P.L. 94-469) and the Resource Conservation and Recovery Act (P.L. 94-580), EPA has jurisdiction over its regulation which, if instituted, could include both solar and other household uses.

It is apparently possible to significantly reduce the cadmium released in the manufacturing process, and it appears that with proper encapsulation CdS cells can be used in residential areas with minimal hazard. Exposure to the material would only occur during fires, accidental breakage, or building demolition. The acceptability of such exposure can be judged better when meaningful standards have been developed (3).

Undisassociated GaAs is harmful but apparently not highly toxic. A lethal dose for an average adult is about one-third kg (0.7 lb). To ingest this much GaAs, a person would have to eat the amount of GaAs in about 20m$^2$ (200 square feet) of flat-plate arrays or the amount of GaAs in a field of concentrating arrays covering about 3 acres, clearly a Herculean task. The coatings and protective encapsulation placed on cells should be able to reduce any hazards associated with normal operation of GaAs devices to acceptable levels (3).

The major danger arises when the material disassociates as a result of a fire or some other accident, since many arsenic compounds are highly toxic and recognized carcinogens. The $As_2O_3$ which would be released in a fire could contaminate land and water near the fire site. Since the cells would most probably be used in connection with high concentrations of sunlight, care must be taken to ensure that the materials do not vaporize and escape if a breakdown in the cell cooling system occurs. The amounts of arsenic used in a concentrating collector, however, would be extremely small. A device with a concentration ratio of 1,000 would, for example, use only about 0.16 g/m$^2$ of collector area. This concentration is 250 to 1,500 times smaller than the concentration of $As_2O_3$ recommended by the U.S. Department of Agriculture for weed control.

# Solar Cell Fabrication

Once a piece of photovoltaic material has been prepared, the fabrication of that piece of material into a solar cell involves physical arrangement of the material, attachment of electrical leads, provision of antireflection coatings and encapsulation for environmental protection. If solar cells are to be successful in cost competitive terrestrial applications, all this must be done on a mass production basis at reasonable cost.

A process for solar cell fabrication developed by *R. Gonsiorawski; U.S. Patent 4,152,824; May 8, 1979; assigned to Mobil Tyco Solar Energy Corporation* involves the steps of:

(1) providing a semiconductor body of a first conductivity type and having first and second opposite surfaces;

(2) forming on the first surface a continuous layer of a material containing a dopant capable upon diffusion into the body of forming a region of a second opposite conductivity type in the body;

(3) removing the layer from selected portions of the surface so as to form a gridlike pattern of apertures defined by intervening nonremoved sections of the layer;

(4) heating the body in an atmosphere containing the dopant at a temperature at which the dopant will diffuse into the body from the layer and the atmosphere so as to form in the body relatively deep diffused regions of opposite conductivity type in line with the apertures and relatively shallow diffused regions of the opposite conductivity type in line with the intervening layer sections, the regions establishing a junction within the body; and,

(5) forming conductive contacts on the surfaces with the contacts on the first surface conforming to and overlying the deep diffused regions.

## FROM SILICON WAFERS

Conventionally, semiconductor device manufacture involves purification and preparation of polycrystalline semiconductor material, and its processing into rods which are then refined into monocrystalline semiconductor material and doped with impurities to achieve specific electrical properties. These doped rods are cut into slices or wafers, which then are mechanically and chemically polished. In a simple manufacturing process, e.g., for production of semiconductor diodes, the slices may typically be subjected to a diffusion process to convert a surface or selected portions of a surface of the slice to opposite conductivity type from the remainder of the slice. At this point, the slice is ready for further processing steps including provision of metal contacts and separation of the slice into individual chips or wafers from which the final semiconductor devices are produced. Such a process, while commercially practicable, does involve wastage of a significant (about 85%) amount of semiconductor material by the time the individual polished slices are obtained and even further wastage before the individual chips or wafers are obtained.

A device developed by *J. Lindmayer; U.S. Patent 3,769,558; October 30, 1973; assigned to Communications Satellite Corporation* is an improved semiconductor device, particularly suitable as a solar cell, which has a surface inversion layer with increased charge density. A P-type semiconductor material such as silicon is covered with a layer comprising oxides of silicon and chromium. The addition of the chromium oxides increases the charge density of the inversion layer by orders of magnitude over that created by the silicon oxide alone. In the fabrication process, a silicon oxide layer is formed first, followed by a layer of elemental chromium. Both layers are oxidized by placing the device in an oxygen atmosphere at temperatures in excess of 800°C.

A process sequence developed by *J. Lindmayer, U.S. Patent 3,895,975; July 22, 1975; assigned to Communications Satellite Corporation* involves diffusing an impurity of a first type conductivity into the front surface of a semiconductor bulk material while simultaneously alloying and diffusing an impurity of a second type conductivity into the back surface of the semiconductor bulk material from a metallic source. During this simultaneous doping, the back surface area of the semiconductor and the second type metallic impurity are in a molten alloy state.

Figure 46 shows, from top to bottom in two parallel sets of diagrams, the process steps (at the left) and the products produced by the various process steps (at the right).

**Figure 46: Steps in Diffusion of Impurities into a Silicon Semiconductor**

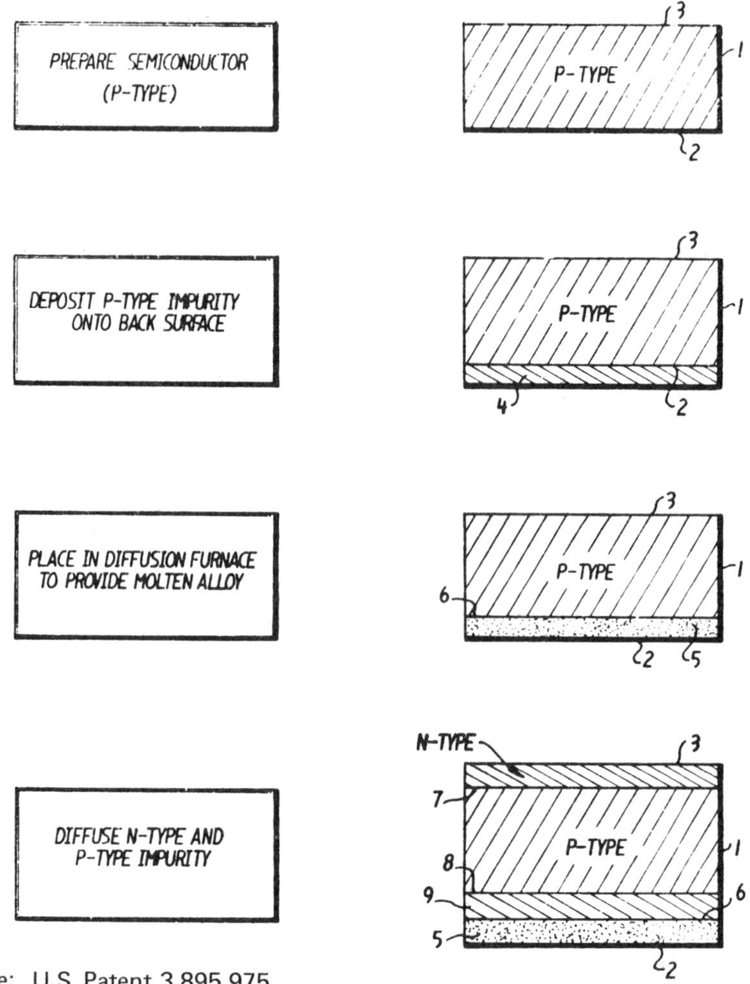


Source: U.S. Patent 3,895,975

In a first step, a wafer of semiconductor bulk material, e.g., P-type silicon, having a back surface 2 and front surface 3 (the surface through which light will enter the solar cell) and dimensions suitable for use as a solar cell, is cleaned and polished in a conventional manner. As a second step, a layer 4 of P-type material, e.g., aluminum, about 5000 to 10,000 angstroms thick, is deposited onto the back surface 2 of the silicon 1. The range of thicknesses is merely representative of a preferred deposit of aluminum. Other thicknesses may be used; however, a layer less than 2000 angstroms may not provide enough stress relief and a layer greater than 10,000 angstroms may result in a rough back surface of aluminum. The P-type layer of aluminum may be deposited onto the back surface of the silicon wafer by means of a standard boat evaporation technique.

As is well known in the art, a boat containing an ingot of the metal to be evaporated is heated to a temperature above the melting point of the metal in a total or partial vacuum. In the preferred embodiment, an aluminum ingot is heated to about 1500°C in a partial vacuum environment including a small amount of oxygen. The aluminum atoms that are evaporated will condense on the back surface of the solar cell that is exposed to the ingot. For the boat evaporation technique, it has been found that the aluminum will form a smoother surface when deposited in a very high vacuum. Other known deposition techniques such as electron beam evaporation, sputtering and plating may also be used.

The silicon wafer having the aluminum layer deposited on the back surface is now placed into the diffusion chamber of a standard diffusion furnace. The wafer will lie on a quartz tray with its coated surface face down and its front surface exposed to the inside of the diffusion furnace chamber. The wafer will remain in the diffusion furnace for a period of about 15 minutes at a temperature of about 800°C. Under these conditions, since the temperature is above the eutectic temperature of the silicon-aluminum combination (577°C) and the melting point of aluminum (660°C), the aluminum layer and adjoining silicon will form a pool of molten silicon-aluminum alloy **5** at the back surface of the silicon wafer. When the coated silicon wafer is first placed into the diffusion furnace the diffusion chamber should have in it only an inert gas, such as nitrogen.

At this stage in the process, a junction **6** is formed which may be characterized as a $P^+$-P junction. That is, the molten silicon-aluminum alloy comprises a very heavily doped P-type region (i.e., $P^+$) while the remaining silicon, which is still crystalline, comprises the original P-type region. The silicon remains because its melting point is above 800°C.

After the silicon-aluminum alloy has been formed in the furnace, the wafer is ready to have an N-type impurity, preferably phosphorus, diffused through the front surface. To enable diffusion of the phosphorus, a diffusion gas comprising $N_2$, $O_2$ and $PH_3$ (1% in argon) may be used. The diffusion gas will flow through the diffusion furnace chamber at a rate of 1,000 cc/min for $N_2$, 75 cc/min for $O_2$, and 550 cc/min for $PH_3$ in a manner well known in the art. The inert gas originally in the chamber will be exhausted by the flow of diffusion gas. Diffusion of the phosphorus is allowed to continue for approximately 10 minutes at a temperature of about 800°C. In this manner a shallow N-P junction **7** is provided at a depth below the front surface of the silicon.

Once the N-type phosphorus has been diffused into the front surface to form the desired junction, the silicon wafer is removed from the furnace and is allowed to cool to room temperature. The molten silicon-aluminum alloy solidifies into the back surface of the silicon wafer. The interface between the aluminum-silicon alloy and the bulk silicon provides what may be described as a $P^+$-P junction **8**. That is, the alloy provides a heavily doped P-type ($P^+$) region **9**. In this manner, an N-P junction **7** and a $P^+$-P junction **8** are simultaneously formed, as shown in the figure. Although some diffusion of the aluminum atoms into the silicon bulk material may take place during the alloying step and form an intermediate junction between the diffused silicon and the alloy, this effect is small in the preferred embodiment and may be neglected.

At this point it would be helpful to review some of the advantages obtained with the present diffusion process. First, it has been found that a more ideal N–P junction and $P^+$–P junction can be obtained. The small pool of molten silicon-aluminum alloy relieves mechanical stresses throughout the whole silicon wafer which would damage the crystal lattice and prevent the uniform formation of a sharp junction. Secondly, the pool of molten alloy prevents any of the phosphorus from diffusing into the back surface of the silicon. Such phosphorus diffusion, if allowed, would tend to contaminate the back surface, thereby producing an undesirable N–P junction near the back surface. Finally, the presence of the $P^+$–P junction will reduce the recombination of carriers generated in the P-type silicon, thereby enhancing the solar cell current and to a smaller degree the voltage output.

Finally, to complete the manufacture of the solar cell, front and back surface photocurrent collecting metallic contacts (not shown) may be applied.

A device developed by *J.G. Besson, T.N. Duy and W. Palz; U.S. Patent 3,928,073; December 23, 1975; assigned to Societe Anonyme de Telecommunications, France* is in the category of known solar cells formed by a P-type doped silicon semiconductor body in whose upper part, which is exposed to the incident light, there is provided a thin N-type doped zone. Such cells are normally provided with metal electrodes, a first electrode covering the lower surface of the P-type doped semiconductor body, a second electrode, in the form of a current collecting grid, being formed on the upper surface of the cell where the semiconductor body is doped N-type. Further, this upper surface is covered with various layers of materials which improve its energy efficiency and/or protect the cell so as to render its quality reliable.

However, it is known that whatever precautions are taken in the course of manufacture of the cell, the N-type doped zone has irregularities or other faults related to the surface conditions. Consequently, the carriers produced in the vicinity of the surface of the cell upon impact of the incident photons manage to recombine in major part on the ionized surface centers of this zone. Therefore, they do not participate in the conversion of energy that the cell must effect. Further, this effect occurs selectively according to the wavelength of the incident photons and is particularly harmful for the components of the solar radiation the wavelength of which is less than 5000 angstroms, and this has an adverse effect on the efficiency of such a solar cell.

An object of the device is to overcome the harmful action of the surface centers for the shorter wavelengths by a controlled compensation of the electric charges they carry. The surface ions then pass to the neutral atom state without affecting the value of the energy efficiency of the cell in which these shorter wavelengths consequently participate, as opposed to what happens in known solar cells.

More specifically, the device is a solar cell comprising a P-type doped semiconductor body in whose upper part, exposed to the incident light, there is provided a thin N-type doped zone, the battery being covered in its lower part by a metal electrode and in its upper part by a metal collector grid, wherein the N-type doped layer between the bars of the collector grid is covered by an insulating layer. The insulating layer is transparent to the incident radiation, being itself covered by a conductive layer which constitutes a third electrode also transparent to the incident radiation and insulated from the collector grid. Such a cell design is shown in Figure 47.

**Figure 47: Section of High-Efficiency Solar Cell Developed by Societe Anonyme de Telecommunications**

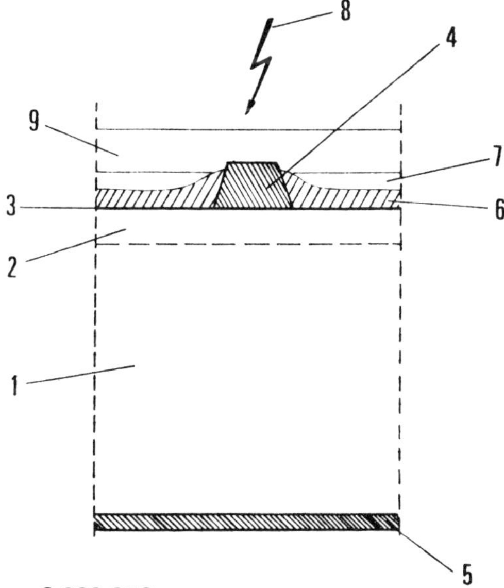


Source: U.S. Patent 3,928,073

The cell comprises a P-type doped zone **1** of the silicon semiconductor body. This zone has, for example, a height of 300 microns. By known processes this zone is surmounted by a N-type doped zone **2** having a height of approximately 0.5 micron. In its upper part, on a height of 0.01 micron, there is a zone **3** comprising recombination centers which are ionized (negatively). The cell is further comprised of a collector grid forming an electrode for zone **2-3**, the grid being constituted by bars **4**, and a solid electrode **5** for zone **1**. These elements are absolutely analogous to and obtained in the same manner as those in an N-P silicon solar cell or battery.

The surface zone **3** is covered with an insulating layer **6** of silicon oxide and a layer **7** of $SnO_2$, or any other body such as indium oxide ($In_2O_3$) which is known to be both an electrical conductor and transparent to incident radiation **8** on the cell. It will be observed that by an appropriate disposition of the layers **6-7**, the layer **7** is insulated by layer **6** from the adjacent element **4** of the collector grid. The assembly may be covered by at least one ordinary and known layer **9** which has antireflection properties and may be used as a protection.

Owing to the fact that it is thin and bearing in mind also that it is sufficiently conductive, zone **3** may be likened to a kind of capacitor plate or electrode brought to a negative potential by the assumed ionic charge of the recombination centers which form it. Further, the conductive layer **7** and the insulating layer **6** may constitute, in effect with zone **3**, a capacitor of which the parts **3** and **7** are the electrodes and part **6** the dielectric. If the electrode is brought to a negative potential, the surface of the dielectric facing the electrode is charged negatively and, by influence, the opposite sun face of the dielectric is charged positively. The whole of this sun face and zone **3** may, if the potential of the electrode is suitably chosen, form a surface assembly having zero resulting charge. Everything operates, therefore, in the present solar cell as if the zone **3** no longer intervened in the efficiency of the cell. In practice, no flow is observed between the electrode **7** and the other electrodes **4** and **5** of the solar cell.

As a solar cell is employed grouped with other cells to constitute a battery, the potential to which the electrode **7** is brought may be achieved from the total voltage of the battery, constituting a solar generator.

A device developed by *A. Yamashita; U.S. Patent 3,928,865; December 23, 1975; assigned to Matsushita Electric Industrial Co., Ltd., Japan* is a photoelectrical transducer element comprising a semiconductor body, a rectifying barrier formed in the semiconductor body, and electrodes formed on the semiconductor body on both sides of the barrier. A deep-level-forming impurity is heavily doped in the neighborhood of the rectifying barrier. Figure 48 shows such a device.

**Figure 48: Matsushita Design for High-Efficiency Solar Cell**

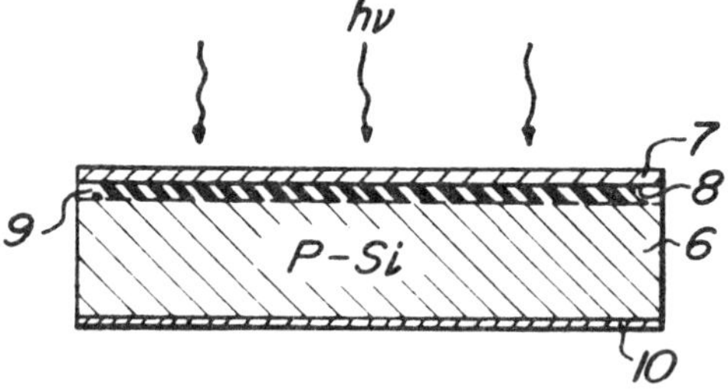


Source: U.S. Patent 3,928,865

In the figure, **6** indicates a P-type silicon, **7** a metal electrode using niobium, **8** a metal semiconductor junction, **9** a region heavily doped with a deep-level-forming impurity, **10** a metal electrode. The element shown may be made as follows:

First a pellet of P-type silicon single crystal is oxidized in an oxidizing atmosphere to form a silicon oxide film on the surface. Then, the silicon oxide film on one side is removed by an etching technique and a deep-level-forming impurity, in this case copper, is vapor deposited on the exposed surface. Then, the pellet is heated in an inert atmosphere to diffuse the impurity. Copper impurities diffuse through the silicon body and these are mostly trapped in the neighborhood of the opposite silicon oxide-silicon interface to form the region **9**. The silicon oxide film is then removed, and as the electrode on the light-receiving side niobium is thinly sputtered, the electrode on the other side gold or gallium is alloyed. The samples thus made but with varying diffusion conditions developed the following open-circuit voltages under constant illumination, at a wavelength of 1,000 m$\mu$. As is evident from the following table, copper diffusion increases the efficiency of photoelectric transformation.

| . . Copper Diffusion. . . | | |
|---|---|---|
| **Temp (°C)** | **Time (min)** | **Open-Circuit Voltage (mV)** |
| No diffusion | | 0.30 |
| 800 | 60 | 0.95 |
| 1000 | 10 | 2.6 |
| 1000 | 30 | 2.6 |
| 1000 | 180 | 6.5 |

A process developed by *J.E. Lawrence and I. Wu; U.S. Patent 4,062,102; Dec. 13, 1977; assigned to Silicon Material, Inc.* is a process for manufacturing a solar cell from a reject semiconductor wafer comprising stripping all external layers from the wafer, etching the surfaces of the wafer so as to effectively remove all P-N junctions without pitting the wafer surface, introducing a layer of dopant to form a P-N junction in the front wafer surface, forming a first patterned conductive electrode over the dopant layer, and forming a second conductive electrode on the back surface of the wafer. In the preferred embodiment a sputtering operation is used to form the conductive electrodes.

The cost of a starting single crystal silicon wafer is more than one-half the total manufacturing cost of a finished solar cell. For example, the cost of a new three-inch wafer is six dollars, whereas the cost of a finished solar cell is about nine dollars. Hence, less expensive sources of single crystal silicon wafers must be developed.

Presently, rejected semiconductor wafers constitute the lowest cost source of single crystal silicon wafers available. Such reject wafers have definite value as future starting material for the fabrication of solar cells (photovoltaic products). This process is designed to take maximum advantage of the availability of such low-cost reject cells.

Figure 49 shows steps in the recovery process.

**Figure 49: Rejected Wafer and Sequential Steps in Wafer Recovery Process**

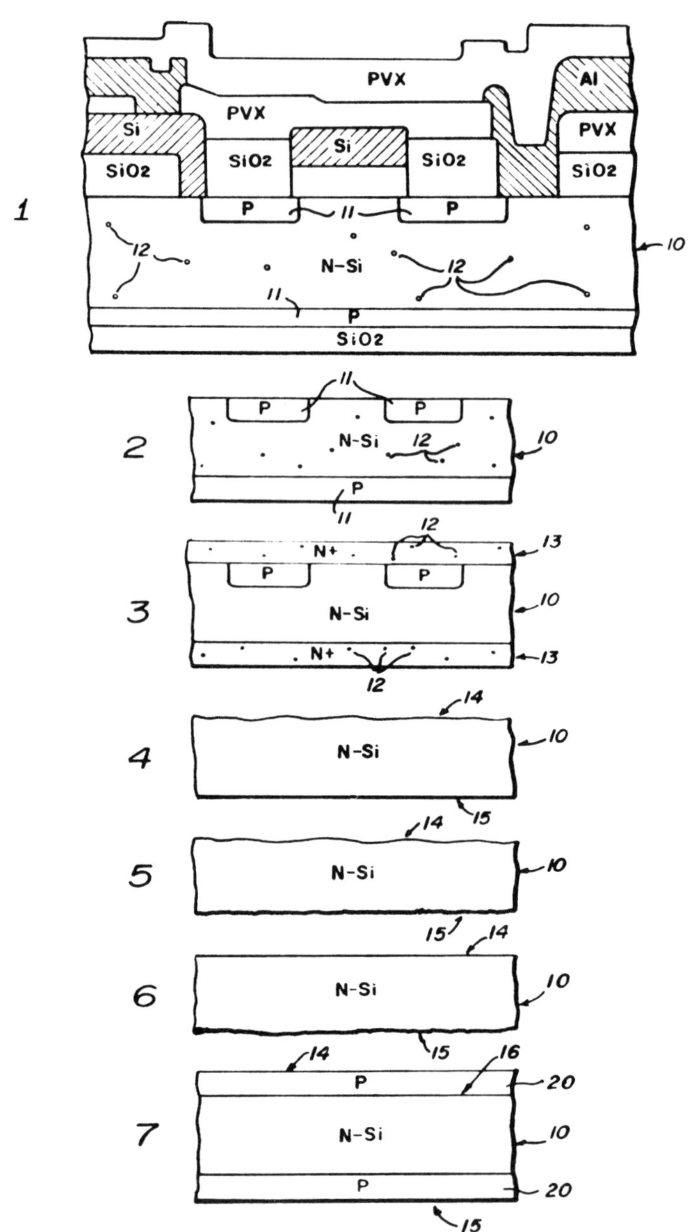


Source: U.S. Patent 4,062,102

Rejected semiconductor wafers, which are to be transformed into solar cells may have a complex structure as shown in Figure 49, Step 1. The bulk silicon wafer **10** has the dopant type, resistivity, and orientation of the original as-grown crystal. The wafer **10** having an N-type conductivity with P-type diffusion regions **11** forming P-N junctions may have epitaxial layers (not shown), and external layers such as silicon dioxide ($SiO_2$), phosphovapox ($Si_3N_4$), silicon nitride (not shown), phosphosilicate glass (PSG, not shown), aluminum (Al), and polysilicon. Although a complex metal-oxide semiconductor (MOS) structure is illustrated, it should be recognized that all reject wafers, even those which were rejected after an initial oxidation step, and also bipolar structures may be processed in accordance with the present process.

The initial required step, i.e., stripping, of the process for transforming a reject semiconductor wafer into a solar cell consists of removing all conducting and insulating material external to the silicon wafer. The as-received semiconductor reject wafers are loaded into acid-resistant carriers, and then immersed in different acids, such as $H_2SO_4$-$H_2O_2$ to remove photoresist; HF to remove $SiO_2$, $Si_3N_4$, PVX and PSG; HCl-$HNO_3$ to remove metals; and HF-$HNO_3$-HAc to remove polysilicon which may be external to the silicon wafer. Other materials external to the wafer may require different solutions to achieve complete removal.

Step 2 illustrates a rejected semiconductor wafer with all external layers removed as a result of this required processing step. Within the stripped wafer, as illustrated in Step 2, many contaminants **12** are embedded. Common contaminants may include oxygen, carbon, and metals (such as copper). Should it be desired to remove most of the contaminants, a gettering step may be used. In the gettering step, a furnace boat containing the stripped wafers is placed on a conveyor and slowly moved into a furnace having a temperature in the range between 970° and 1150°C. When the temperature of the wafers is approximately that of the furnace, a phosphorus impurity is carried in a gas stream to the wafers whereupon the phosphorus is diffused into the surfaces of the wafers. The source of the phosphorus is functionally infinite throughout the furnace cycle and is preferably $P_2O_5$, although $POCl_3$ and other gaseous sources of phosphorus may also be used.

The phosphorus is diffused from the infinite source to a depth of about 2 microns into the semiconductor wafer. This generally takes 20 to 40 minutes, depending on the furnace temperature. After the predetermined period, the wafers are withdrawn and the diffusant source is removed. Then the wafers are transferred into a cool zone and allowed to cool to a temperature suitable for handling.

With reference to Step 3, the diffused phosphorus is illustrated by **13**. At the completion of this furnace gettering step, the wafers are immersed in HF to remove the phosphorus glass.

A key to the successful application of furnace gettering steps is to generate the greatest possible surface lattice damage as a result of the phosphorus diffusion. Phosphorus atoms are smaller than the silicon substitutional lattice site they occupy, thus high concentrations of diffused phosphorus will generate more crystal strain, i.e., lattice damage. Highly mobile contaminant impurities are drawn into the silicon surface lattice damage by the Cottrell getter mechanism.

Next, with reference to Step 4, the semiconductor wafer is silicon etched to chemically dissolve the silicon lattice which contains dopant atoms introduced after crystal growth, while leaving the remaining silicon surfaces relatively flat. (See Step 4, where numerals **14** and **15** identify the front and back wafer surfaces, respectively.) In this step, the silicon wafers are loaded into a carrier which holds the wafers apart while they are immersed in a silicon etching solution. Preferably, the etchant comprises hydrofluoric acid, nitric acid, acetic acid and iodine in accordance with the following formula: $1HF\text{-}3HNO_3\text{-}4HAc\text{-}I_2$ (saturated). This etchant provides an etching rate near 12 microns per minute per side at 25°C. While the wafer carrier is immersed in the etchant, continuous agitation of either the carrier or the container should be used. The desired nonpreferential etch and nonstaining features of this chemical polishing step are enhanced by having the freshest available solution in contact with the silicon wafer at all times.

After removing 12 microns of silicon from each outer surface of about 100 silicon wafers having a 3-inch diameter, it has been found that the above-identified etchant tends to heat up and subsequently would cause a preferential etching of additional wafers immersed therein. Such an occurrence is undesirable since preferential etching introduces excessive topographical roughness to the wafer surfaces. Thus, when using 1-liter quantities, the etchant is changed after 100 wafers have been immersed therein.

Alternatively, other etching solutions that remove silicon at a constant rate regardless of the impurity type, impurity concentration, crystalline orientation, and lattice strain, as well as etching to retard the formation of strains and preferential etching may be used.

Typically, the diffusion depth of the P-N junction is known from the history of the reject silicon wafer. Generally, it is about 3 microns on PMOS wafers and 25 microns on LIC wafers. Consequently, the time required to remove the silicon varies in accordance with such a depth. In the preferred embodiment, in order to insure that all traces of the P-N junctions are removed, the wafer is immersed in the etchant for a time sufficient to remove about 0.5 mil of material more than the depth of the P-N junctions that have been diffused into the wafer.

It should be apparent that after the removal of the outer regions of the silicon wafer, the resultant wafer is relatively thin compared to a virgin wafer and in many cases may be broken.

In an alternative embodiment a back-grinding step follows the silicon etch. In this step the silicon wafer front surface is placed on a vacuum chuck. The wafer is then rotated and an abrasive material is drawn across the back surface, thus introducing a controlled amount and distribution of back wafer surface lattice damage. The wafer's back surface is continually cooled and lubricated by running deionized water preceding, during and following the abrading. Step 5 illustrates a wafer with back surface wafer lattice damage. Alternatively, the lattice damage on the back surface may be accomplished by lapping, ion implantation, or bombarding the surface with hard material.

A key to the successful application of this process is to generate a large amount of lattice damage in a controlled way such that the newly formed lattice damage does not propagate through the wafer during the subsequent furnace treatments which will be used to form the solar cell's P-N junction. The large amount of lattice damage on the back wafer surface is important for such damage will attract, and thus electrically neutralize, all highly mobile contaminants in the wafer during the subsequent furnace treatment used to form the solar cell's P-N junction. It has been found that solar cells in which this process step is employed tend to develop a longer minority carrier lift time and have a lower P-N junction leakage current, and hence exhibit an increased efficiency.

Referring now to Step 6, in another alternative embodiment the front surface **14** of the silicon wafer is polished to a mirrorlike finish. Such a finish is achieved by the conventional chemical-mechanical polishing procedure given new silicon wafers. In this step, silicon wafers are mounted on a plate using a waxlike glycolthylate. The plate is mounted on a polishing machine. The plates are pressed against a polishing cloth while the cloth and plates are rotated counter to each other to produce a polishing action. A slurry is added to the cloth/wafer interface while the polishing action takes place. Preferably, the slurry is colloidal silica in which the pH is increased by the addition of sodium hydroxides. Approximately 10 to 20 microns of silicon are removed by this front wafer surface polishing process. After the silicon surface is polished to a mirrorlike finish, the slurry is replaced by deionized water in a rinsing action. The wafers are then demounted and the wax is removed from the wafer back surface by immersing the wafers in trichloroethylene, then acetone, then deionized water.

Next, it is required to diffuse a dopant layer **20** into the front surface **14** to form a P-N junction **16** as shown in Step 7. For N-type silicon wafers, the P-N junction is formed by the diffusion of boron. This is commonly achieved by placing 50 to 300 wafers on a furnace boat. The boat is then inserted to a furnace of a temperature between 1000° and 1200°C. Boron then passes from a $BCl_3$, $BBr_3$, diborane, or boron nitride source to the wafer. The furnace cycle times and ambients are set to provide the silicon wafer with a P-N junction having a depth of about 0.3 micron (3000 A) beneath the silicon wafer surface and with a boron diffused region having a sheet resistivity of 60±30 ohms per square. The furnace doping process described will diffuse boron into all surfaces of the wafer, i.e., front, edge, and back.

Solar cells require the P-N junction only on the front wafer surface. The back wafer surface should be free of P-N junctions such that metal electrodes placed on the front and back wafer surfaces, in a later processing step, will be electrically isolated. However, as shown, the boron is also diffused into the back wafer surface. A wafer mesa etching technique for selectively removing the P-N junction from the wafer edge and back surface is carried out in a later processing step.

In an alternative embodiment, the P-N junction is formed by ion implantation. Accordingly, only the front surface facing the ion source is doped with the boron. Hence, the subsequent removal of the P-N junction from the edge and back surface, as will be described, is not required.

After the front surface has been bombarded with the ions, an annealing step is required. During annealing, the wafer is placed in a furnace filled with nitrogen and having a temperature between 900° and 950°C for a period of about 15 to 30 minutes. This allows the boron dopant to stabilize within the silicon lattice and also serves to cure radiation damage caused by the bombardment of the silicon lattice.

In yet another alternative embodiment, the boron dopant may be spun onto the front surface and then the wafer is heated to about 1000°C in the diffusion furnace. In such a manner, the dopant is restricted to the front surface of the wafer. Accordingly, as in the ion implantation step, the subsequent removal of the back surface is not required.

Still another alternative embodiment begins with a P-type starting wafer. The P-N junction is formed by the diffusion or implantation of an N-type dopant such as phosphorus or antimony. Such N-type dopants are introduced by furnace sources, ion implantation, or spun on just like the boron described above. The solar cell's P-N junction must be localized to the front wafer surface.

All silicon wafers with newly formed P-N junctions, regardless of the dopant source, should be immersed in hydrofluoric acid to remove the dopant glass phase from the wafer surfaces. Such a treatment is not identified in the figure.

Figure 50 shows an alternative process sequence. It is an alternative since it is not required if by the preceding steps the wafer's P-N junction is localized to the front wafer surface. As shown in Step 8, the back surface **15** of a doped silicon wafer is placed onto a hot plate. The wafer is then heated to about 100° to 150°C. Next, an acid-resistant wax (such as apiezon wax) **19** is placed on the top surface of the boron layer **20** of the wafer. The wax has a melting temperature between 100° and 150°C and thus is caused to melt in a covering relationship over the top surface without covering the side surfaces or the back surface of the wafer. In alternative embodiments, carnuba wax or beeswax may be used.

The structure is removed from the hot plate **18**, cooled to room temperature and then immersed in a container **17** filled with a silicon etching solution **21**, similar to that previously described which serves to remove the P-N junction from the portions of the wafer that are not covered by the wax. Such an assembly is shown in Step 9.

Next, the wafer with its wax mask is quenched in water to stop the reaction of the solution **21**, dried and thereafter immersed in trichloroethylene (TCE) solution to dissolve the apiezon wax. Afterwards, the wafer may be immersed in acetone, followed by deionized water. The P-N body **22** shown in Step **10** is the result of the processing steps applied thus far.

**Figure 50: Mesa Etch Technique as Alternative to Process of Figure 49**

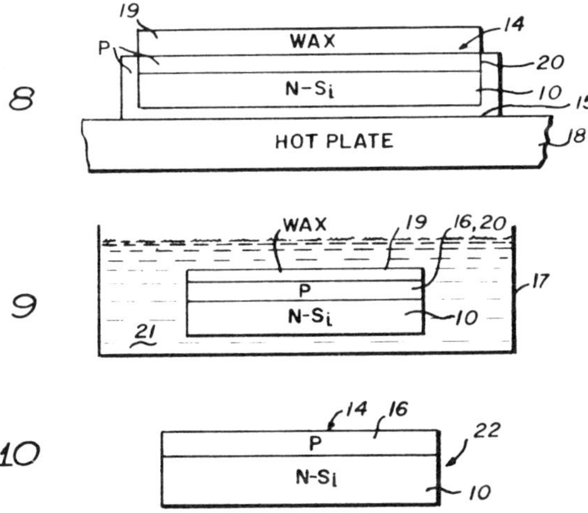


Source: U.S. Patent 4,062,102

As was previously explained, the steps illustrated in the above figure and commonly referred to as the mesa/back surface etch are not required if the P-N junction dopant is introduced by ion implantation or by spin-on techniques.

Alternatively, the P-N junction may be removed from the back surface by lapping, grinding, or bombarding the back surface with a hard material. However, if the alternative techniques are used, care must be taken not to damage the P-N junction on the front surface.

A sputtering operation may be carried out using the apparatus shown on the following page in Figure 51a to selectively deposit conductive electrodes on the front and back surfaces of the body **22**. The sputtering apparatus **23** includes a vacuum chamber **24**, a cathode electrode **25**, a shutter **26**, an anode electrode or platform **27** mounted to a shaft **28**, a motor **29** for rotatably driving the shaft, an evacuation system **30** for evacuating the chamber, an electrical energizing system **31** for applying radio frequency (RF) and dc energy to the anode and cathode electrodes, a switch **32** for selectively connecting the system **31** to target portions **33-36** of the cathode **25**, an argon source **37**, conduit **38** for connecting the argon source to the vacuum chamber, and a pressure regulator **39** for controlling the pressure of the argon introduced to the chamber.

The electrical energizing system includes an RF generator, an RF impedance matching network and a dc power supply. In the preferred embodiment, the target **33** is titanium, target **34** is silver, and target **36** is silicon monoxide. Since the apparatus is commercially available (Perkin Elmer Ultek Model 2400-85A), details of the construction and operation of the apparatus are not described.

A sputter mask processing step is then utilized to effect direct printing of the front electrode pattern on the body. The objective of this step is to position the bodies and their masks **40** (Figure 51b) such that direct metal electrode pattern formation can occur during subsequent steps. In this step the silicon wafers are placed onto the sputter apparatus platform **27** such that the front surface of the wafer is facing upward.

**Figure 51: Sputtering Apparatus for Electrode Deposition on Recovered Wafers**

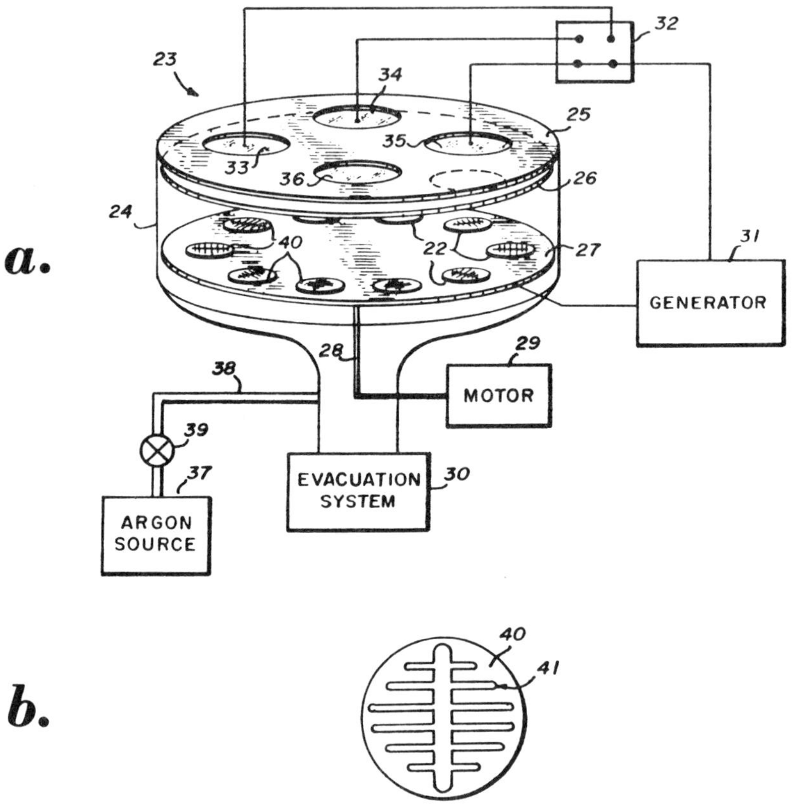


Source: U.S. Patent 4,062,102

The metal mask is placed over the wafer. The mask consists of a sheet of metal with an etched pattern of slots **41**. The etched pattern will allow for the restricted passage of metal through the slots during a subsequent step of metal deposition, for direct metal pattern formation. In the preferred embodiment, the mask is formed from full-hard stainless steel having a thickness of about 8 mils and having extremely flat outer surfaces.

Steps 11 through 16 in Figure 52 complete the production of the solar cell. The assembly of the mask **40**, wafer **22**, and platform **27** is shown in Step 11. Following the masking step, a front metallization step is required to clean and then deposit a metal pattern on the wafer's front surface. This step is illustrated in 12 and 13 in the figure and utilizes the vacuum system **30** to draw a pressure of $10^{-6}$ torr in the sputter chamber **24** (Figure 51). (Vacuum pressures of $10^{-5}$ to $10^{-7}$ torr may be utilized.) After the desired vacuum pressure is drawn, argon **37** is then released to the chamber such that a pressure near $2 \times 10^{-2}$ torr is obtained and controlled by the regulator **39**. Next, the power generator **31** is energized to apply an electrically reverse RF power flow between the cathode **25** and anode **27**. In this manner the surface of the wafer exposed through the slots in the mask is bombarded with ions. This action removes small amounts of silicon surface contamination (as indicated in Step 12) which would otherwise increase silicon-metal contact resistance.

After the sputter cleaning action is completed, the cathode target **33** comprised of a first metal to be deposited on the wafers is energized. Initial deposition occurs onto the shutter **26** to minimize the possibility of transferring cathode target surface contamination onto the silicon wafers. After this initial deposition, the shutter is rotated to a position that allows the cathode target material to be transferred as a metal layer **42** onto the assembly comprising the silicon wafers **22**, the metal masks **40**, and the platform **27**. This activity is illustrated in Step 13. The initial metal that comprises the target and is deposited as the layer is titanium.

After the deposition of 1500 A of titanium, the source is deenergized, and the second metal cathode target **34** is energized. In a manner similar to that previously described, the cathode surface contamination is deposited on the shutter and the shutter is rotated to allow a second metal layer **44** to be deposited on the first metal layer that covers the wafers, the metal mask and the platform. The second metal that comprises the target **34** and is deposited as layer **42** is silver. After the deposition of 50,000 A of silver on the first metal layer, the source is deenergized. The deposition of the front metal electrode on the solar cell is thus completed and the chamber is allowed to return to atmospheric pressure. The metal masks are lifted from the silicon wafers.

The next required processing steps, i.e., back metallization, are illustrated in Steps 14 and 15. In these steps, the wafers are turned over such that the silver layer **44** rests on the platform. Accordingly, the back surface **15** now faces upward toward the shutter **26** and the associated targets of the sputter apparatus **23**. The sputter chamber **24** is closed, a vacuum is drawn and a sequence of ac-

Figure 52: Final Steps in Solar Cell Production from Rejected Semiconductor Wafers

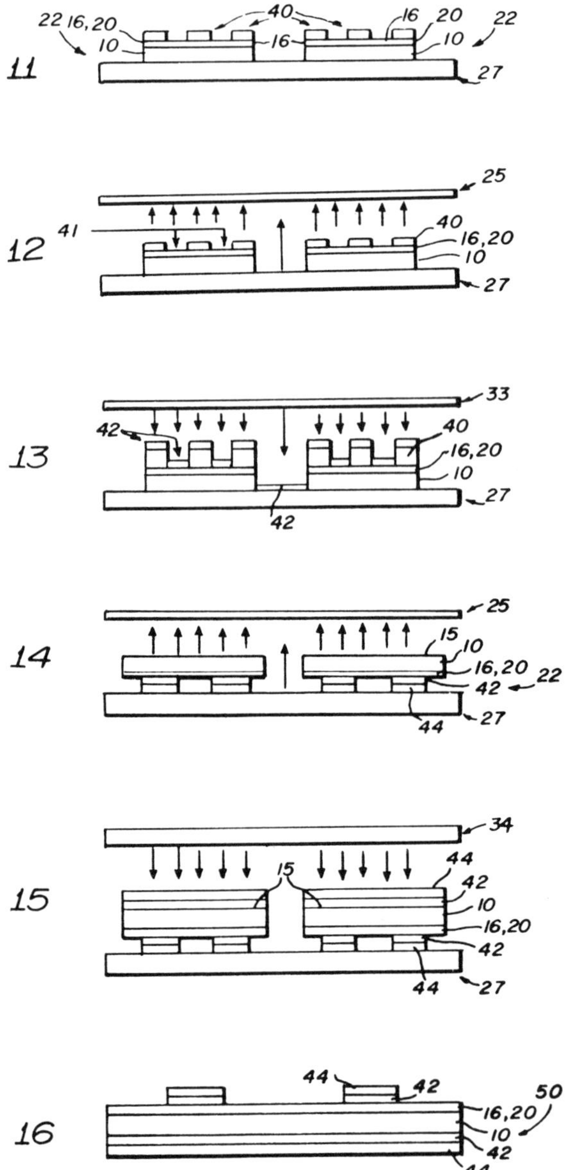


Source: U.S. Patent 4,062,102

tions similar to those described in the front metallization processing step is followed. However, in this step no metal mask is used to selectively cover the back wafer surface. Step 14 illustrates the cleaning action of the back surface and is similar to that shown in Step 12. Step 15 illustrates the deposition action (similar to that shown in Step 13) by which two metal layers are deposited on the back surface. As before, upon the completion of second metal deposition, the sputter chamber is allowed to rise to atmospheric pressure.

With the described process a solar cell **50**, as shown in Step 16, has been formed. The solar cell includes the silicon wafer having a front surface P-N junction, a titanium layer and a silver layer.

Relative to the metal layers, it has been found that the layers described above, i.e., 1500 A titanium covered by 50,000 A silver, provide the resulting solar cell with very favorable high conductivity electrodes. However, it should be recognized that the metal layer thicknesses can vary and the types of metal can be changed, to provide similarly favorable results in solar cell product performance.

In fabricating the solar cell, various alternatives are possible. For example, a sintering step can be used to reduce silicon-metal contact resistance and thus increase solar cell performance efficiency. In the sintering step the solar cell wafers are loaded on a furnace boat. The boat is then inserted into a furnace having a temperature of about 500°C (for titanium-silver metallization layers) and a nitrogen or nitrogen-hydrogen mixture of gasses as the furnace ambient. The wafers are left in the furnace for a period of approximately 15 minutes. After this period, the wafers are withdrawn from the furnace and allowed to cool to room temperature.

In many of the applications to which solar cells have been put, an important utilization is to continuously apply a "trickle charge" to a battery that powers a device located far from more conventional sources of electricity. Thus, solar panels using a multiplicity of cells have found employment on sea-going oil and gas drilling rigs, and for recharging battery-powered microwave relay stations located at the peaks of mountains.

Whether the battery being recharged is small or large, the possibility of overcharging the battery can present a serious hazard. While it is possible to use a voltage regulator to prevent overcharging, such a regulator increases expenses, may reduce efficiency of the overall system, and is simply another piece of equipment subject to corrosion and, ultimately, to failure. Without the voltage regulator, the battery can be overcharged to the point of failure, indeed even to explosion. In any case, failure of the battery leads to inoperativeness of the load devices, e.g., the light on a buoy, or the operation of a solar powered watch or flashlight or pocket calculator.

In terrestrial uses of solar cells, therefore, it will be apparent that it is often desirable for a solar cell to maintain a substantially uniform voltage output, since there will be wide variations in illumination present under ambient conditions.

The purpose of a cell design developed by *J. Lindmayer; U.S. Patent 4,137,095; January 30, 1979; assigned to Solarex Corporation* is to provide a solar energy cell in which, once the desired voltage has been reached, exposure to significantly increased illumination will not significantly increase the voltage generated by the cell to a level such that a device or battery powered by the cell will become inoperative. This object is accomplished not by a regulator circuit, but through the inherent electrical characteristics of the cell, which are thereby preserved even at low light levels.

While there are other methods of forming a silicon solar energy cell having a substantially constant voltage output under varying conditions of illumination, the present process involves the use of gallium as an alloying agent to form a back junction in the cell.

The use of gallium does produce a cell that has a different function than some aluminum-alloyed cells: it results in a solar cell having a substantially constant voltage output.

Solar cell efficiency is enhanced if the reflectivity of the silicon surface is reduced in order to increase the amount of light absorbed by the cell. Prior art researchers have recognized the desirability of altering silicon surfaces, preferably by chemical etching.

In a process developed by *W.L. Bailey, M.G. Coleman, C.B. Harris and I.A. Lesk; U.S. Patent 4,137,123; January 30, 1979; assigned to Motorola, Inc.* a surface etchant for silicon comprising an anisotropic etchant containing silicon is disclosed. The etchant provides a textured surface of randomly spaced and sized pyramids on a silicon surface. It is particularly useful in reducing the reflectivity of solar cell surfaces.

The following is one specific example of the conduct of the present process. An etchant solution was made by dissolving 7.0 g of silicon in a solution of 500 ml $H_2O$, 25 g potassium hydroxide and 50 ml of ethylene glycol. The solution was covered and heated to 80°C. Silicon wafers having a (100)-type orientation were immersed in the heated etchant solution. After 1 hour of immersion in the heated solution, the wafers were removed, rinsed in water and dried. The wafers were examined under an electron microscope and found to have textured surfaces. Reflectance measurements before and after etching indicated that the etching step reduced reflectance from about 30 to 9%. The wafers were then processed into solar cells which were tested and found to be operable.

A process developed by *J.W. Yerkes and J.E. Avery; U.S. Patent 4,141,811; February 27, 1979; assigned to Atlantic Richfield Company* is a process for manufacturing solar cells in which silicon wafers are assembled in a holding jig such as a diffusion boat in pairs with adjacent surfaces of each of the pairs in contact. The thus-assembled wafers are subjected to a chemical vapor deposition diffusion step during which a phosphorus pentoxide glass layer is formed predominantly on the exposed surfaces and side edges of the wafers along with a P–N junction thereunder. A small amount of the phosphorus pentoxide glass is also formed on the surfaces which are in contact with each other, particularly at the outer edges. The wafers are then reassembled in such a manner that the surfaces having the phosphorus pentoxide layer thereon are placed in contact with each other and again the wafers are placed in a holding jig. The thus-assembled wafers are then placed in a plasma etching reactor for removal of the phosphorus pentoxide glass layer and the underlying P-N junction from the side edges and the surfaces opposite the surfaces exposed during the previous diffusion step.

Figure 53 shows a typical diffusion jig or boat **10** constructed, for example, of quartz. The diffusion boat has a plurality of slots, shown at **12** formed normally to receive a single wafer with appropriate space between wafers to allow the desired diffusion gases to totally surround the wafer during the diffusion process. The wafers are assembled into the boat in pairs, such as shown at **14** and **16**. Through appropriate sizing of the slots or by providing other mechanical means, the individual wafers comprising each of the pairs are assembled in such a fashion that adjacent first surfaces are maintained in close and intimate contact.

**Figure 53: Apparatus for Chemical Vapor Deposition Diffusion Step in Solar Cell Manufacture**

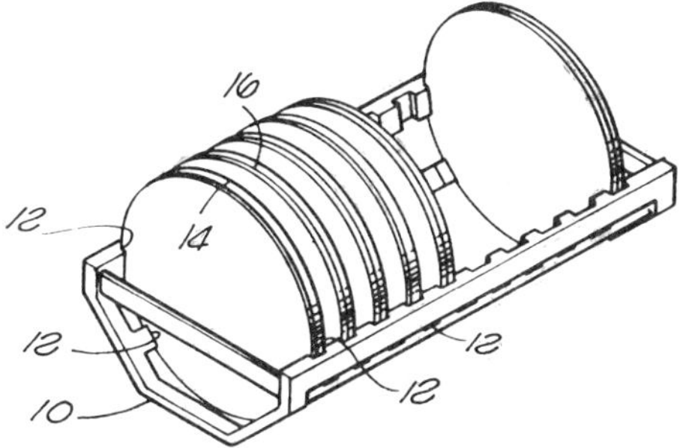


Source: U.S. Patent 4,141,811

Then Steps (a) through (e) in Figure 54, shown on the following page, illustrate the device in various stages of the overall process.

In Step (a) there are illustrated a pair of wafers **20** and **22** having first surfaces **24** and **26** in contact with each other. With the wafers thus assembled the contact between the surfaces functions as a diffusion mask. Although masking does occur in that diffusion gases do not penetrate to provide an effective deposit of the diffusion material, some of the material does penetrate between the surfaces and thus there is some diffusion of active impurities into those surfaces (mostly at the outer edges). However, when compared to the amount of the phosphorus pentoxide glass deposited on the exposed surfaces **28** and **30** and the exposed edges **32** and **34**, there is very little diffusion into the surfaces **24** and **26**.

With the wafers arranged in pairs as described above, the boat **10** is then inserted into the diffusion furnace and a diffusant such as phosphorus oxychloride is caused to pass over the heated wafers as is represented by the arrows **36** in Step (a). The diffusion of active impurities into semiconductor wafers is well known to the art and a detailed description thereof is not required herein.

**Figure 54: Steps in Solar Cell Production by Atlantic Richfield Process**

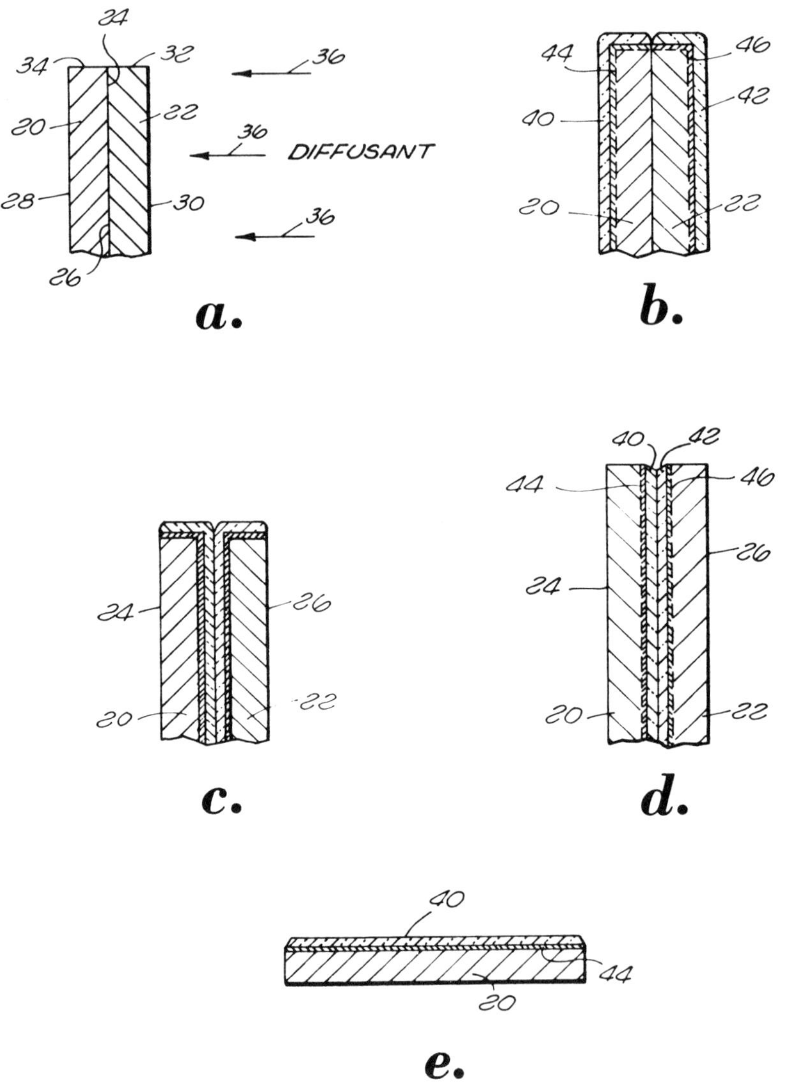


Source: U.S. Patent 4,141,811

After sufficient time has expired, the wafers will then appear as illustrated in Step (b). As is shown, the wafers **20** and **22** have a layer of phosphorus pentoxide glass **40** and **42**, respectively, thereon. The layers of glass extend over the exposed surfaces **28** and **30** as illustrated in Step (a) and the exposed peripheral edges **32** and **34**. It is noted, however, that little or no glass is formed on the adjacent contacting surfaces **24** and **26**. Thus, as above pointed out the arrangement of the pairs of wafers with the adjacent first surfaces in contact effectively functions as a mask to preclude substantial growth of the phosphorus pentoxide glass and subsequent junction formation in the contacting surfaces.

By spacing the pairs of wafers apart, as illustrated in Figure 53, the gaseous diffusant material is able to easily penetrate the spaces between the adjacent pairs of wafers to provide an adequate source of material to cover the exposed surfaces of the pairs of wafers to thereby provide the desired formation of the phosphorus pentoxide layers such as shown at **40** and **42** of Step (b).

Subsequent to formation of the layers of phosphorus pentoxide glass and diffusion of the active impurity phosphorus therefrom into the wafers to form a P-N junction as shown, for example, at **44** and **46**, the wafers are removed from the boat. As illustrated in Step (c), the wafers are then reassembled by reversing the faces of contact with each other. Thus the faces of the wafers **20** and **22** bearing the phosphorus pentoxide layer **40** and **42** are placed in contact with each other. The first surfaces previously in contact with each other, as shown at **24** and **26** of Step (a), are now exposed as illustrated in Step (c). The pairs of wafers thus rearranged are again inserted into the boat and, as indicated above through arrangement of appropriate sized slots or with mechanical devices, the adjacent faces are brought into close and intimate contact. The boat with the wafers thus arranged is placed in the reaction chamber of a plasma etching apparatus.

Figure 55, on the following page, shows the plasma etching apparatus.The apparatus includes a reactor chamber **52** typically made of quartz, having a cover **54** and a gas inlet manifold **56**. The side of the reactor has been partially broken away to better illustrate gas diffusion tubes **57** disposed therein and being externally connected to the manifold.

The pressurized supply **58** of a binary gaseous mixture comprised of oxygen and a halocarbon gas is connected through a pressure-regulating valve **60**, a three-way solenoid valve **62**, and a flow meter **64** to the manifold. A vacuum gauge **66** provides an indication of total reaction pressure in the reactor. At any time, and prior to introduction of the gas mixture to the manifold, the corresponding flow lines are constantly evacuated through the three-way solenoid valve leading to the mechanical vacuum pump **68**. Such is also accomplished where air and atmospheric pressure prevail in the reactor through the utilization of a three-way isolation valve **80**. A source of radio frequency power **82** provides exciting energy through a matching network **84** to coil **86** which surrounds the reaction chamber.

The halocarbon utilized should be selected from the group of organohalides having not more than 2 carbon atoms per molecule and in which the carbon atoms are attached to a predominance of fluorine atoms. The preferred gaseous mixture is produced from a mixture containing approximately 8.5% by volume of oxygen and 91.5% tetrafluoromethane gas.

Through the generation in the reaction chamber the fluorocarbon-based plasma reacts with the exposed phosphorus pentoxide and silicon material to combine therewith to thus etch such material.

**Figure 55: Plasma Etching Apparatus**

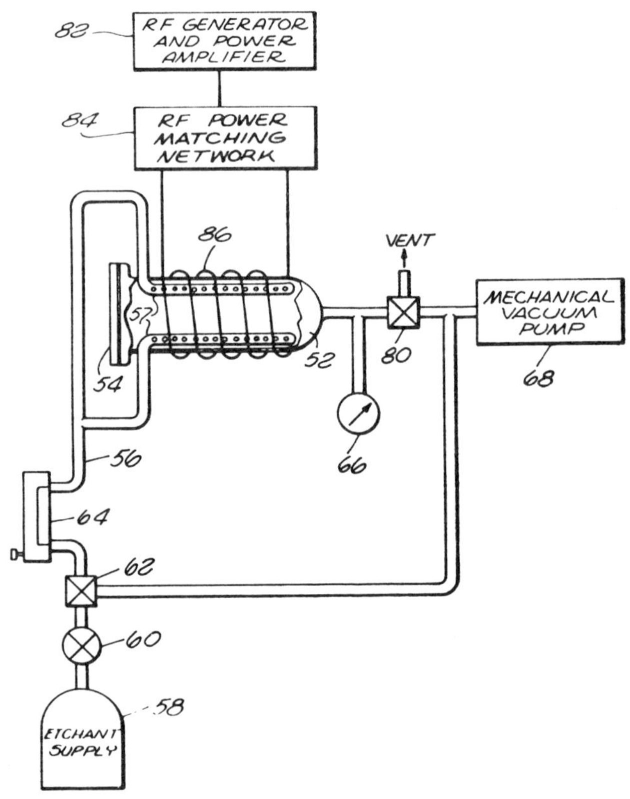


Source: U.S. Patent 4,141,811

While in the reaction chamber of the plasma etching apparatus the peripheral edges of the wafers with the phosphorus pentoxide layers formed thereon are etched more rapidly than is the exposed surface of the wafers. As pointed out above, such is even enhanced by the exothermic reaction which further heats the wafer. There is clearly a nonuniform etching of the wafer with the more rapid etching occurring on the outer edges than on the flat surface. It has further been found that the plasma etching of semiconductor wafers is such that the phosphorus doped silicon dioxide will be etched more quickly than the boron doped silicon dioxide. As a result, the nonuniform etching of the silicon wafer is further enhanced. To speed up the etching process on the wafers, the boat bearing the wafers arranged as illustrated in Step (c) of Figure 54 is placed at the center of the reaction chamber without any shielding, thereby being subjected to the maximum concentration of the plasma, and particularly the ions and free radicals contained therein.

By increasing the pressure within the reaction chamber to approximately 20 torr the etching process will be increased, thereby increasing the through-put of the plasma etching apparatus.

Upon completion of the plasma etching process the boat, constructed of plastic, glass or aluminum, is removed from the reaction chamber and the wafers will appear as illustrated in Step (d) of Figure 54. As shown, the wafers have the phosphorus pentoxide coating removed from the peripheral edges and surfaces **24** and **26**. Also removed from the surfaces and peripheral edges is any converted semiconductor material. However, beneath the layers of phosphorus pentoxide coated on the surfaces **28** and **30** the P-N junction remains as shown at **44** and **46** and the layers **40** and **42** of phosphorus pentoxide also remain.

Upon removal and disassembly the completed and etched device is as illustrated in Step (e) of Figure 54 and includes the original boron-doped P-type wafer **20** having the coating of phosphorus pentoxide **40** thereon with the P-N junction **44** inbetween. The device thus formed is now ready for further and appropriate processing to apply the desired contacts to each of the surfaces for utilization as a solar cell.

It can thus be seen that by utilizing what has previously been considered to be deficiencies in the plasma etching process there results a number of distinct advantages, particularly when manufacturing solar cells in production quantities through utilization of the surfaces of the wafers as selfmasks. It is readily apparent that twice as many wafers can be diffused at a time than has previously been possible. This obviously reduces the amount of diffusant source material required by a substantial degree, thus further decreasing the cost of formation of the cell. Through utilization of the plasma etching the previous wet etching process is eliminated, thus eliminating additional steps of wetting and drying from the process and also eliminating the difficulty in handling the acid etch residue and pollutant material.

Again, through utilization of the selfmasking twice as many semiconductor wafers may be etched at a time in the plasma etching apparatus as compared to the previous utilization of the apparatus without any increase in surface area to be etched. Through utilization of the previously considered deficiencies in the plasma etching, the plasma etch times can be made much shorter and thus the through-put of the plasma etching apparatus can become higher. As a result of the shorter etching times and the plasma etch apparatus, less damage is done to the surfaces of the wafers in which the P–N junction has been formed.

It can thus be seen that through the utilization of the selfmasking and dry-etching techniques in the formation of solar cells in accordance with the present process, not only is there a dramatic cost reduction per unit but in addition there is a dramatic reduction in the formation of pollutants through utilization of the wet-acid-water-etching techniques previously in existence.

The two most common methods for doping crystalline silicon to form P–N rectifying junctions are ion implantation and high-temperature diffusion. Both of these techniques are subject to disadvantages. The chief disadvantage of ion-implantation is that the energetic dopant ions produce displacement damage in the silicon crystal, necessitating an additional, carefully controlled annealing operation. The chief disadvantage of high-temperature diffusion is that a dense region of dopant precipitates may be formed very near the surface of the crystal, resulting in a dead layer characterized by an extremely short carrier lifetime. In solar cells, this layer limits the short-wavelength photovoltaic response.

A more recently investigated method for the forming of P–N junctions in silicon bodies is the use of laser pulses to induce diffusion of a layer of dopant into the silicon substrate.

A process developed by *J. Narayan and R.T. Young; U.S. Patent 4,147,563; April 3, 1979; assigned to the U.S. Department of Energy* is one in which high-quality junctions are prepared by effecting laser-diffusion of a selected dopant into silicon by means of laser pulses having a wavelength of from 0.3 to 1.1 $\mu$m, an energy area density from 1.0 to 2.0 J/cm$^2$, and a duration of from 20 to 60 nanoseconds. Initially, the dopant is deposited on the silicon as a superficial layer, preferably one having a thickness in the range of from 50 to 100 A. Depending on the application, the values for the abovementioned pulse parameters are selected to produce melting of silicon to depths in the range from 1000 A to 1 $\mu$m. The process has been used to produce solar cells having a one-sun conversion efficiency of 10.6%, these cells having no antireflective coating or back-surface fields.

## THIN FILM DEVICES

Pros and cons of thin film solar cell designs have been reviewed (with an excellent bibliography) by Hovel (2). He describes thin film cells as the fourth major category of cells (in addition to silicon, cadmium sulfide cells and concentrator cells). Of course there is an overlap in this categorization since cadmium sulfide cells often fall into the thin film category. Hovel goes on to point out the advantages of thin film devices as:

(1) Little semiconductor material is needed.

(2) Low-cost substrates can be used.

(3) Continuous processing of large sheets can be visualized.

The disadvantage to-date primarily lies in low cell efficiency because of:

(1) Grain boundary effects.

(2) Poor quality of semiconductor material often resulting from growth on foreign substrates.

Thin films span a range of perfection extending from single crystals through large crystallites to highly disordered or amorphous materials. Many fabrication techniques for producing them (e.g., vapor phase epitaxy, chemical transport, and molecular beam deposition) are well adapted for making single crystalline, electronically sophisticated, multilayered structures. Some of these same techniques, as well as others (e.g., spray pyrolysis), are also applicable to inexpensive production of films of far less perfection.

Highly disordered thin films are typically characterized by large defect concentrations and short lifetimes compared to those found in single crystals. Despite this fact, there are cells with efficiencies near 10% (e.g., $CdS/Cu_2S$).

Such thin film cells have the advantage of using little active material because they can absorb sunlight more efficiently than silicon. However, they are fragile, can exhibit chemical degradation over long periods, are difficult to manufacture reproducibly, and are frequently inefficient energy transducers. The requisite service life has not yet been demonstrated. Many of the highly disordered thin film systems are sufficiently complicated, both structurally and chemically, that the basic understanding of their operation is severely limited. The same considerations concerning stability and service life may well apply to sophisticated multilayer devices having greater crystalline perfection.

Many of these difficulties may be resolved either by improving the technology or by lowering manufacturing costs to a point where considerations such as longevity are less important. Thus, thin films could form the basis of an inexpensive cell technology. Work to develop this potential for low-cost power generation will include the search for new materials and research exploring basic phenomena in thin film and prototype crystalline systems (1).

Since no thin film system has as yet been proven satisfactory, other potentially useful thin film systems should be sought. Materials with strong optical absorption and band gaps matched to the solar spectrum need to be identified and characterized. Systematic trends in the density and spectrum of traps in disordered layers must be delineated. The problem of making stable contacts to such fragile materials needs to be explored. Much of this work might be purely empirical; however, research concerned with the basic phenomena in thin films also requires emphasis. While work on prototype crystalline systems may be useful, the unique complications associated with thin films, which are quite different from those in single crystals, should be primary targets of the investigations.

As in other flat plate systems, effective application of inexpensive, presumably low efficiency cells requires innovation in other than cell related costs. Low cost encapsulants, support structures, and installation that maintain the structural integrity and chemical stability of the cell are required (1).

A 3% cell made at, say, the cost of aluminum foil would not be usable as a source of grid electricity because of the other area costs associated with building a reliable and durable field of arrays. Although the possibility of a high efficiency thin film cell that is cheap to fabricate cannot be ruled out, it is likely that low cost mass production techniques will yield cells with performance inferior to single crystal cells for reasons of materials purity and quality control. The reduction of noncell area costs is thus critical to the success of very inexpensive cells and should be regarded as an integral part of thin film development.

A process developed by *F.A. Shirland; U.S. Patent 3,713,893; January 30, 1973; assigned to Gould, Inc.* is one in which vacuum evaporated cadmium sulfide, thin film solar cells are employed to form a high voltage solar battery quickly and economically that can transfer useful amounts of power to an external useful device with a high degree of reliability.

A plurality of small areas of cadmium sulfide, CdS, or other suitable semiconductive material is vacuum evaporated through a suitably shaped evaporation mask onto a substrate on which an appropriate pattern of negative electrometallic leads has been vacuum evaporated, or otherwise formed, so that each area of CdS will partially cover one lead. A barrier is formed on the upper surface of each CdS area by applying a cuprous sulfide ($Cu_2S$) slurry, or by other means known in the art. However, the slurry barrier is kept separated from the negative electrode leads. The slurry is heated to form barriers.

Then the cells so formed are connected by superimposing a thin flexible insulating layer in a suitably laid down pattern (by vacuum evaporation or other means) of positive electrode leads so arranged geometrically that each lead on the upper insulating barrier bridges from the barrier of one cell to the negative electrode lead which projects from the side (or edge) of the previous cell. Prior to the attachment of the upper insulating layer, however, an insulating stripe is applied over the edges of the cells to keep the positive electrodes from shorting across the edge of cadmium sulfide where the barrier does not extend. Either the substrate or the upper insulating layer may be translucent, but one of them must be, so that either a front wall or back wall cell is possible.

The number of cadmium sulfide, or other suitable semiconductor, areas is chosen to yield the total voltage desired, for the area of each cadmium sulfide area is made only as large as needed to yield the desired output current.

A process developed by *A.G. Milnes and D.L. Feucht; U.S. Patent 3,993,533; November 23, 1976; assigned to Carnegie-Mellon University* involves producing thin semiconductor film for use in solar cells. The desired semiconductor is grown epitaxially on a second semiconductor film which may be epitaxial on a third semiconductor. The second semiconductor has a lower melting point than the desired semiconductor. The temperature of the second semiconductor is increased. This creates a molten state in the second semiconductor and the desired semiconductor is stripped away from the second semiconductor. The desired film may be detached by dissolving the second semiconductor with a chemical agent that dissolves the second semiconductor.

More particularly, this process describes a different means of fabricating single crystal films of silicon or other semiconductors in the range of 1 to 50 micrometers for use in solar cells designated as "peeled film technology" (PFT). Such films, produced at low cost, offer the possibility of providing N-P junctions or Schottky barrier metal-N semiconductor junctions for cheap photovoltaic solar cells for the direct generation of electrical energy from sunlight. Figure 56 shows the fundamentals of thin film production by this process.

**Figure 56: Thin-Film Solar Cell Construction—Carnegie-Mellon Process**

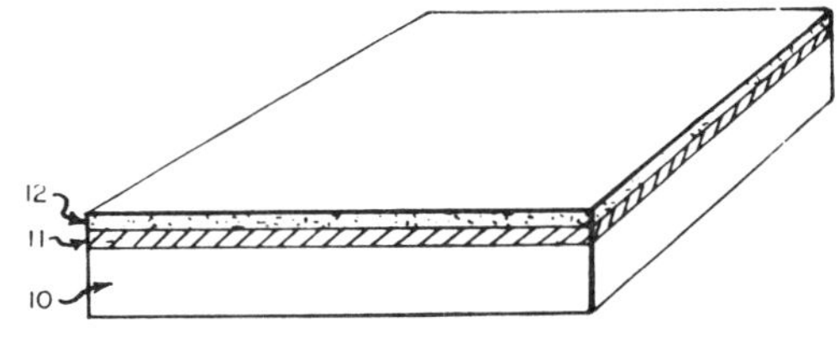


Source: U.S. Patent 3,993,533

The figure shows a slab of single-crystal silicon **10** rectangular in shape and having dimensions of 20 cm long by 6 cm wide by 0.5 cm thick. On the top face of the slab a single-crystal, or nearly single-crystal, layer of germanium or $Si_xGe_{1-x}$ alloy (where x may have values from 0 to 0.8) **11** has been grown having a thickness from 0.05 to 5 $\mu$m, but preferably of about 0.5 $\mu$m. This layer is single crystal because it has been nucleated by the clean, oxide-free, single-crystal surface of the silicon that matches the lattice constant and crystalline structure (cubic-face-centered diamond structure) of the silicon-germanium to within 4%. Under such conditions of close lattice match it is well known that single-crystal epitaxy occurs. Up to 80% Si may be included in the $Si_xGe_{1-x}$ layer to improve lattice match conditions and to allow a high temperature of deposition of the Si layer **12** to permit good single-crystal epitaxy.

This single-crystal layer may have been produced in one of many ways that are known in the art for the production of germanium or silicon-germanium layers on germanium or silicon seeds. Preferred techniques include the thermal decomposition of $GeH_4$, $GeHCl_3$, $GeCl_4$, $GeH_2Cl_2$, $GeH_3Cl$, $SiH_4$, $SiHCl_3$, $SiCl_4$, $SiH_2Cl_2$, and $SiH_3Cl$, in a stream of gas, such as $H_2$ or He or Ar, at temperatures in the range of 700° to 1300°C. The growth of $Si_xGe_{1-x}$ single-crystal layers on Si seeds from a mixed stream of $GeCl_4$, $SiCl_4$ and $H_2$ has been described by Miller and Grieco (*J. Electrochemical Soc.* 109, 1962). Also, there has been further work by Lyutovich, et al (*Sov. Physics—Crystallography* 15, 1971) and others over a substantial range of values of x.

Some techniques involve the growth of germanium and silicon from iodine or bromine compounds of Ge and Si such as $GeI_2$ by disproportionation into $GeI_4$ and Ge at temperatures in the range of 400° to 700°C. Yet other techniques include the passage of the Si slab through a molten bath of Ge or Ge-Si, or a bath of Ge dissolved in Sn or Ga or Pb, or such a bath containing both Ge and Si. Yet other techniques involve evaporation or sputtering of germanium or germanium and silicon.

On top of the Ge or $Si_xGe_{1-x}$ layer has been grown a single, or substantially single, crystal layer of silicon having a thickness from 1 to 50 $\mu$m. The layer is single crystal because it has been epitaxially grown on the single-crystal $Si_xGe_{1-x}$ layer. This single-crystal Si layer may have been produced in one of many ways that are known in the art for the production of silicon layers on silicon seed crystals. The preferred techniques include the thermal decomposition of $SiHCl_3$ or $SiH_4$ or $SiCl_4$, or $SiH_2Cl_2$ or $SiH_3Cl$ in a stream of gas such as $H_2$ or He or Ar at temperatures in the range of 700° to 1200°C, but preferably below the melting point of $Si_xGe_{1-x}$ alloy. Other techniques include the thermal decomposition of the analogous iodine and bromine compounds of silicon with hydrogen. Other techniques of depositing the Si can involve evaporation of Si or sputtering.

The silicon layer can be removed from the structure by raising the temperature for a short period above the melting point of the intermediate layer **11**. The silicon does not melt because its melting point is 1420°C. Since the silicon-germanium alloy is molten at some temperature in the range above 960° to 1350°C, the silicon layer can be lifted (detached) off by a vacuum pick-off or by any attachment that may have been made to it. It may be pushed off by a probe or it may even slip off under its own weight if the slab is tilted and if the intermediate layer is sufficiently liquid and of low viscosity.

During the growth of the $Si_xGe_{1-x}$ and Si layers, dopants may be added to the deposition process so that the semiconductor properties are convenient for subsequent fabrication of a solar cell from the Si layer. For instance, in the deposition of the $Si_xGe_{1-x}$ and Si layers diborane can be added to the gas stream to produce P-type boron-doped layers and phosphine or arsine to produce N-type phosphorus or arsenic-doped layers. Thus, one structure would consist of the $Si_xGe_{1-x}$ layer **11** heavily P-doped and the Si layer **12** lightly P-doped, except for the final top surface of the layer (0.1 to 3 $\mu$m deep) which is N-doped to form a P-N junction in the silicon **12**. The heavily P-doped layer of $Si_xGe_{1-x}$

that is residual on the underside of the P-Si film after peeling is helpful in assuring good ohmic contact to the Si film. Another structure would consist of a heavily doped N-$Si_xGe_{1-x}$ layer on which N-Si is followed by a thin P-Si layer, or by a thin Schottky barrier of Pt or Au.

Although the process has been described with reference to the growth and detaching of a Si film, it is equally appropriate for the growth and detaching of single-crystal or near-single-crystal films of GaAs or $GaAs_xP_{1-x}$ or GaP, or $Al_yGa_{1-y}As$. The lattice match of GaAs is almost identical with that of Ge and the melting point of GaAs is 1240°C so that Ge can be made molten (MP 960°C) for the detaching process. The GaAs film could be grown by interacting Ga, $GaCl_3$ and $AsH_3$ or $AsCl_3$, or by flash evaporation or any other known art process that does not damage the Ge film. It may be observed that Ge is a very slow diffuser in GaAs and, therefore, will not seriously dope the GaAs film during growth. The seed slab for the growth of a GaAs film could, for example, be GaAs or Si single-crystal material. If the slab is GaAs the epitaxy tends to be more perfect than if it is Si.

In the same way, using Ge or $Si_xGe_{1-x}$ as a release, single-crystal films of GaP, $GaAs_xP_{1-x}$, and $Al_xGa_{1-x}As$ can be grown. All these semiconductors have lattice constants within about 4% of that of Ge and melting points above that of the release agent.

The method is limited in materials only by the requirement that layer **11** shall have a lower melting point than layer **12** on the slab **10**; and that preferably materials **10, 11** and **12** shall have closely similar lattice constants so that the epitaxy by known-art methods shall be good and single crystal or nearly so. For example, InP could be used for either the third semiconductor or the desired semiconductor film with an intermediate layer of lower melting point such as Ge or InAs.

In an alternate version of the process, the desired thin semiconductor film is detached by dissolving the second semiconductor film in a chemical agent that dissolves it more readily than the desired film and the third semiconductor.

A process developed by *R.J. Janowiecki, M.C. Willson and D.H. Harris; U.S. Patent 4,003,770; January 18, 1977; assigned to Monsanto Research Corporation* is one in which polycrystalline silicon films useful in preparing solar cells primarily for terrestrial application are prepared by a plasma spraying process. A doped silicon powder is injected into a high temperature ionized gas (plasma) to become molten and to be sprayed onto a low-cost substrate. Upon cooling, a dense polycrystalline silicon film is obtained. A P-N junction is formed on the sprayed film by spray deposition, diffusion or ion implantation. A sprayed junction is produced by plasma spraying a thin layer of silicon of opposite polarity or type over the initially deposited doped film. In forming a diffused junction, dopant is applied over the surface of the initial plasma-sprayed film usually from the vapor phase and heat is used to cause the dopant to diffuse into the film to form a shallow layer of opposite polarity to that in the original film.

A junction is also formed by implanting dopant ions in the surface of the originally deposited film by the use of electrical fields. When used in conjunction with ohmic contacts and electrical conductors, the P-N junctions produced using plasma-sprayed polycrystalline silicon films are formed into solar cells which are useful for directly converting sunlight into electricity by means of the photovoltaic effect.

In the prior art, polycrystalline silicon films have been produced by chemical vapor deposition and by evaporation. Both processes are slow, requiring much time to produce films of useful thickness, and both processes are conducted in a vacuum environment. The plasma spraying process described herein is a fast deposition process, conducted at essentially atmospheric pressure, capable of the large-scale production of low-cost polycrystalline silicon photovoltaic cells in large-area form on an automated, continuous basis. The process is capable of depositing not only polycrystalline silicon but also electrical contacts, antireflective or protective coatings, and other semiconductor materials.

The plasma generators used in this process are well known in the art and commercially available. They are electrical devices used for continuously generating a high temperature gas which is partially or completely ionized (a plasma). The high temperature gas is used for the melting of injected particles and the spraying and deposition of molten particles onto a substrate to solidify and form a polycrystalline film. By injecting doped silicon particles into the hot gas, doped polycrystalline silicon films can be produced. Films can be deposited on an electrically-conducting substrate such as a metal or graphite which can then serve as ohmic contact in the solar cell. Metallized nonconductors can also be used as substrates. Free-standing silicon films can be produced by depositing onto removable substrates.

Figure 57 schematically shows various techniques for P-N junction formation in solar cell manufacture. Using a doped polycrystalline silicon film formed by melting powder with a direct current or an RF plasma generator, a P-N junction can be formed by plasma-spraying a polycrystalline silicon layer of opposite polarity or type over the initial film or by diffusing dopant atoms or implanting dopant ions into the surface of the initial film.

Figure 58 shows the type of apparatus which may be used for generating continuous plasma streams and utilizing them to produce polycrystalline silicon films. In the dc arc plasma generator system, an arc is formed and sustained between a cathode and an anode and a plasma is formed by passing a nonoxidizing plasma-forming gas through the arc. Doped silicon particles are contained in a feeder and carried in a separate gas stream and injected into the torch for transport through the plasma stream to be melted and deposited on a substrate.

In the ac radio frequency induction plasma generator system, an arc is produced by a high frequency ac induction coil and used to form a plasma stream from a nonoxidizing gas. Silicon particles may be injected axially or in other ways into the plasma stream for melting and deposition on a substrate to form a polycrystalline silicon film.

Figure 57: Various Techniques for P-N Junction Formation in Solar Cell Manufacture

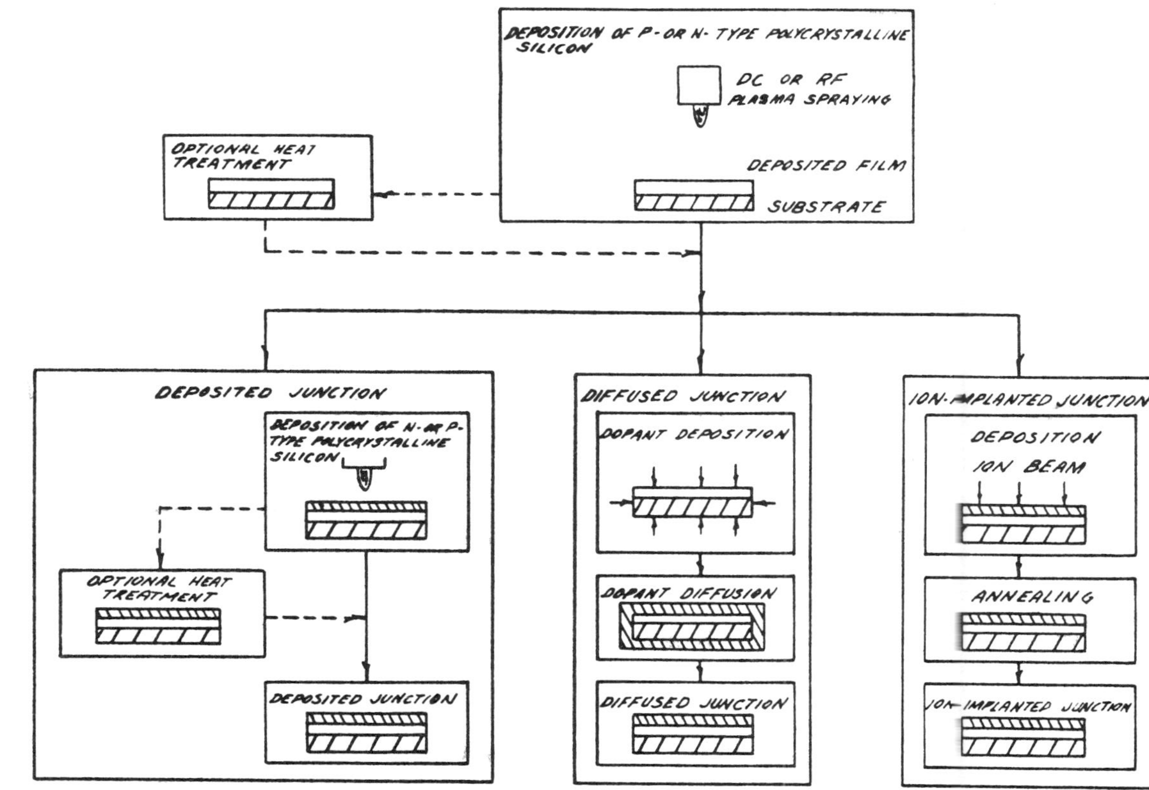


Source: U.S. Patent 4,003,770

**Figure 58: Plasma Spraying Process for Preparing Polycrystalline Solar Cells**

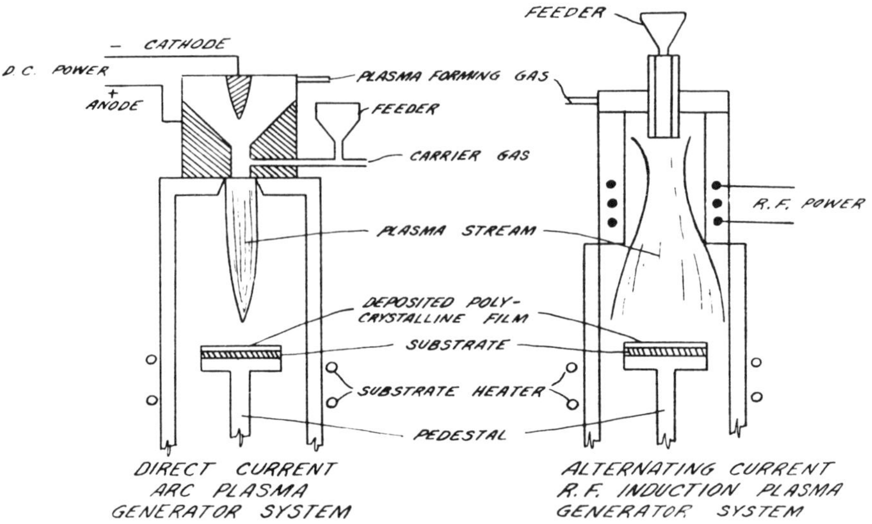

Source: U.S. Patent 4,003,770

A process developed by *F. Forrat; U.S. Patent 4,053,326; October 11, 1977; assigned to Commissariat a l'Energie Atomique, France* is one in which large, high-quality crystalline films can be produced in an inexpensive manner. This result is obtained by using a metal having a low melting point which covers the metal support and which, during the manufacture of the cell, is raised to a temperature such that it is in a liquid state, whereupon the crystal grows by epitaxy on the liquid metal.

Advantageously the metal support is a steel- or iron-silicon or iron-nickel plate. The metal having a low melting point is advantageously tin or tin alloy, e.g., tin alloyed with lead. The thickness of the tin layer is preferably less than 5 $\mu$. The semiconductive material is preferably silicon.

Some known photovoltaic cells are made from silicon discs or tape, or from a metal substrate on which a silicon film is deposited. Advantageously, in the latter case, a steel substrate is used, but precautions then have to be taken to obtain silicon having large monocrystalline grains (for obtaining high conversion yields) and to prevent the silicon from being doped by the substrate during the manufacture of the photovoltaic cell.

Figure 59 shows the present cell design which comprises a tin-plated steel plate **2** the central part of which bears a semiconductive layer **4** comprising, e.g., a P-silicon layer **8** and an N-silicon layer **10**. On the top surface, connections **12** are connected to a collector **13** connected to an external connection (not shown); the other connection **15** which has the opposite polarity from connection **12**, is connected to plate **2**. The cell can be closed at its top by a glass plate **16** brazed to the plate at its ends **18** and **20**; insulating support **22** completes the assembly.

**Figure 59: Solar Cell Design**

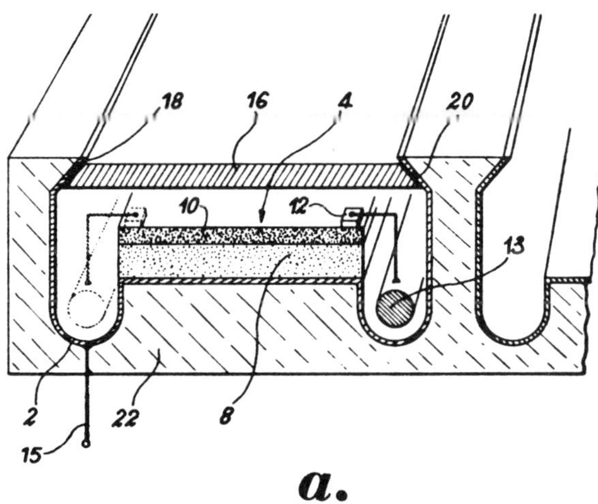


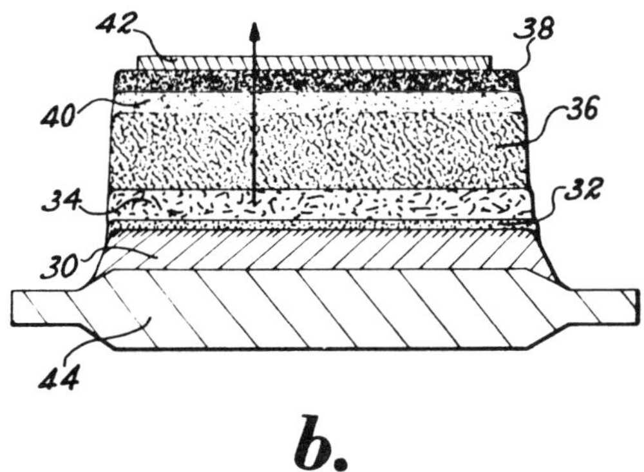


(a) Section of photocell featuring steel substrate.
(b) Detail of cell structure and substrate.

Source: U.S. Patent 4,053,326

Referring to Figure 59, the tin deposited on the steel support is adapted to facilitate the crystalline growth of silicon during manufacture and to produce a chemical barrier against iron atoms from the sheet, to prevent too much silicon dissolving in the iron, to anchor the silicon film on the steel sheet and to match the thermal expansion of silicon and iron. In addition, the tin can be used to braze the glass plate to the steel plate.

Advantageously, the steel plate can have a textured structure obtained, e.g., by heat treatment [orientation (110) or (100)]. The steel plate can be tin-plated by electrolysis or by liquation; the electrolysis operation can be followed by heat treatment at approximately 800°C under hydrogen, thus facilitating adhesion between the tin and the support. The plate can be made of iron-silicon steel and be, for example, a few tenths of a millimeter thick. By way of example, a photovoltaic cell of this type can deliver a current of 10 $mA/cm^2$ at a voltage of 500 mV in normal sunlight (approximately 1 $kW/m^2$).

Figure 59b is a detail diagram of the structure of the semiconductive layer deposited on the metal support. For the sake of illustration the dimensions are not, of course, to scale. A metal surface **30** having a low melting point is, for example, a film of tin or tin-based alloy. During manufacture, the surface is liquid but during normal operating conditions it is solid. Layer **30** is covered with a layer **32** of $P^+$ doped silicon and then in succession with a $P^+$ doped layer **34**, a P silicon layer **36**, and an $N^+$ type layer **38**. The juxtaposition of P-type layer **36** and N-type layer **38** results in a P–N junction **40**.

Of course, the signs of the doping operations can be inverted, in which case layers **32, 34** and **36** will have N and $N^+$ doping. The entire structure is covered by a transparent metal comb or grid-shaped electrode **42**, made, for example, of aluminum having a thickness from 2 to 2000 A.

Figure 60, on the following page, shows the manufacturing process. As shown, a chamber **102** is provided with heating means **104** and an inlet **106** for injecting a gaseous compound into chamber **108**, the compound containing the substance to be deposited. A support **110** surmounted by a metal substrate **112** is disposed in the chamber. The temperature of the substrate is raised by the heating means to such that the gaseous compound in chamber **108** is decomposed, producing a material for forming the crystal and such that the metal substrate is liquefied at least on its surface. After becoming liquid, the substrate produces nuclei **114** at its surface, resulting in the formation of a thin layer **116** which in turn forms a nucleus on which crystal **118** can grow by epitaxy in direction **120**.

If the crystal-forming material is silicon, the thin layer comprises a compact hexagonal arrangement, as in the case of nontextured steel. As soon as the silicon atoms are deposited, they gather around the first deposited atoms. As soon as the first layer is deposited on the liquid metal, growth occurs by conventional epitaxy on the base assembly. In order to increase the crystal deposition rate, the operating temperature can be increased above the decomposition point of silane.

Figure 60: Process Apparatus for Solar Cell Manufacture on Steel Substrate

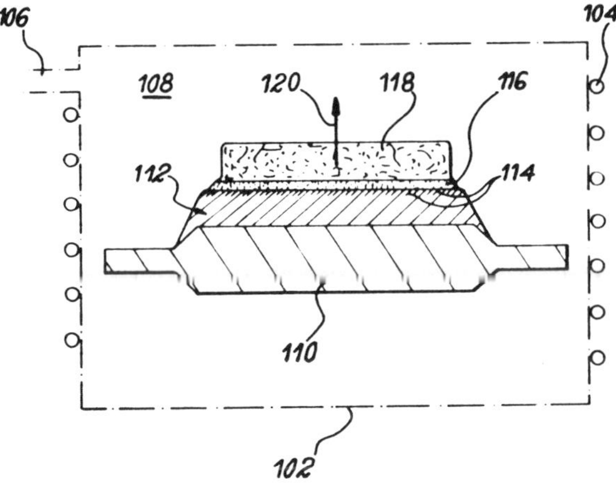


Source: U.S. Patent 4,053,326

The silicon crystal can be doped by mixing the silane with doping agents, such as boron, to obtain type $P^+$ or P doping ($P^+$ doping corresponds to a concentration of about $10^{16}$ atoms of doping agent per $cm^3$, and type P doping to about $10^{15}$ atoms per $cm^3$). To obtain type N doping, the silane can be mixed with phosphorus.

Advantageously, the metal support is a liquid film of a metal having a Curie point corresponding to the highest temperature range in the method of operation, so as to ensure automatic control. For example, the metal support can be made of a steel sheet having a Curie point between 600° and 1000°C.

A process developed by *J. Lindmayer; U.S. Patent 4,058,418; November 15, 1977; assigned to Solarex Corporation* is one in which a thin film of silicon is deposited on a metallic foil.

In order to force the deposition into an orderly structure the following general conditions must be met. The carrying substrate should have a melting point above the silicon-depositing temperature. The substrate should be coated with a material that is liquid at the deposit temperature. The coating material should be or should contain a desired dopant. For example, for an N-P cell the coating metal should provide a $P^+$ region in the silicon. For a P-N cell the coating material should provide an $N^+$ region in the silicon. The substrate and the depositing silicon material should move in relation to one another. For example, when the deposition beam is stationary, the substrate must be moving in the deposition chamber. Silicon may be deposited through a suitable aperture so that the crystallization begins at a seed point, subsequently spreading over the width of the substrate and then continuing to grow laterally over the liquid phase substrate. The conditions described above will provide for the growth of an ordered silicon film.

When this silicon film is deposited on a solid surface by any known method, it will grow epitaxially, or form a polycrystalline or amorphous layer, neither of which is suitable for the formation of an efficient and inexpensive solar cell. In the technique described here, however, upon entering into the deposition area, the coating metal on the substrate will melt and therefore will provide a floating base. This way the silicon arriving through an aperture will deposit on a liquid surface which has no crystalline structure and no orientation. The problem with such material is that when silicon is deposited on this surface it may deposit itself in an amorphous form. It was found, however, that silicon can be deposited on some liquid surfaces or on surfaces which have no crystalline structure, and it was also found that single crystallization silicon may be deposited by forcing the silicon material deposited from gas, vapor or from evaporation, sputtering or other known techniques into a crystalline structure by starting its crystallization at one point and moving the liquid underneath so that the deposited silicon is gradually building up from the point of crystallization in a lateral direction.

As the silicon arrives through an aperture and the base is moving, the newly arriving silicon may find it convenient to orient itself laterally to the region already deposited. In this fashion the layer leaving the deposit chamber is likely to be laterally oriented and will be highly ordered. The system described above will provide the required low defect state density large crystallites of $10^{18}$ $cm^{-3}$ or less. If these conditions are met, then a thin film silicon solar cell with suitable efficiency can be created.

By governing the speed at which the liquid coated carrier substrate is moving under the silicon vapor, the crystallization and the thickness of the layer can be adjusted. This speed will have an effect on the size of the crystallites on the doping and also on the thickness of the layer. The amount of silicon vapor which means the density of the silicon-containing vapor above the carrier as well as the slit size and shape will also have a definite effect on the above. Other effects will come from the carrier metal utilized, as well as from the coating material. All of these variables have a very large range where, by setting one parameter, the others have to be changed empirically to meet the conditions to obtain the desired large or single-crystal thin film silicon material.

Clearly, the substrate material must be inexpensive and must be selected to have a higher melting point than the temperature where the deposition takes place. Such material may be selected from a large variety. It may be a tantalum foil, steel, aluminum, or even low-grade metallurgical silicon. There is no other restriction on the material except its melting point. The coating material has to be selected to be liquid at the silicon deposition temperature. It may be aluminum applied to the carrier material by any of the known coating or plating techniques. If the deposition temperature is over 660°C, the aluminum layer will melt and form an alloy with the deposited silicon.

This system has an eutectic temperature of 577°C. The alloyed region of silicon remains in the liquid phase until it cools below 577°C. Upon recrystallization the bottom part of the film automatically becomes heavily P-type.

The importance of such a region should be heavily emphasized. When the silicon is thin, many of the photogenerated carriers will reach the rear surface; the formation of a $P^+$ junction will provide a retarding field which, in turn, will keep away the carriers from the rear interface. If such a field is not provided, heavy internal recombination will occur, causing a great loss in the photocurrent. If the cell is made with aluminum as the coating metal, the cell must be the N–P type. For the P–N type cell Sb, Bi or other materials with N-type character must be used. Many other materials may also be used as coatings: indium, tin, or lead are metals which may be used with dopants of arsenic or boron or a combination of these or alloys. Many of these materials are generally used for semiconductor doping and those materials may also be prepared as a coating on the substrate.

Figure 61, on the following page, shows an overall fabrication process flow diagram. The following steps are utilized in this fabrication process. The carrier film, which may be supplied from a roll or from strips, is coated with a coating material, the carrier film being a material the melting point of which is above the deposition temperature as described. This carrier film will be coated in the coating application step by the coating material. This coated carrier film is then introduced in a gas or a vacuum environment.

Before the silicon deposition is started, an initial seeding step is required. The seeding can be performed either by applying previously prepared silicon single-crystal seed or by a mechanical shutter system or a combination of these. The seeding step may take place either in the gas or vacuum environment or before the coated carrier enters this gas or vacuum environment.

After the seeded or coated carrier is introduced in the gas or vacuum environment, a heating step is taking place. During the heating step the coating metal melts and subsequently the silicon deposition step takes place.

The next step is the cooling process in which the coating, and on top of it the silicon, are solidifying.

The next step is the front junction formation, followed by a front contact formation, which completes the process and the solar cell is ready. The front contact may be a conductive pattern or film.

The conductive material may be a metal, combination of metals, or a conductive transparent film. The metals found to be suitable are aluminum, chromium, tantalum, silver, gold, copper, titanium, nickel, or others used in making contacts on silicon material. The transparent film may be made by the deposition of oxides of tin or indium.

Figure 61: Steps in Solarex Process for Thin Film Silicon Solar Cell Production

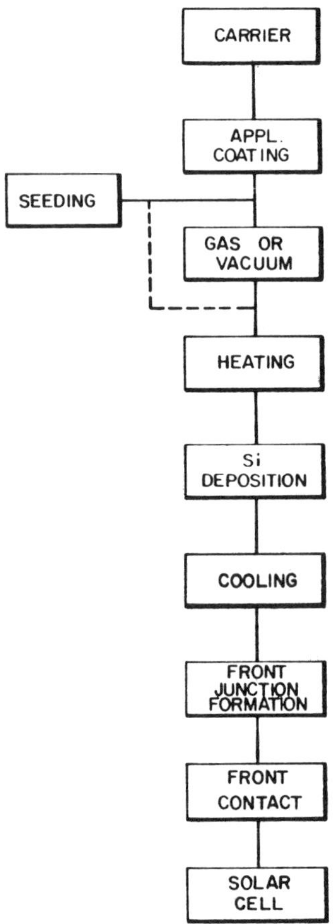


Source: U.S. Patent 4,058,418

A cell design developed by *M.P. Selders; U.S. Patent 4,101,341; July 18, 1978; assigned to Battelle Development Corporation* is a photovoltaic cell for converting solar energy into electrical power and which has, in sandwich construction, two different polycrystalline semiconductor layers in intimate contact and disposed on a metal or metal-coated substrate. The cell is provided with a light-transmissive or grating-shaped electrode on the side toward the light. The cell layers comprise the semiconducting selenides of cadmium and tin.

Such a photovoltaic cell can be fabricated with low-cost deposition techniques because of the good crystallographic match between semiconductive material layers. The good match also allows thin polycrystalline layers of less than about 3 microns to be used.

The construction of such a cell is shown in Figure 62. It consists of polycrystalline semiconductor layers of N-type cadmium selenide **1** and P-type tin selenide **2**, on a conductive substrate **3**. The second electrode **5**, in a grating pattern, is disposed on the tin selenide layer. Incident radiation is represented by arrows **4** striking the upper surface of the tin selenide layer and the second electrode. Leads **6** are connected to the substrate and the second electrode for tapping electricity from the cell.

**Figure 62: Photovoltaic Cell**

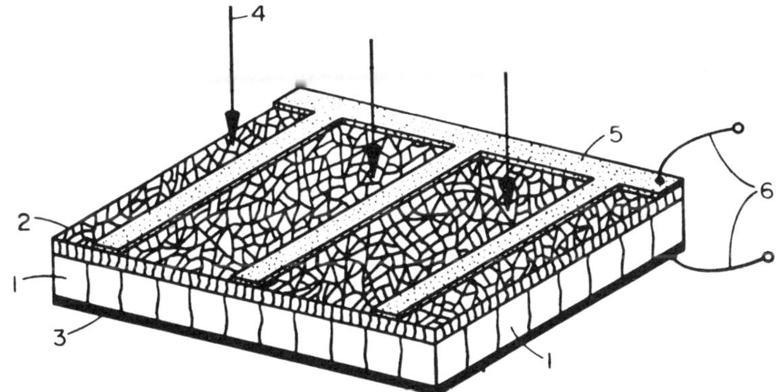


Source: U.S. Patent 4,101,341

To produce such a photovoltaic cell, the two selenide layers are advantageously vapor-deposited in the desired sequence on the heated substrate at a pressure of about $10^{-5}$ torr. Preferably, the first deposited layer is about 3 $\mu$m in thickness, and the second layer is about 0.3 to 1 $\mu$m thick. The electrode on the side which is towards the light is thereafter also vapor-deposited.

In another production method, a thin cadmium selenide layer is vapor-deposited on the substrate and then partially converted into a tin selenide layer by immersion in a tin ion-containing solution. Optionally, the tin selenide may be first deposited, followed by a cadmium for tin ion-exchange at the exposed surface.

In a third production method, the cadmium selenide and the tin selenide layers of the cell, according to the process, can also be produced in either order by spraying and thermally decomposing cadmium- and selenium-containing or tin- and selenium-containing solutions, respectively, on a substrate which is heated to about 100° to 300°C.

A type of construction developed by *A.G. Gulko and J.H. Steinberg; U.S. Patent 4,101,923; July 18, 1978* is one in which a solar cell is provided which comprises a film of crystalline silicon doped with boron on at least one surface thereof and having the other surface doped with vanadium.

A process developed by *J.F. Jordan and C. Lampkin; U.S. Patent 4,104,420; August 1, 1978; assigned to Photon Power, Inc.* is a large photovoltaic cell comprising a layer of multicrystalline cadmium sulfide, about 1 to 2 microns thick, formed by simultaneously spraying two suitably selected compounds on a uniformly heated plate of Nesa glass, thereafter forming a coating of $Cu_2S$ over the cadmium sulfide layer while the latter is heated, to form a photovoltaic heterojunction, applying thereover a layer of $CuSO_4$, and applying electrodes of Cu and Zn, respectively, to separated areas of the layer of $CuSO_4$, and heating the cell to form a cuprous oxide rectifying junction under the copper electrode by reaction of the Cu electrode with the $CuSO_4$, while diffusing the zinc through the body of the cell.

The diffusion of the zinc provides a negative electrode coplanar with the positive copper electrode, eliminating any need for introducing mechanically complex provision for making a connection to the Nesa glass, while the use of a $CuSO_4$-Cu combination enables use of a layer of CdS only 1 to 2 microns thick, rather than the usual 20 microns, despite the fact that such thin layers tend to have pinholes, which in the prior art render the cells inoperative, but in the present case do not.

Very similar ground is covered by the same inventors in *U.S. Patent Reissue 29,812; October 24, 1978; assigned to Photon Power, Inc.*

The reader is referred back to the section on Materials for Solar Cells and to the subsection on Group II Compounds for discussion of the manufacture of such thin film cells as described there in connection with U.S. Patent 3,902,920.

A cell design developed by *F.A. Shirland; U.S. Patent 4,120,705; October 17, 1978; assigned to Westinghouse Electric Corp.* is one comprised of (1) a $Cu_2S$ thin film evaporated on a conductive substrate at an elevated temperature thereby growing a polycrystalline film of preferred orientation; and (2) an outer CdS layer grown epitaxially on the $Cu_2S$ film.

Some effort has been made in the past to produce a backwall CdS-$Cu_2S$ thin film solar cell employing a transparent substrate anticipating advantages over conventional frontwall CdS-$Cu_2S$ solar cells. For the purposes of this discussion, the term frontwall refers to devices illuminated through the $Cu_2S$ layer to the junction, while backwall refers to devices illuminated through the CdS layer to the junction. In a frontwall arrangement, which is the standard type of cell construction, a layer of CdS is deposited on a metallic substrate and then a layer of $Cu_2S$ is formed on the CdS. A metallic grid is then applied to the surface of the $Cu_2S$, and, finally, transparent protective coating is applied.

In a particular configuration of a backwall CdS-$Cu_2S$ thin film solar cell, the $Cu_2S$ is adjacent to the substrate and the CdS is then deposited on the $Cu_2S$. In this backwall structure a grid is also attached, but in this case it is attached to the thicker, more conductive CdS layer which permits a coarser, less expensive grid to be used. Furthermore, in the backwall structure a protective coating is not needed since the $Cu_2S$ layer is not vulnerable to the atmosphere as a result of it being sandwiched between the substrate and the stable CdS layer.

Unfortunately, these and other hoped-for advantages of the backwall structure described above have not been fully realized since past attempts to produce such a backwall structure failed to reproduce the essential epitaxially-formed $Cu_2S$-CdS heterojunction.

The present process overcomes these problems and produces a useful epitaxial heterojunction device by vacuum depositing a first crystallized compound on a substrate, then abruptly shifting to vacuum deposition of a second crystalline compound epitaxially on the first crystalline compound. One version of such a device is shown in Figure 63.

**Figure 63: CdS-$Cu_2S$ Heterojunction Solar Cell**

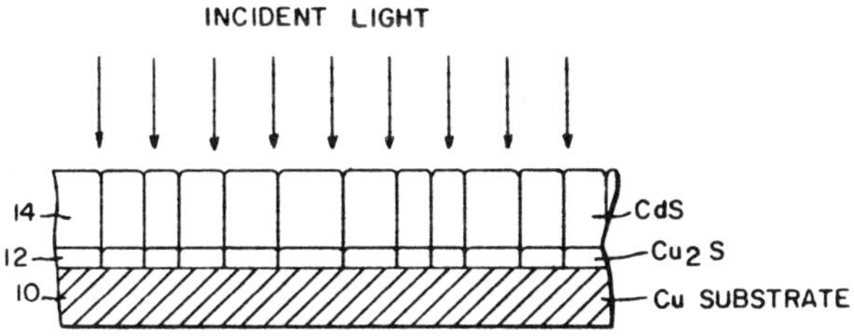


Source: U.S. Patent 4,120,705

As shown in the above figure, a $Cu_2S$ layer **12** is disposed upon a metallic substrate **10** and a CdS layer **14** is disposed upon the $Cu_2S$ layer. In the manufacturing process, $Cu_2S$ starting material is held close to the stoichiometric ratio of 2 copper atoms to 1 sulfur atom and is vacuum evaporated onto the metallic substrate held in the temperature range of 200° to 435°C, thereby producing a polycrystalline film deposit **12** of $Cu_2S$ in the hexagonal crystalline form. The hexagonal $Cu_2S$ crystals assume a preferred orientation on the substrate, whereby the C-axis of each crystal is generally perpendicular to the substrate and the three A-axes of each crystal are randomly oriented in directions generally parallel to the plane of the substrate.

When a sufficient thickness of $Cu_2S$ is achieved, the $Cu_2S$ evaporation is started abruptly while being at the elevated temperature. For example, a fast-action shutter arrangement in a vacuum chamber is preferably used so that CdS crystallites **14** grow epitaxially on those of the previously deposited layer **12** of $Cu_2S$. The resulting CdS layer is made many times as thick as the $Cu_2S$ layer so that when the whole structure is cooled to room temperature the normally hexagonal crystalline form of CdS will dominate and hold the $Cu_2S$ in the same structural form. A preferred thickness for the $Cu_2S$ layer is from about 0.1 to 1.0 micron, while a preferred thickness for the CdS layer is from about 1.0 to 10.0 microns.

In this manner a clean abrupt epitaxial heterojunction is produced between the $Cu_2S$ layer and the CdS layer providing an efficient device for solar energy conversion. It will be apparent to those skilled in the art that the present process provides an all-vacuum-deposited epitaxial CdS-$Cu_2S$ thin film solar cell.

Furthermore, such a structure has several marked advantages over the conventional frontwall CdS-$Cu_2S$ thin film solar cell:

> First, a very fine mesh grid contact is no longer necessary; the thinner $Cu_2S$ layer **12** is now backed up by a continuous metallic contact **10**. Any grid (not shown) on the CdS side of the junction can be much coarser, and thus much lower in cost because the CdS layer is thicker and more conductive than the $Cu_2S$ layer.
>
> Second, a separate passivation layer is no longer needed to protect the $Cu_2S$ layer from the atmosphere. The metallic substrate can readily be made impermeable to air and water vapor, and several microns or more of the CdS film on the other side of the $Cu_2S$ layer will be equally impermeable. Thus, only the edges of the cell need to be protected.
>
> Third, the $Cu_2S$ layer can now be made somewhat thicker to absorb more of the incident light in a planar film without sacrificing collection efficiency, since the light is entering from the junction side rather than from the side opposite the junction, as indicated by the direction of the arrows. Also, with the CdS film only a few microns thick, higher output is obtained since more of the shorter wavelength photons that are absorbed in the CdS will be absorbed closer to the junction than would normally be the case for a prior art backwall cell.
>
> Fourth, with this process several steps in the normal cell processing are eliminated and the key steps are completed in the same vacuum chamber with the same pump-down, thus providing maximum production efficiency.

Figure 64, on the following page, shows how this method may be employed to produce another frontwall CdS-$Cu_2S$ solar cell. In place of the copper substrate of Figure 63, the process of Figure 64 employs a transparent glass support member **20** which is coated with a thin conductive layer **21** in a known manner. For example, the member **20** is preferably coated with from about 40 to 100 A of $SnO_2$.

The method of producing layers **22** and **24** of Figure 64 is the same as the method of producing layers **12** and **14** of Figure 63. That is, layer **22** is produced by vacuum deposition of $Cu_2S$ at a temperature between 200° and 435°C with the $Cu_2S$ starting material being held in the stoichiometric ratio of 2 copper atoms to 1 sulfur atom. When a thickness of about 0.1 to 1.0 micron of $Cu_2S$ is achieved, the $Cu_2S$ evaporation is abruptly halted and the CdS evaporation is abruptly started, preferably by means of fast shutters.

The resulting structure (Figure 64) is capable of being operated either as a frontwall or backwall solar cell. With the incident light entering the device through the transparent support **20** as shown in Figure 64, the device is being operated in the frontwall mode.

It will be apparent to those skilled in the art that the arrangement of Figure 64 achieves an optimum combination of frontwall operation with many of the structural advantages of a backwall device. For example, an advantage of the embodiment of Figure 64 over that of Figure 63 is that photons having wavelengths shorter than 5200 A are absorbed in the CdS layer **14** of Figure 63 without reaching the junction, whereas such short wavelength photons entering through the $Cu_2S$ layer **22** of Figure 64 are absorbed at or near the junction thereby achieving more efficient solar energy conversion.

**Figure 64: Alternative Construction of CdS-$Cu_2S$ Heterojunction Solar Cell**

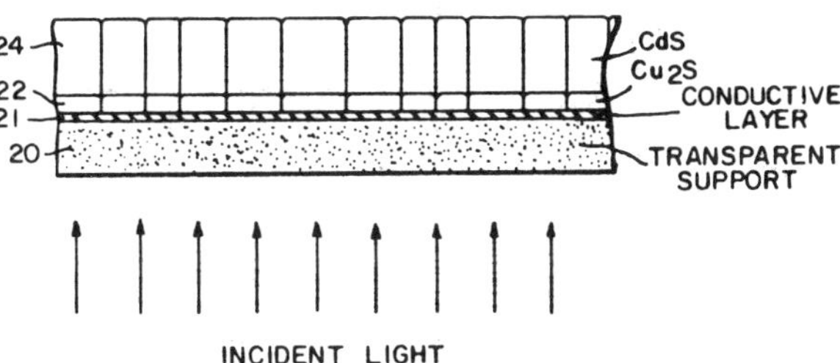


Source: U.S. Patent 4,120,705

A solar cell developed by *L.R. Ullery, Jr.; U.S. Patent 4,127,424; November 28, 1978; assigned to Ses, Incorporated* comprises a substrate, a bottom electrode, a first layer of cadmium sulfide, a second layer of cuprous sulfide forming a barrier junction with the first layer and a top electrode; an improvement is that the substrate is an insulative ceramic material and the bottom electrode is a conductive ceramic layer fused to the substrate. The conductive layer is optionally coated with a metal having a high electrical conductivity.

A process developed by *H.B. Maier; U.S. Patent 4,137,096; January 30, 1979* is one in which low cost solar cells are made by electrodeposition of a metallic coating on a conductive substrate, followed by cathodic conversion of the coating in a thiosulfate containing bath to form metallic sulfides. Electrical contacts are made to portions of the sulfide layer and the coated substrate is dried to a point sufficient to yield a photoresponsive device.

A type of construction developed by *J.A. Duisman; U.S. Patent 4,143,235; March 6, 1979; assigned to Chevron Research Company* yields a cadmium sulfide photovoltaic cell of improved efficiency comprising a barrier layer and a cadmium sulfide-containing bilayer. The bilayer is formed by depositing at a first temperature an initial layer of cadmium sulfide in interfacial contact with the substrate and then depositing a subsequent layer of cadmium sulfide at a second temperature which is at least 20°C below the first temperature.

## VERTICAL MULTIJUNCTION DEVICES

The edge-illuminated vertical multijunction cell shown in Figure 65 consists of a stack of silicon homojunction devices qualitatively similar to standard cells, illuminated in such a way that the light enters the cells parallel to the junctions. Several recent calculations have indicated that these devices have the potential of achieving efficiencies as high as 30%, although very little work has been done on designing and testing optimized cells. The maximum measured efficiency to date is 9.6% for a device without an antireflective coating.

**Figure 65: The Vertical Multijunction Cell (Edge Illuminated)**

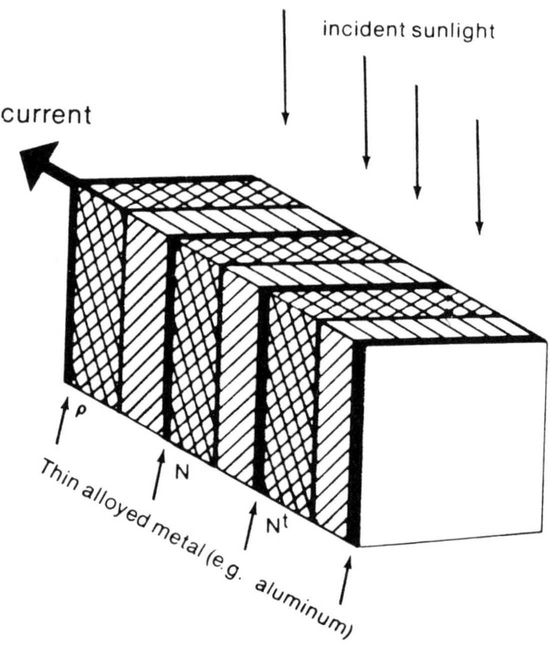


Source: Reference (3)

The high potential efficiency results from several features of the device which are as follows (3):

> Since the vertical multijunction devices are connected in series, they produce higher voltages and lower currents than other concentrator cells with the same power output. The low currents reduce the resistance losses. Series connections also mean, however, that considerable care must be taken to ensure that all of the cell elements are illuminated since an unilluminated unit will act like a large series resistance.
>
> As is the case with the interdigitated cell, the vertical multijunction devices do not require contacts on the surface (the aluminum connections can cover less than 1% of the surface area) and thus front surface reflections are minimized.
>
> The multijunction device should be able to make more efficient use of light with relatively long and short wavelengths.
>
> The multijunction devices should be able to perform better than conventional silicon cells at high temperatures, and their performance at high temperatures improves in high light concentrations; the cells should, for example, be only half as sensitive to temperature at 1,000 suns as they are in unconcentrated sunlight.

The extra manufacturing steps required to fabricate these devices will make them somewhat more expensive than conventional silicon cells, but this difficulty would be offset if the high efficiencies are realized.

A cell designed by *J.F. Wise; U.S. Patent 3,690,953; September 12, 1972; assigned to the U.S. Secretary of the Air Force* is a solar cell constructed from a slab of epitaxially grown silicon containing a plurality of very thin alternate N and P zones, cut, lapped, and polished to a thickness of approximately 0.010 inch, having a common deposited aluminum contact connecting on the back side of the cell to each of the N zones and another common deposited aluminum contact on the back side of the cell connecting to the P zones, and the cell oriented such that the direction of impingement of the solar energy on the front surface of the cell is in a direction generally parallel to the alternate zones of N and P material and provides a high efficiency hardened solar cell.

This design provides a solar cell, primarily for use in outer space, having optimum conversion efficiency by having all charge carriers formed in relatively close proximity to a junction. This results in 15%, or better, initial efficiency compared to approximately 11% initial efficiency for conventional cells. The construction disclosed also provides a cell that is much more radiation resistant because degradation of minority carrier diffusion length will affect conversion efficiency much less than in previous cell construction.

Typically cells constructed in this manner have a degradation of approximately 5% after exposure to high radiation, whereas conventionally constructed cells exhibit approximately a 35% degradation. The combination of these two features results in cells that have approximately 100% greater output than prior cells after exposure to severe nuclear or natural radiation. Due to the specific configuration of the cell, all contacts on the top of the cell are eliminated, thus no dissimilar material interfaces are exposed to the radiation and survival of the cell is only dependent upon the survival of the silicon material. The combined results obtained from the disclosed structure are a cell that has a much longer operating life in outer space and an improved efficiency such that for output equivalent to prior devices approximately only one-half the area, weight, and volume of existing similar devices is required.

A process sequence developed by *D.L. Kendall, F.A. Padovani, K.E. Bean and W.T. Matzen; U.S. Patent 3,969,746; July 13, 1976; assigned to Texas Instruments Incorporated* involves fabricating a vertical multijunction solar cell utilizing an orientation-dependent etch to selectively provide parallel grooves in a monocrystalline silicon body to provide P–N junctions. In some instances the grooves are filled with silicon of the same conductivity type as the silicon body.

A cell developed by *W.P. Rahilly; U.S. Patent 3,985,579; October 12, 1976; assigned to the U.S. Secretary of the Air Force* is a vertical multijunction solar cell fabricated with the junction channels perpendicular to a ribbed electrical grid structure which provides an improved cell having increased mechanical strength and decreased electrical resistance.

A type of construction developed by *R.E. House, R.A. Irvin and D.F. Kane; U.S. Patent 4,082,570; April 4, 1978; assigned to Semicon, Inc.* comprises a cell formed from a plurality of integrally interconnected P–N junction-containing semiconductor wafers. The wafers are stacked end-to-end in the cell so that the respective junctions in each wafer are parallel to each other. The efficiency and performance of the cell is improved, particularly upon exposure to concentrated sunlight, by imposing various conditions on the cell fabrication and design. Improvements result, for example, by selecting a high-resistivity semiconductor as the starting material in the fabrication of the cell, controlling the diffusion process to optimize the junction gradient and minimize the thickness of the base region in each wafer, orienting the wafers in the cell so that they are illuminated at a small angle relative to the plane of the respective junctions therein, and treating the exposed surfaces of the wafer to reduce reflectivity and surface recombination velocities.

Such a cell is shown on the following page in Figure 66. The cell comprises a plurality of semiconductor diodes **10** which may be of silicon into which impurities are diffused to form a P–N junction **10A**. The plane of the junction in each diode is parallel to the end faces of the diode. The diodes are stacked end-to-end and fused together with a plurality of electrically conductive interface layers **12** interposed between successive diodes. Electrically conductive contact layers **14** are formed on the end surfaces of each of the two end diodes in the cell. Electrical leads **16** attach to each of the contact layers.

**Figure 66: Semicon, Inc. Design for Vertical Multijunction Solar Cell**

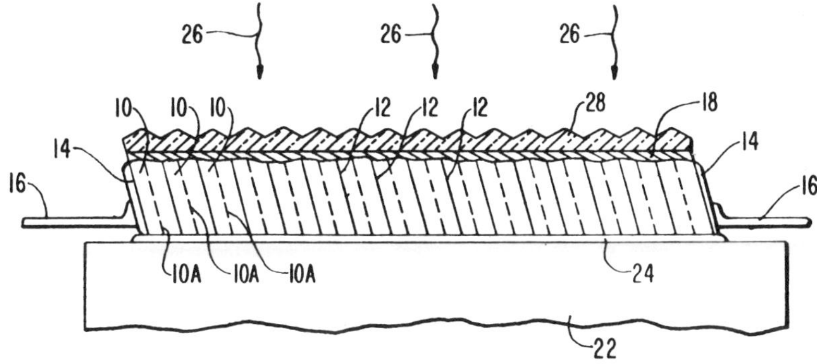


Source: U.S. Patent 4,082,570

The cell is coated with a protective outer coating **18** which is also transparent to radiant energy. The cell is fixed to a substrate **22** by a layer of adhesive **24**. The substrate may in turn be mounted on a suitable heat sink (not shown).

The arrows **26** indicate the direction of incidence of the radiant energy to be converted by the cell, which is illustratively solar energy. A multiple plano-convex lens **28** (i.e., essentially flat on the cell side and convex on the opposite side over each junction) is placed on the top surface of the cell to concentrate and direct the radiant energy to the regions of the junction in each diode.

In operation, the radiant energy absorbed within each diode generates electron and hole pairs therein. The junction establishes a potential gradient within the diode which urges the electrons to collect on one side of the junction and the holes to collect on the opposite side of the junction. Provided the lifetimes of the photogenerated carriers are sufficiently long, a net voltage will appear across the junction in each diode. As the individual diodes in the cell are electrically connected in series, the cell voltage across the leads is the sum of the individual voltages generated by the diodes. The cell voltage is thus available for electrical utilization in a load. Typically, a plurality of cells like the cell shown are interconnected in series and/or parallel to provide an electrical output of the desired characteristics.

A cell design proposed by *H. Diepers; U.S. Patent 4,099,986; July 11, 1978; assigned to Siemens AG, Germany* provides a semiconductor body consisting of single crystal semiconductor whiskers which are grown on a substrate surface permitting relatively inexpensive manufacture and high efficiency of the resulting solar cell.

Such a cell design is shown schematically in Figure 67. The solar cell, which is shown in a partial cross section in the figure, comprises a substrate surface **2**, on which stand a multiplicity of single crystal semiconductor whiskers. In the figure only eight whiskers **4** of equal size, arranged parallel side-by-side, are shown. Their heights and their diameters, which are, for instance, in the order of **100** or several hundred microns, may be different, however. In addition, the whiskers may have a cross section which varies over their height.

With respect to the incident sunlight radiation which is indicated by individual arrows **5**, the solar cell is oriented so that its whiskers are directed substantially against the direction of incidence of this radiation. Thus, the radiation can be absorbed almost completely by the whiskers with this structure.

**Figure 67: Vertical Multijunction Solar Cell Comprising Semiconductive Whiskers**

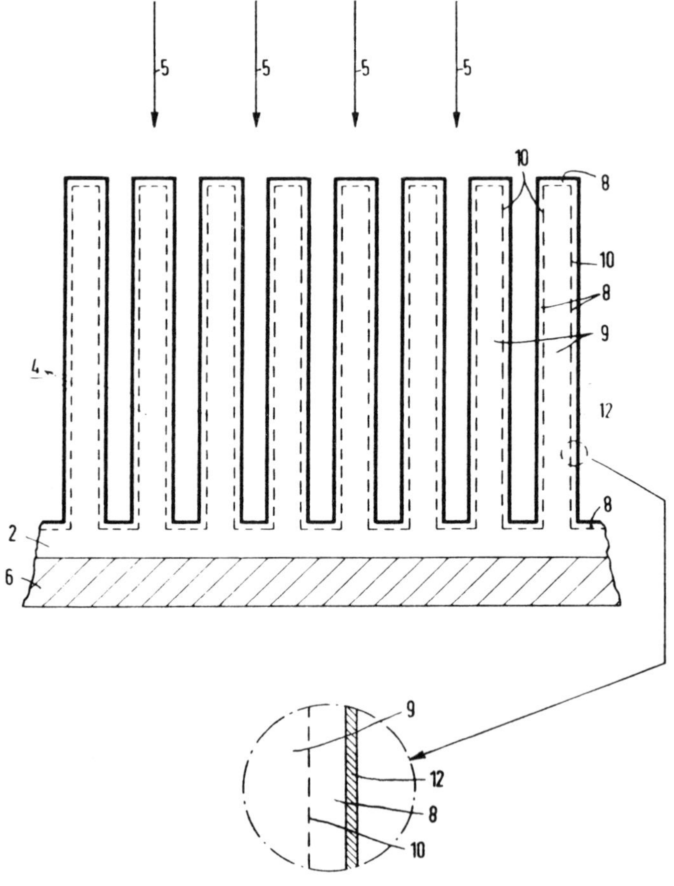


Source: U.S. Patent 4,099,986

Due to the spectral composition of the sunlight, the optimum band gap of the whisker material used for the solar cell, as measured in electron volts (eV), should be about 1.5 eV. The band gap of silicon is approximately 1.1 eV, so that the output voltage produced by a silicon solar cell is correspondingly small and the efficiency of the energy conversion of such a cell is in the order of about 11%. One will, therefore, attempt to use materials with larger band gaps for solar cells. Such materials are, for instance, certain semiconductive III-V compounds, or also ternary alloys with such compounds. Thus, gallium arsenide, for instance, has a band gap of approximately 1.4 eV.

For growing the semiconductor whiskers of a solar cell according to the present process, the so-called vapor-liquid solid mechanism (VLS mechanism) can advantageously be used, which is known from the journal *Transactions of the Metallurgical Society of AIME*, vol 233, pp 1053-1064 (June 1965). According to this crystal-growing mechanism, the material to be crystallized is absorbed in a predetermined amount of a metal which is placed on the substrate and in which the material to be crystallized is soluble, and which is called the agent. At a suitable predetermined temperature, an alloy is formed during the precipitation with the material to be crystallized, which is saturated upon further precipitation of this material. Thus, supersaturation and precipitation of the material on the substrate comes about and, finally, growth of the whiskers with the liquid agent at the tips of the former.

The crystal growth which occurs is heavily anisotropic, i.e., it takes place nearly in a direction perpendicular to the substrate surface, since the absorption of the crystallizing material or its components takes place preferentially at the free surface of the liquid metal phase, while the precipitation from the liquid metal phase occurs only at the boundary surface between the drops and the substrate. With the abovementioned method, a large area can advantageously be provided with whiskers.

Referring to the figure, the whiskers are advantageously grown on the substrate in accordance with the known VLS mechanism. The method is applicable, for instance, to Si and particularly also to GaAs, GaP and $Ga(As_{1-x}P_x)$. While in the case of silicon, Au, Pt, Pd, Ni, Cu or Ag can be used as the agent, the Ga itself advantageously serves as the agent in the case of the gallium compounds mentioned. Other highly effective compounds with large band gaps are InP, CdTe, AlSb and CdS, to which the VLS method is likewise applicable. With these compounds also a foreign material agent is not necessary, so that the first compound partner can serve as the agent, i.e., In, Cd, Al or Cd, respectively. Also, Ge whiskers can be grown by the known method, likewise using Au as the agent.

In the known VLS method, the growth conditions for the whiskers in a reaction chamber provided for this purpose are heavily dependent on the temperature of the substrate. A similarly strong influence is also exerted by the vapor deposition rate or the degree of supersaturation of the vapor in the reaction chamber. The whisker diameter depends substantially on the particle size of the reagent material and the temperature. Thus, increasing temperatures lead to larger whisker diameters due to better wetting of the substrate surface. The agent material can be applied, for instance, through masks to specific points on the substrate surface or may also simply be vapor-deposited on the substrate. During the vapor deposition or the heating of the substrate, small droplets then form on the substrate surface. The size of the droplets depends, for instance, on the layer thickness of the vapor-deposited material.

With the known method, whisker densities of $10^4$ to $10^6$ per square centimeter can be obtained. This corresponds to a mean whisker width of 100 to 10 microns if the arrangement is rectangular.

Any substance favoring whisker growth or whisker germination can be used as material for the substrate. Thus, single or polycrystalline silicon substrate can be provided for growing silicon whiskers. As is shown in Figure 67, such electrically nonconducting substrates are advantageously placed on an electrically conducting carrier body **6**, which acts at the same time as an electrode. Advantageously, electrically conducting metal strips can also be provided as the substrate body and at the same time as the electrode. In the case of silicon whisker growing, such a strip can consist of carbon-free steel.

Doping of the whiskers grown by the VLS method can be carried out in accordance with known techniques. Thus, P-doping of silicon whiskers can take place after they are grown or, in some cases, while they are being grown, with boron or aluminum. Subsequently, the surface of this now P-conducting whisker is given a doping of the opposite type for forming an N-conducting border zone **8**, for instance, by diffusing phosphorus from the gaseous phase into the surface up to a depth which approximately corresponds to the diffusion length. The remaining P-conducting layers of the whiskers are designated **9** in the figure. The P-N junction formed between the N- and P-conduction zones **8** and **9**, respectively, is indicated in the figure by a dashed line **10**. The position in depth of this P-N junction can be adjusted in a manner known per se by the diffusion conditions, e.g., the diffusion time, the diffusion temperature or the gas flow.

Although N-doping of the border zone **8** near the surface and P-doping in the underlying zone **9** has been assumed, the doping of the two zones can just as well be arranged in the opposite manner, as is also known.

For developing an electrode facing the incident light for the solar cell, the surface of the whiskers is coated with a layer of transparent material **12**, which is at the same time electrically conductive. Advantageously, materials which absorb only a small fraction of the energy of the incident radiation are used. Such materials are, for instance, tin oxide doped with antimony, $SnO_2$(Sb) or indium oxide doped with tin, $In_2O_3$(Sn). A suitable technique for applying these layers is the cathode sputtering method. The layers can also be vapor-deposited or applied by means of ion plating, where the materials are vapor-deposited, the vapor is partially ionized by a plasma discharge and the ionized portion in the vapor is precipitated electrostatically with the neutral vapor.

*R. Kaplow and R.I. Frank; U.S. Patent 4,128,732; December 5, 1978; assigned to Massachusetts Institute of Technology* developed a solar cell design which is comprised of a plurality of series connected unit solar cells formed from a common substrate of semiconductor material. Each unit solar cell has spaced elongate sidewalls, and a "dead space" area between adjoining sidewalls of adjacent units is made substantially smaller than an active, light-receiving area, extending between the opposite sidewalls of each individual unit. In addition, the width of the active area is concisely limited to ensure that radiation incident on the active area is incident at a point which is spaced from the P-N junction of each unit by no more than a predetermined optimum distance. Reducing the dead space area while concisely limiting the width of the active area provides improved solar cell performance without requiring focusing lenses.

Similar devices are described by *R. Kaplow, R.I. Frank and J.L. Goodrich; U.S. Patent 4,129,458; December 12, 1978; assigned to Massachusetts Institute of Technology.*

## HORIZONTAL MULTIJUNCTION CELLS

There have also been proposals for using horizontal multijunction devices using silicon or GaAs. It may be possible to design a cell array capable of producing relatively high voltages on a single chip, thereby reducing the cost of interconnecting devices and reducing series resistance in connections. The inherent efficiencies of these devices are approximately the same as conventional cells, but it may be easier to use the approach to develop practical cells which can more nearly approximate the potential of the materials (3).

One alternative is the use of a vertical stack of photovoltaic cells arranged so that upper layers absorb only high energy (i.e., short wavelength) photons, allowing the remaining photons to reach lower cell levels. Little experimental work has been done on these devices, but theoretical analysis has predicted that the light-filter devices could achieve efficiencies as high as 46% if two separate cells are used and 52% if three cells are used. Multilayer cells using Ge, Si, GaAs, and other materials with two or more layers may also be able to achieve efficiencies above 40% (3).

A cell design developed by *M. Ettenberg and H. Kressel; U.S. Patent 3,990,101; November 2, 1976; assigned to RCA Corporation* is a solar cell having two heterojunctions.

At the incident surface and in the body is a first region having a band gap energy greater than 2.1 eV which thus is substantially transparent to solar radiation. Spaced from the first region and at the opposite surface is a second region which is of a material having a band gap energy in the range of 1.5 to 1.9 eV. Between and in contact with both first and second regions is a third region of a material having a band gap energy less than either the first or second regions. The third region is the most active region of the device, and the second region is substantially transparent to solar radiation not absorbed by the third region. The junctions between the third region and each of the first and second regions are heterojunctions. On the opposite surface of the body is an electrode capable of reflecting back into the body of the device solar radiation passing through the second region which was not absorbed by the active region.

Such a design is shown in Figure 68. The solar cell device **10** comprises a body **12** of semiconductor materials having the capability of converting solar energy into electrical power, such as the III–V semiconductor materials and their alloys.

**Figure 68: Solar Cell Device Having Two Heterojunctions**

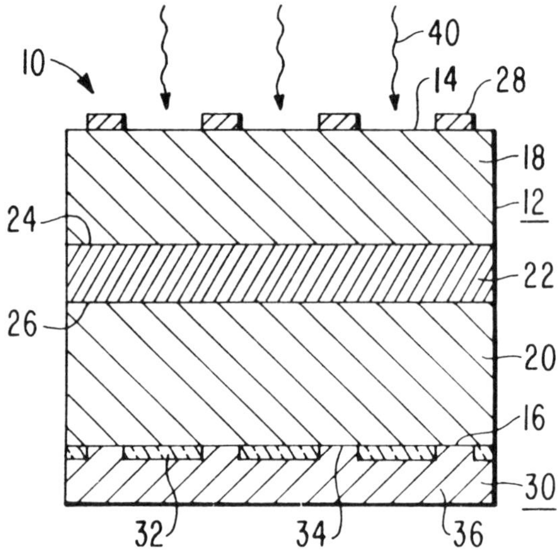


Source: U.S. Patent 3,990,101

A portion of surface **14** of the body **12** is exposed so as to allow solar radiation **40** to enter the semiconductor device. Opposite the incident surface **14** of the body is a surface **16**. The body includes a first region **18** of one conductivity type at the solar radiation incident surface **14** and a second region **20** of an opposite conductivity type spaced from the first region and at the opposite surface **16**. Between and contiguous to both the first region and the second region is a third region **22**. The third region is of the same conductivity type as the second region and of a material substantially uniform in composition. Thus, the junction between the first region and the third region is a P–N junction **24**. The first region is of N-type conductivity and the second and third regions are of P-type conductivity.

The semiconductor material of the first and second regions differs in composition from the semiconductor material of the third region. Therefore, the P–N junction **24** and a junction **26** between the third region and the second region are heterojunctions. The first region is of a semiconductor material, preferably of the III-V Group, which is substantially transparent to solar radiation. Since the solar radiation which is incident to the device is typically in the range of 4000 to 9000 A, the semiconductor material of the first region will be of a high band gap energy, i.e., greater than 2.1 eV. The third region is the region where substantially most of the solar radiation absorption in the device occurs. Thus, the third region is of a semiconductor material having a band gap energy such that it will absorb most of the incident radiation. It is well known to those in the solar cell art that the most efficient absorption of solar radiation occurs in semiconductor materials having a band gap energy in the range of 1.4 eV, such as GaAs. As for the second region, it must be substantially transparent to the solar radiation not absorbed by the third region, but of a band gap energy such that a potential barrier is formed at the heterojunction. It is desirable that the second region be of a semiconductor material having a band gap energy in the range of 1.5 to 1.9 eV.

It is preferred that the third region be of gallium arsenide, GaAs. It is most desirable that the material of the first region and the second region be lattice matching with the material of the third region. Lattice matching between the third region and each of the first and second regions is an important factor in preventing strains and dislocations from forming in the body. Thus, it is preferable that the first region and the second region be of aluminum gallium arsenide, but of different aluminum concentrations. Since the first region is required to have a higher band gap energy than the second region, typically the first region will be of $Al_xGa_{1-x}As$ wherein x is greater than 0.5 but less than 1, and the second region will be of $Al_yGa_{1-y}As$ wherein y is in the range of 0.1 to 0.4. Other semiconductor materials which are lattice matching to GaAs and could be utilized are indium gallium arsenide phosphide, or indium aluminum arsenide phosphide.

On the incident surface of the body is an electrode **28**. Typically the electrode is in the form of a grid and covers no more than about 5 to 10% of the surface area of the incident surface. Keeping the surface area, on which the grid electrode is on the incident surface, at a minimum allows as much solar radiation to enter the solar cell device as possible.

At the opposite surface is a solar radiation reflecting electrode **30**. Typically, the reflecting electrode functions both as a collector of current and a reflector of solar radiation not absorbed by the third region. The reflecting electrode includes a noncontinuous layer **32** of an insulating material such as silicon dioxide on the opposite surface. The noncontinuous layer is noncontinuous because of openings **34** which are in the form of a grid pattern.

In the openings at the opposite surface and on the noncontinuous insulating layer is a metallic layer **36**. The metallic layer is of a metal or metals having good electrical conduction properties and reflective to solar radiation. As an example, the metallic layer may be of a first layer of chromium directly on the opposite surface and noncontinuous insulating layer with a second layer of a metal such as gold on the layer of chromium. While the gold is an excellent electrical conductor, the chromium aids in the adhesion of the gold to the device.

In the operation of the solar cell device solar radiation first strikes the device at the incident surface and enters the first region. As previously stated, the first region is of a material having a band gap energy greater than 2.1 eV, thus being substantially transparent to the solar radiation which is predominantly in the frequency range of 4000 to 9000 A.

After the vast majority of the solar radiation passes through the first region, it enters the third region, which is preferably GaAs. Gallium arsenide has a band gap energy of approximately 1.4 eV and is capable of absorbing solar radiation below 9000 A. As the solar radiation passes through the third region, it is substantially absorbed and electron-hole pairs are formed.

As is well known in the art, there exists an electric field at the P-N junction. The minority carrier generated in the third region, in this case electrons, which are in the vicinity of the P-N junction are swept by the electric field across the P-N junction. Preferably the third region is no thicker from the P-N heterojunction **24** to the heterojunction **26** than the diffusion length of minority carrier in the third region. The displacement of these electrons across the junction **24** results in the generation of an electric current. Since the junction **26** is a heterojunction having a potential barrier as a result of the second region possessing a higher band gap energy than the third region, minority carriers generated in the third region and near the heterojunction **26** are repelled toward the P-N junction **24**. This repelling of the minority carriers at the heterojunction **26** increases the minority carrier collection efficiency at the P-N junction **24**.

In addition, as a result of the second region having a band gap energy higher than that of the third region, solar radiation not absorbed by the third region will probably not be absorbed by the second region. This unabsorbed radiation passing through the second region will next strike the reflecting electrode and be reflected back into the third region for possible absorption. The unabsorbed solar radiation will pass through the noncontinuous insulating layer, which is transparent to solar radiation, but it is, upon striking, the metallic layer which reflects the radiation back into the body. If the metallic layer were in intimate contact with the second region instead of the insulating layer in the deposition of the metallic layer, a thin alloy layer would form between the layer and the semiconductor material of the second region. Such an alloy layer will usually itself absorb solar radiation incident onto it, resulting in the loss of solar radiation which is not absorbed within the semiconductor material of the body.

Wherever the insulating layer is an intermediate between the second region and the metallic layer, no alloy layer (absorbing solar radiation) is formed. Therefore, the alloy layer will only be formed where metallic layer **36** is in openings **34** and on the second region, which accounts for only a small portion of the second region at the opposite surface **16**.

It is also conceivable that solar radiation not absorbed by the third region on this second pass may strike the incident surface at an angle greater than the critical angle and again be reflected back into the body.

The solar cell device may be made by epitaxially depositing the regions on a substrate, such as GaAs, with the second region being deposited first then the third region on the second region and finally the first region on the third region. The regions are preferably deposited by liquid phase epitaxy, although they may be deposited by vapor epitaxial techniques well known in the art. The regions may be sequentially deposited on the substrate by using the method and apparatus described in U.S. Patent 3,897,281.

In the method described in U.S. Patent 3,897,281, charges of the semiconductor material and conductivity modifiers to be deposited are placed in the wells of a refractory furnace boat. A substrate is placed in a slide which extends longitudinally through the boat and across the bottom of the wells. The boat and its contents are heated to a predetermined temperature at which the charges become molten. After reaching the predetermined temperature, the slide is moved into a first well and the boat and its contents are cooled. During the cooling of the molten charge, the semiconductor material is deposited onto the substrate. The substrate is sequentially moved into the remaining wells while the furnace contents are cooled for the further deposition of the semiconductor material in the wells.

After the substrate with the epitaxially grown regions is removed from the furnace the substrate is mechanically or chemically removed by techniques such as grinding or etching, leaving only the body. Next, the grid electrode **28** is formed by evaporating a metallic layer onto the incident surface **14** of the body and by etching techniques well known in the art, a grid pattern is formed into the metallic layer. A layer of insulating material, such as silicon dioxide, is then deposited on the opposite surface **16** of the body, and openings in the form of a grid pattern are made in the insulating layer by well-known masking and photoresist techniques, thereby forming the noncontinuous insulating layer **32**. In the openings and on the noncontinuous layer is evaporated the metallic layer **36**.

A cell design developed by *L.W. James; U.S. Patent 4,017,332; April 12, 1977; assigned to Varian Associates* is one whose objects are to provide (1) a radiant energy or photovoltaic conversion cell of greatly increased overall efficiency; (2) a practical photovoltaic cell which can convert several photon wavelengths or energy ranges; and, (3) an energy conversion cell in which plural conversion cells can be stacked efficiently, economically and practically and in a specific manner.

A device developed by *Y. Tarui, T. Sakamoto and Y. Komiya; U.S. Patent 4,046,594; September 6, 1977; assigned to Agency of Industrial Science and Technology, Japan* comprises a pile of active layers, each being composed of a P-N junction semiconductor element and a coextensive electrode, the semiconductor elements and electrodes being so shaped and dimensioned as to expose the side of each layer to the light falling on the multilaminated structure, thus converting the solar energy to electric power at an increased efficiency.

A device developed by *H. Kressel, R.V. D'Aiello and P.H. Robinson; U.S. Patent 4,070,206; January 24, 1978; assigned to RCA Corporation* is a three-layer device. The third layer is of a conductivity type opposite that of the first and second layers so as to form first and second P-N junctions, respectively, therebetween. The thickness of the third layer is at least twice the minority carrier diffusion length of the semiconductor material, so that carriers generated within the third layer have a high probability of being collected by one of the P-N junctions. The body includes means for electrically connecting the first and second P-N junctions and means for transferring the carriers collected at the first P-N junction to a portion of the first surface.

A type of construction developed by *A.G. Milnes; U.S. Patent 4,094,704; June 13, 1978* provides composite solar cells of improved efficiency which comprise two cells of different characteristics arranged in optical series but electrically insulated from each other. Preferably, each cell is of larger crystal grain size than its substrate, which grain size is achieved by growing the cell semiconductor on a molten intermediate rheotaxy layer of a suitable semiconductor which solidifies at a temperature below the melting temperature of the solar cell semiconductor. The substrate and the intermediate rheotaxy layer of the overlying cell are transparent to that fraction of sunlight which is utilized by the underlying cell. Various configurations of overlying and underlying cells are described.

A device developed by *H.W. Brandhorst, Jr.; U.S. Patent 4,131,486; Dec. 26, 1978; assigned to U.S. National Aeronautics and Space Administration* is a back-well cell, for example, a solar cell which comprises a first semiconductor material of one conductivity type with one face having the same conductivity type but more heavily doped to form a field region arranged to receive the radiant energy to be converted to electrical energy; and a layer of a second semiconductor material, preferably highly doped, of opposite conductivity type on the first semiconductor material adjacent the first semiconductor material at an interface remote from the heavily doped field region.

Instead of the opposite conductivity layer, one may employ a metallic layer to form a Schottky diode. If the metallic Schottky diode layer is used, no additional back contact is needed. A contact, such as a gridded contact, pervious to the radiant energy may be applied to the heavily doped field region of the more heavily doped same conductivity material for its contact.

A cell design developed by *K. Morimoto; U.S. Patent 4,140,610; February 20, 1979; assigned to Futaba Denshi Kogyo KK, Japan* has a plurality of alternate P-type and N-type semiconductor layers formed parallel with the plane on which the rays of light are incident. Therefore, all the electron-hole pairs produced in the vicinity of the light-receiving surface and at deep positions distant therefrom can move across the P-N junctions without being subject to recombination. Accordingly, the present solar battery can have a photoelectric conversion ability over a wide frequency range of incident light and, thereby, can remarkably improve the conversion efficiency.

Such a device is shown in Figure 69. Reference numeral **21** designates a substrate which acts as the base plate of a solar battery. The substrate may be made of silicon, sapphire, glass, ceramics, etc.

A plurality of N-type semiconductor layers **22** and P-type semiconductor layers **23** are alternately deposited and formed on the upper surface of the substrate in a laminated manner and in parallel with a plane on which light rays **A** are incident; thus a plurality of P–N junctions **24** is formed.

The outside edges of the N-type semiconductor layers and P-type semiconductor layers are staggered relative to each other in the opposite directions. Connection ears **25** and **26**, which are made of the same semiconductor materials as the foregoing N-type and P-type layers, connect together the abovementioned staggered outside edges of the N- and P-type semiconductor layers.

Terminal electrodes **29** and **30**, which have leads **27** and **28**, respectively, are mounted on a part of the connection ear of the N-type semiconductor layer and part of the connection ear of the P-type semiconductor layer by evaporation or the like so that an ohmic contact may be obtained therebetween.

The terminal electrode **29** formed on the connection ear of the N-type semiconductor layer may be made of an alloy containing components for forming N-type material, for example, gold antimonide, or aluminum, etc.; the terminal electrode **30** may be made of an alloy containing components for forming P-type material, for example, indium alloy, or indium itself, aluminum, or the like.

On the upper surface of the topmost one of the P-type semiconductor layers there is provided an antireflection layer **31** for preventing the reflection of incident rays of light and efficiently directing these rays into the semiconductor layers. Thus, a light-receiving surface **32** is formed on the top of the antireflection layer.

**Figure 69: Horizontal Multijunction Cell Design**

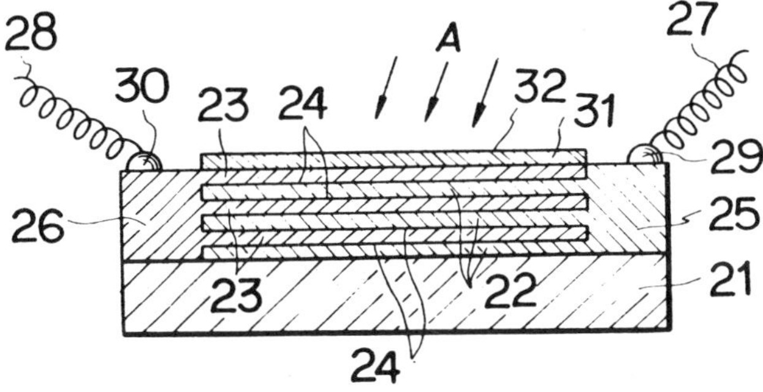


Source: U.S. Patent 4,140,610

## OTHER CONFIGURATIONS TO IMPROVE CELL EFFICIENCY

Based upon a semiempirical formulation, the theoretical maximum efficiency for silicon solar cells was determined by P. Rappaport, *RCA Review*, 20 (1959), as approximately 22%. Then, a theoretical limit of approximately 30% was derived by W. Shockley and H.J. Queisser, *Journal of Applied Physics*, 32 (1961), by assuming that only radiative recombination mechanisms are active, as required by the principle of detailed balance. In practice, some of the recombination processes will be nonradiative and the true practical limit is probably between the two limits, perhaps 25%.

Since the best commercial cells are 14% with good production lots being 8 to 10%, considerable improvement in efficiency can be expected with improved fabrication techniques. This is true even if the lower theoretical minimum efficiency is assumed. Another goal, directed towards reducing the cost per kilowatt, is the development of a production technique yielding high efficiency cells without an extensive selection procedure.

Table 1, in the introduction to this volume, summarizes some "probable maximum achievable efficiency" values for various types of photovoltaic cells, as well as the maximum measured efficiency for prototypes of such cells built to date (3).

Apart from the question of efficiency related to materials choice, there is the difficulty of providing means for efficiently converting sunlight photons incident over a given area to electrical current. This difficulty arises because incident photons from sunlight cover a relatively wide frequency/wavelength spectrum or energy range distribution. A single solar cell, which is sensitive to a limited range of solar wavelengths or photon energies, is able to convert photons of such range to electrical energy with relatively high efficiency. However, solar photons with energies above such range will not be converted efficiently by such a cell, and photons with energies below such range will not be converted at all, so that the resulting overall solar energy conversion efficiency of the cell may be poor, as pointed out by *L.W. James; U.S. Patent 4,017,332; April 12, 1977; assigned to Varian Associates.*

According to *W.J. King and S.J. Solomon; U.S. Patent 3,496,029; February 17, 1970; assigned to Ion Physics Corporation* the ion implantation technique for P-N junction formation is capable of achieving higher efficiency cells in production quantities. Through exact control of fabrication variables a semiempirical optimization of the important cell parameters is obtained. The major technical advantage of the ion implantation technique lies in the capability of accurately producing the required junction profile. The ability of accelerated particle techniques to accomplish this precise tailoring of the characteristics of a semiconductor layer lies in the property that energetic heavy ions have very discrete ranges in matter. By the use of a particle accelerator to produce these ions, it is possible to obtain not only discrete ranges but also highly accurate control of the ion energy, ion current, integrated beam current and beam distribution.

In addition, ions with sufficient energy for implementation by this technique may be readily obtained.

A cell design developed by *E.R. Streed; U.S. Patent 3,591,420; July 6, 1971; assigned to U.S. National Aeronautics and Space Administration* utilizes phosphors in the cover glass which are excited to fluorescence by solar ultraviolet radiation and particulate radiation. This fluorescent energy passes through the interference filter for utilization in the solar cell, whereas the ultraviolet and other radiation would not normally be converted to electrical energy because the wavelength is not within the spectral response limits of the solar cell.

According to *A.I. Bennett; U.S. Patent 3,682,708; August 8, 1972; assigned to Westinghouse Electric Corporation* a solar cell with improved efficiency is provided with a convoluted P-N junction whereby a higher proportion of carriers produced by exposure of the solar cell to a source of radiation will be collected by the P-N junction rather than being lost by recombination. Such a solar cell is also claimed to have an increased resistance to radiation damage. Figure 70 shows such a cell design.

**Figure 70: Convoluted Junction Solar Cell Design for Improved Efficiency**

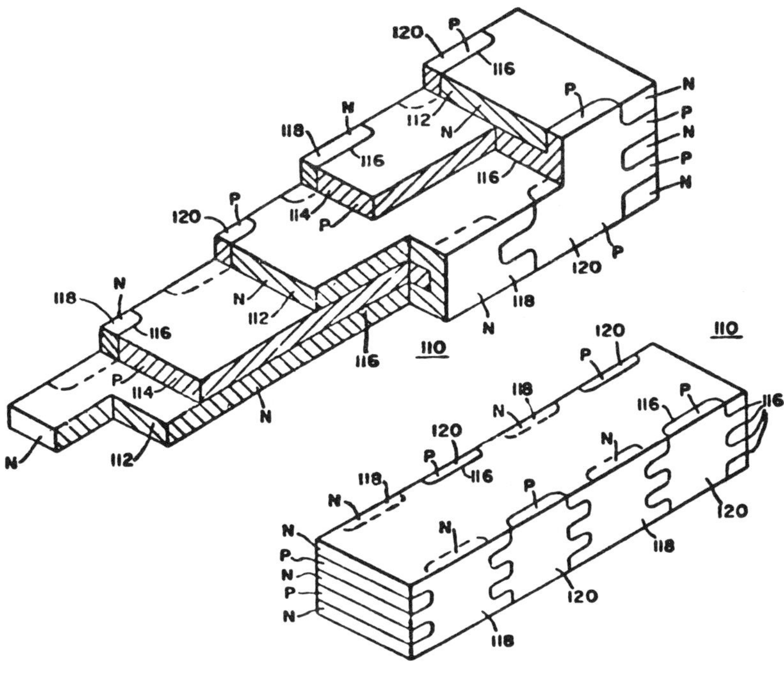


Source: U.S. Patent 3,682,708

Generally, this improved solar cell **110** comprises an N-region and a P-region interconvoluted or interconnected so as to present to a passing photon of light a plurality of substantially parallel alternately and successively disposed N-type and P-type layers of semiconductor material. Each of the layers has a preferred thickness of less than the diffusion length of the minority carriers generated within the cell when it is activated by exposure to solar radiation. The top layer is very thin, preferably from 0.3 to 0.5 micron in thickness, and has a high level of impurity concentration to reduce the electrical resistance within the layer. The bottom layer may be of the same thickness and doping concentration as the top layer, though it may have a thickness greater than that of the top layer but less than that of the intermediate layers. For some purposes the bottom layer may also have a level of doping impurity concentration less than that of the top layer.

When initially prepared, a P–N junction **116** exists as several separate P–N junctions formed by the contiguous surfaces of the adjacent layers **112** and **114** of different type semiconductivity. Employing suitable means, for example, as diffusion through masks, N-type regions **118** and P-type regions **120** may be produced in any suitable arrangement about the sides and ends of the cell. The N-type regions connect the N-type layers together and the P-type regions connect the P-type layers together whereby only common regions **112** and **114** exist. Each contiguous surface of each region **118** with a layer **114** forms a P–N junction which is a segment of the convoluted P–N junction **116.** In the same manner, each contiguous surface of region **120** and a layer **112** forms a P–N junction which is a segment of the convoluted P–N junction **116.**

A process developed by *H. Fischer and W. Pschunder; U.S. Patent 3,772,768; November 20, 1973; assigned to Licentia Patent-Verwaltungs-GmbH, Germany* provides an improved method of producing a solar cell wherein a liquid metal-semiconductor phase is produced on a semiconductor body which has been previously provided with a P–N junction. The liquid metal-semiconductor phase is allowed to remain for a certain time before cooling it and then electrodes are attached to the semiconductor body.

This process ensures that the efficiency of the solar cells is distinctly improved in comparison with those solar cells produced by conventional methods. In the usual cells with wafers about 300 μm thick, an increase in efficiency of about 10% has been observed as a result of the use of this technique. With thinner cells having wafers about 150 μm thick, such as have been used recently in generators, the gain in power is still more pronounced and is about 13%.

A technique developed by *W. Pschunder; U.S. Patent 3,802,924; April 9, 1974; assigned to Licentia Patent-Verwaltungs-GmbH, Germany* is designed to increase the power/weight ratio of solar cells beyond the present value. The solar cells are provided with a recess on the reverse face, wherein the remaining thicker rim portions are narrow, compared with the dimensions of the remaining thinner zones of the recess. Thus, the solar cells according to the process have a U-shaped cross section.

Since the major part of the light incident on the solar cells is absorbed within a region of thickness about 100 μm from the surface, the cell has in the recess a thickness of preferably about 100 μm, while the rim of the cell has a thickness of about 200 to 300 μm.

Such a solar cell design provides good electrical properties along with good mechanical strength and a minimum weight. The connection between the cells is made at a point where the cells have a thickness of 200 to 300 μm. The high pressure which may occur during the welding of the cells does not result in damage to the cell due to the strength of the cells.

With large area solar cells, the frame-like thicker rim is preferably reinforced by intermediate webs with the same thickness so that the reverse face has several recesses, separated from each other by intermediate webs.

A process developed by *R. Epple; U.S. Patent 3,905,836; September 16, 1975; assigned to Telefunken Patentverwertungsgesellschaft mbH, Germany* is a process for making a P–N junction photoelectric semiconductor device from a semiconductor body with particular emphasis on the step of producing recombination centers in the interior of the body, whereby the maximum spectral sensitivity of the device may be chosen as a function of the number of recombination centers produced.

According to *J. Lindmayer; U.S. Patent 3,907,595; September 23, 1975; assigned to Communications Satellite Corporation (COMSAT),* the efficiency of a solar cell is improved by incorporating a metal layer beneath the P–N junction of the solar cell. The metal layer reflects back towards the junction light which would otherwise penetrate too deeply to contribute to the voltage and current output. The metal also produces charge carriers which are photo-excited up to the conduction band of the semiconductor by light having photon energy less than the semiconductor band gap. The metal and the semiconductor material forming the solar cell are both highly ordered, thereby forming a barrier contact at the metal-semiconductor interface.

A cell design developed by *E. Kittl; U.S. Patent 3,929,510; December 30, 1975; assigned to the U.S. Secretary of the Army* includes solar radiation conversion means integral with or spaced from a silicon cell. The solar radiation conversion means is characterized by a band-emission spectrum that provides a good spectral match with the spectral response of a silicon cell.

The specific improvement involves including rare-earth compounds in the radiation conversion means. More particularly, the radiation conversion means includes ytterbium oxide and small amounts of at least one other rare earth oxide from the group praseodymium oxide, neodymium oxide, holmium oxide, erbium oxide, and thulium oxide. Ytterbium oxide is used as the major component in the radiation conversion means because the thermal emission spectrum of ytterbium oxide is unique in that it exhibits a single strong emission band at 0.97 micrometer. In this instance, a maximum amount of energy is concentrated in a single band and the ratio of band radiation to total solar radiation is a maximum and comes closest to the ideal case for the silicon cell illuminated with monochromatic light of 1.0 μm. Small amounts of other rare earth oxides can be included for energy conservation and to sensitize the ytterbium oxide. This is accomplished by doping the crystal lattice of ytterbium oxide with another rare earth ion having higher lying energy levels that are specifically suited for adsorption of the incident solar radiation input. The schematic arrangement of such a cell design is shown in Figure 71.

**Figure 71: U.S. Army-Developed Improved Solar Cell System Design Featuring Rare Earth Filter**

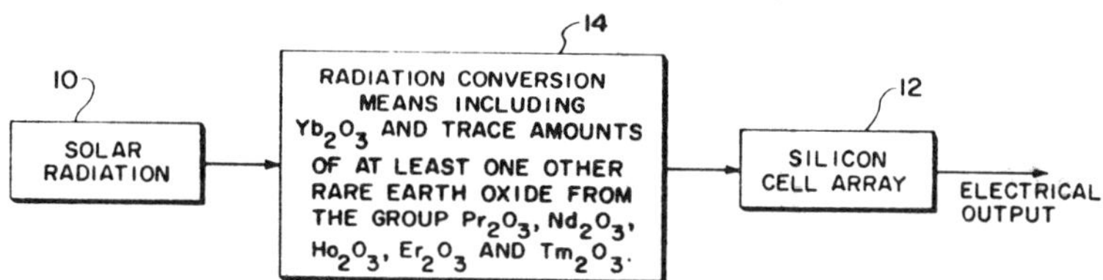


Source: U.S. Patent 3,929,510

According to *T.R. Anthony, H.E. Cline and D.M. Winegar; U.S. Patent 3,936,319; February 3, 1976; assigned to General Electric Company* a solar cell with improved efficiency is provided with a convoluted P-N junction whereby a higher proportion of carriers produced by exposure of the solar cell to a source of radiation will be collected by the P-N junction rather than being lost by recombination. The solar cell has an increased resistance to radiation damage. The solar cell is made from a body of semiconductor material in which two regions of opposite type conductivity are formed. The material of one region is substantially the same as the body and the material of the other region is recrystallized material of the first region having solid solubility of a metal therein to impart a selective type conductivity and resistivity thereto.

It is interesting to compare this convoluted P-N junction design with that described earlier in this section under U.S. Patent 3,682,708.

A device developed by *M.G. Coleman, L.A. Grenon, R.A. Pryor and F. Restrepo; U.S. Patent 4,144,094; March 13, 1979; assigned to Motorola, Inc.* has as its object the provision of a more efficient radiation responsive current generating cell, particularly solar, wherein the amount and efficiency of minority carrier flow across the cell P-N junction is increased.

In this device, this efficiency improvement is provided by a layer of material having charges disposed over the thin layer of semiconductor material used to form one side of the current generating P-N junction. The layer of material having charges possesses a charge of the same polarity as the minority carriers associated with the thin layer of semiconductor material for aiding minority carrier flow across the P-N junction and for minimizing surface recombination of minority carriers. Such a device is shown in Figure 72.

The figure illustrates a silicon solar radiation responsive cell **40** comprising a first P-conductivity type region **42** and an overlying N-conductivity type region **44**. A back side metallized layer **46** and upper surface metal contacts **48** to the N region **44** (only one shown) complete electrical connections to a P-N junction **50** formed by regions **42** and **44**, respectively. A third region or layer of material **52** having charges is disposed over the thin upper region **44**. In this embodiment, the layer **52** possesses dopants having charges of the same polarity as the minority carriers associated with region **44**. In the case of an N-type conductivity upper layer **44**, the charges in layer **52** are positive as indicated generally at **54**.

Operationally, solar radiation passing through the upper layer **52** and into the regions **42** and **44** creates electron-hole pairs. Due to the positive charge contained in layer **52**, minority carriers in layer **44** will be aided in traveling downwardly across the P-N junction **50** so as to recombine in region **42** in order to generate usable output current. Conversely, the positive charges in layer **52** oppose the movement of minority carriers upwardly and thus assist in preventing surface recombination of the minority carriers existing in layer **44**.

The layer **52** can be constituted by a layer of appropriate antireflective optical thickness which is substantially transparent to the incident radiation and made of a material which is intrinsically or impurity doped to contain the desired charges. Examples of intrinsically doped materials are P- or N-type zinc sulfides, P- or N-type tin oxides, etc. Impurity doped materials can be constituted by oxides, nitrides or oxynitrides of silicon-containing impurity ions. Negative or positive ions can be created with the introduction of chlorine and sodium, respectively. Additionally, suitable compounds other than oxides or nitrides of silicon, can be selectively doped, such as oxides of titanium, tantalum, cerium, etc. Appropriately doped layers also can be disposed over both upper and lower surfaces of the cell in order to increase current generation.

**Figure 72: Solar Cell Design Using Charged Surface Layer to Improve Efficiency**

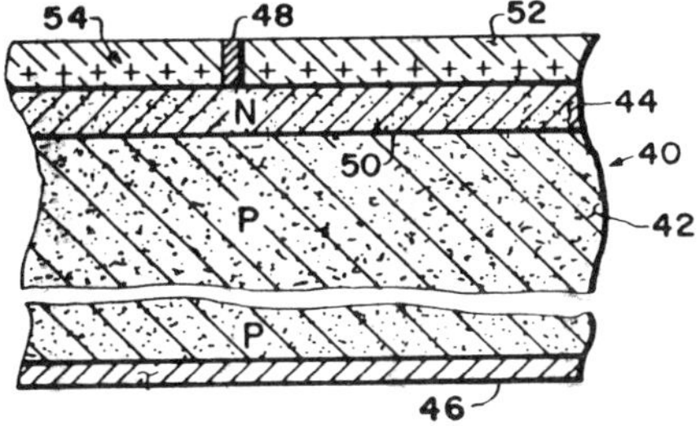


Source: U.S. Patent 4,144,094

## ATTACHMENT OF ELECTRICAL LEADS

The business of attaching leads to solar cells sounds like an insignificant problem compared to the manufacture of semiconductor materials, doping to form P-N junctions, etc. As a matter of fact however, the design and construction of electrical lead systems to maximize the efficiency of operation of solar cell arrays is quite a complex business which has been the subject of much research and development work.

Thus, as pointed out by *K.S. Tameja and V.A. Rossi in U.S. Patent 3,589,946; June 29, 1971; assigned to Westinghouse Electric Corp.*, one of the most important parameters that limit the efficiency of the solar cell is its series resistance. Various methods have been devised to reduce the series resistance among which the use of electrical grid contacts on the diffused surface gave most satisfactory results. One of the serious disadvantages of electrical grid structures on solar cells is that part of the active area of the cell is lost.

Various electrical grid structure patterns have been used in which the entire electrical grid structure is located on the upper surface of the solar cell exposed to a light source. The electrical grid structure consists of a number of metallic strips joined by a common collector strip. In the conventional design of the solar cells, the collector strip also is located on the active surface of the solar cell. Since the collector strip has to be wide enough to carry the current density, a further appreciable loss in active area results because of the location of the collector strip on the active surface. Consequently, from 10 to 12% of the active area of a cell is lost because the active area is covered by the electrical grid structure.

Additionally, these electrical grid structure patterns are not capable of efficiently collecting the carriers released in solar cells having a graded, or drift field, region of semiconductivity and a shallow P-N junction. Solar cells embodying a drift field region of semiconductivity and a shallow P-N junction are desirable for space applications where the solar cells are subject to the effects of radiation. These electrical grid structure patterns do not enable one to achieve the high efficiencies expected from solar cells having a shallow semiconductivity transition region.

One attributed reason for failure to obtain these expected high efficiencies is that the electrical grid structures do not provide an efficient means for collecting the carriers present in the region of semiconductivity to which the grid is affixed. Consequently, the radiation experienced by the solar cell extinguishes the activity of a greater number of carriers than anticipated with the resultant decrease in the expected power output for the solar cell.

Also, as pointed out by *J. Epstein; U.S. Patent 3,664,874; May 23, 1972; assigned to the U.S. National Aeronautics and Space Administration* the contact electrodes frequently must have the ability to withstand extreme environmental operating conditions of temperature, humidity, vacuum and radiation without adversely affecting the original characteristics of the semiconductor device. For many applications, it is also necessary for these contact electrodes to be ohmic, i.e., nonrectifying, and to have low contact-to-surface resistance. Furthermore, the process of making the contact electrodes should not degrade the characteristics of the semiconductor device.

Heretofore, many semiconductor contact electrodes, particularly contact electrodes on silicon devices, have not adequately possessed these properties with resulting deleterious effects on the operation of the devices.

One of the main causes of contact electrode failure occurs as a result of the type of surface adhesive bonding employed to bond the metal electrode to the semiconductor surface. One surface bonding technique involves providing an adhesive layer or substance between the contact electrode material and the supporting semiconductor, while another employs spring-type means to force the contact electrode against the semiconductor surface. Due to the different physical characteristics of the contact electrode and semiconductor materials, extremes in temperature, for example, cause different expansion and/or contraction of the contact electrode material with respect to the semiconductor material. This unequal flexing results in the physical separation of the contact electrode from the semiconductor surface, and a consequent increase in contact-to-surface resistance which causes the dissipation of useful electrical energy as heat or, in extreme cases, a completely open circuit.

An early prior art attempt to provide an acceptable contact electrode for semiconductor devices involved the use of a buffer layer physically located between the semiconductor material and the metal contact electrode. The buffer material has a temperature coefficient of expansion intermediate the temperature coefficient of the contact electrode material and the semiconductor material.

The buffer layer provides expansion-contraction compliance between the semiconductor material and the contact electrode material, and provides some degree of success in preventing contact electrode separation from the semiconductor surface in high and low temperature environments. However, in some special applications involving environmental extremes, this buffer layer still does not completely solve the contact electrode separation problem.

Another solution involves the use of a laminated type of contact electrode including a buffer layer, wherein a heat induced sintered region is formed between the contact electrode material and the semiconductor material. The sintered regions are formed by a process of solid state recrystallization, that is, the contact electrode material is placed on the semiconductor material and both are heated to cause a softening at a common interface. Heating causes atoms in the crystal structure of each to loosen so that when the heat is removed some of the atoms of both the contact electrode material and the semiconductor material physically commingle with the atoms of the other. Thus, upon recrystallization of the contact electrode and the semiconductor materials, a surface adhesive type contact is formed by virtue of the commingling of the atoms of each. This type of laminated contact electrode structure involving the buffer layer and the sintered region does not completely give the desired result because it still provides no more than a surface adhesive type bond between the contact electrode material and the semiconductor material and is therefore susceptible to the abovementioned deficiencies of that type of contact electrode.

Another main cause of contact electrode failure has been the failure to use contact electrode materials, which are, when adhered to the semiconductor device, able to withstand the extreme environmental conditions to which the contact electrodes will be subjected. One material which will withstand the extreme environmental conditions of temperature, humidity, vacuum and radiation, particularly x-rays, is tungsten. Elemental tungsten, a refractory metal, has a relatively high melting point, an acceptable low electrical resistivity, and a sufficient density to enable it to perform as a high Z material, i.e., an x-ray shield. A further property of tungsten, useful in contact electrodes, is that it combines with silicon to form stable compounds having low electrical resistance and high resistance to radiation.

A major drawback in the prior uses of tungsten as a contact electrode has been the fact that tungsten, as a refractory metal, is extremely susceptible to oxidation, and will form oxides at relatively low temperatures, i.e., below 450°C. Such oxide layers are difficult to work with, and must be removed from such contact electrodes when leads are made to the tungsten electrode. Prior art processes for attaching tungsten electrodes to semiconductor devices have necessitated passivation steps with the danger of oxidizing the tungsten contact. Such oxidation layers are insulators and undesirable in contact electrodes.

As pointed out by *H. Kressel and V.L. Dalal; U.S. Patent 3,988,167; October 26, 1976; assigned to RCA Corp.* another problem encountered in solar cells is having electrical contacts on the solar cell which do not interfere with solar radiation incident onto the cell. Solar radiation falling onto a contact will probably be reflected away from the cell, lowering the solar cell's collection efficiency.

While it is important to prevent the loss of solar radiation collection, it is also important to have electrical contacts on the solar cell which are conveniently located for the current generated anywhere in the cells' active region. If an electrical contact is only around the periphery of a solar cell it will not interfere with solar radiation incident onto the cell, but current generated within the solar cell will have to travel a greater distance to the contact than current generated at the cells' periphery. The farther current must travel through the semiconductor material to the contact, the larger the resistance it will encounter, and thus the lower the power conversion efficiency.

Therefore, it would be most desirable in the field of solar cells to provide (1) maximization of solar radiation collection without the cost of thick active regions, and (2) electrical contacts not interfering with incident solar radiation but yet are conveniently located to current generated anywhere in the active region.

As pointed out by *J.I. Pankove in U.S. Patent 4,139,858; February 13, 1979; assigned to RCA Corp.* the standard grid ohmic contact consists of a six fingered geometric pattern of a metal such as gold, chromium, silver, etc. However, a grid system suffers from the shadowing effect of the metal fingers, which reduces the total area of the solar cell which can be exposed to solar radiation, and thereby reduces the total collectable photocurrent.

Transparent metals or metal oxides, such as indium tin oxide, tin oxide, or copper oxide, do not shadow the solar cells due to their transparency; however, the metal oxides have higher resistivities than the metal fingers. This limits the usable voltage available from the photocell. In addition, the bandgap of the oxides may be sufficiently low to shield or filter out some of the solar radiation. Furthermore, tin oxide and indium tin oxide tend to chip or flake when applied in a sufficient thickness to reduce the resistance of the coating.

Thus, it would be desirable to have a transparent metal or semiconductor coating with a low resistivity, which is transparent to visible and infrared solar radiation, and has a high thermal conductivity and stability to elevated temperatures for use in solar cells with solar concentrators.

A scheme developed by *A.E. Mann and E.L. Ralph; U.S. Patent 3,489,615; January 13, 1970; assigned to Spectrolab* is one in which a solar cell for use in a solar cell array is provided with a top electrode on its photosensitive surface extending around a side of the solar cell to overlap a second electrode on the bottom surface of the cell. The overlapped portions are held electrically separate by suitable insulating means. By using insulation means, a large bottom electrode may be used to cover a larger area of the bottom surface of the cell thereby decreasing its resistance. Further, freedom in the arrangement of the electrodes on the bottom surface is realized so that flexible interconnecting tabs for an array of cells can all lie in substantially the bottom plane of the cells.

A type of construction developed by *R.K. Yasui; U.S. Patent 3,493,437; Feb. 3, 1970; assigned to U.S. National Aeronautics and Space Administration* is one in which a multicell submodule is provided, in which a first busbar has portions thereof electrically connected to a first terminal of each of the cells in the submodule, with raised portions of the first busbar being present between the portions thereof which are in contact with the cell's first terminals. A second busbar, which is electrically connected to the second terminals of the various cells in the submodule, includes a plurality of tabs which extend from the second busbar. These tabs are adapted to be connected to the raised portions of a first busbar of another submodule in order to form a multisubmodule cell matrix.

A design developed by *A.E. Mann; U.S. Patent 3,502,507; March 24, 1970; assigned to Textron, Inc.* is one in which a solar cell is provided with a top solar sensitive surface wrapping around at least one edge portion of the cell to lie in close electrically separated relationship with a second electrode means on the bottom surface of the cell. The geometry of the electrodes is such that interconnecting tab or strip-like conductors can extend across the bottom electrode to effect pairs of positive and negative connection points, the connections in each pair being in close relationship to each other and the pairs themselves being at spaced portions of the cell remote from each other. With this arrangement, the risk of circuit discontinuity should the cell crack is minimized. Further, the interconnecting means lies substantially in the plane of the bottom of the cell thereby enabling the cells to be packed relatively closely to each other.

To be efficient, the contacts to the P and N regions should have low resistance to the flow of output current. The contact on the side of the P-N wafer nearest the source of radiant energy must not block the illumination of the wafer while still providing low resistance. For this reason the contact is fabricated in the form of fingers which geometry is an attempt to provide low radiant energy-blocking and low resistance.

In addition, the surface of the wafer upon which the radiant energy impinges should be treated by film evaporation or a similar technique to provide good transmission of the radiant energy into the interior of the wafer. It is difficult to provide the desired surface treatment where provision must also be made for the contact "fingers".

A type of construction developed by *B.D. Wedlock; U.S. Patent 3,549,960; December 22, 1970; assigned to Massachusetts Institute of Technology* is one which is efficient in the sense of having low resistance to the flow of current and in having a surface upon which the radiant energy impinges which is free of obstructing "finger" contacts.

A device developed by *F.A. Shirland; U.S. Patent 3,571,915; March 23, 1971; assigned to Clevite Corporation* provides an integral battery of serially connected photovoltaic cells on a single insulating substrate. Metallized areas are formed on the substrate with semiconductive film such as cadmium sulfide vacuum evaporated upon each of the metallized areas. Barrier layers are formed on the cadmium sulfide films to produce P-N junctions. Electrode leads extend from each metallized area under the semiconductor film to a top surface of the barrier layer of an adjacent semiconductor film.

A design developed by *A.E. Mann; U.S. Patent 3,575,721; April 20, 1971; assigned to Textron, Inc.* consists of an electrical connector arrangement for interconnecting an array of solar cells includes an elongated, flexible strip formed with laterally projecting tabs attached to adjacent rows of solar cells. The tabs are dimensioned and attached so as to preserve electrical continuity between various cells in the event that cell cracking occurs that otherwise would disrupt electrical continuity. The flexible strip incorporates short tabs projecting from one side for connection with a top electrode of one cell and at least one extended tab laterally projecting from the other strip side for connection with the bottom electrode of an adjacent cell. The extended tab bridges across the major portion of the bottom of the adjacent cell and is connected thereto only at points beyond the midpoint of the adjacent cell. Cracks splitting the cell in half are then prevented from completely incapacitating the solar cell.

A device developed by *K.S. Tarneja and V.A. Rossi; U.S. Patent 3,589,946; June 29, 1971; assigned to Westinghouse Electric Corp.* is a solar cell with an electrical grid structure which when affixed to a surface of the solar cell leaves exposed at least 95% of the surface for exposure to a radiant energy source.

A process developed by *W.A. Hasbach; U.S. Patent 3,616,528; November 2, 1971; assigned to U.S. National Aeronautics and Space Administration* provides an electrically-connected matrix of discrete solar cell blanks. Electrode contact receiving areas are provided on the light-sensitive surface of each blank and discrete continuous conducting layers are directly attached to the contact areas and extend between at least two of the blanks to form integral electrode contacts on and interconnections between the blanks. The cell blanks are disposed in separated, adjacent, side-by-side arrangements with their light-sensitive faces lying in substantially the same plane. Bridges for supporting the interconnection are placed in the separations. Portions of the blanks are masked and metal is deposited through the mask onto the bridges and surfaces to form integral electrode contacts on and interconnections between the separated blanks.

Processing time for production of matrices of solar elements is substantially reduced by forming both electrode contacts on the same surface of the silicon blank although there is some loss in light-gathering efficiency because of the areas rendered opaque by the deposited metal. The continuous nature of the electrode contacts, grid lines and interconnections and their direct adherence to the silicon surface without any intervening metallic layer provides improved collection of solar energy as compared to the conventional arrangement utilizing previously deposited contacts and subsequently soldered leads. As compared to the prior art, this approach does not place a plurality of dissimilar materials into contact with the silicon surface and does not produce resistance-creating interfaces.

A technique developed by *J. Epstein; U.S. Patent 3,664,874; May 23, 1972; assigned to the U.S. National Aeronautics and Space Administration* is one whereby tungsten contact electrodes are formed on doped silicon substrates, such as in solar cells. The method comprises depositing elemental tungsten on contact areas of doped silicon semiconductor devices, and thereafter processing the silicon devices to induce a chemical reaction between the tungsten and the silicon. The resulting tungsten contact electrodes thereby include a transition region at the interface of the tungsten electrode and silicon device comprising tungsten-silicon compounds which are physically and chemically bonded to, and form an integral part of, both the tungsten electrode material and the silicon semiconductor material. A lead is bonded directly to the tungsten electrode.

A type of construction developed by *R. Gereth and H. Fischer; U.S. Patent 3,686,036; August 22, 1972; assigned to Telefunken Patentverwertungsgesellschaft mbH, Germany* involves the provision of an electrical contact which contains silver as a first layer, has titanium, chromium, molybdenum or tantalum as a second layer and a precious metal as the third layer. The precious metal is contained in the second layer or is disposed in the form of an intermediate layer between the second layer and the silver. If the precious metal is contained in the second layer, the precious metal is admixed with the second layer or alloyed therewith. This contact is very resistant to temperature and corrosion and, in addition, can be subjected to temperature cycles with considerable fluctuations in temperature.

Such a cell design is shown in Figure 73. Referring now to the drawing, there is shown the construction of a so-called N-type on P-type solar cell which consists of a semiconductor body **1** of silicon of P-type conductivity in one surface of which an N-type region **2** is indiffused. The P-type doping of the silicon substrate is obtained, for example, by the introduction of boron, while the N-type region **2** may be produced, for example, by the indiffusion of phosphorus. The N-type region **2** may have a thickness of 0.3 $\mu$ for example.

Between the N-type region **2** and the portion of the semiconductor substrate excluded from the diffusion there is formed the P-N junction **3** necessary for the solar cell. The dimensions of the silicon body **1** may amount, for example, to 2 x 2 x 0.03 cm. The technique may likewise be used for so-called P-type on N-type solar cells wherein the P-type region is produced by diffusion instead of the N-type region.

As the drawing further shows, contact is made to the two semiconductor regions **1** and **2** forming the P-N junction by means of electrodes. Thus one electrode is provided at the front and one at the back of the solar cell, and the electrode which is at the front, which makes contact to the N-type region **2** is termed the front contact **4**, while the electrode at the back of the semiconductor body, which is provided on the semiconductor substrate and hence on the P-type region **1** is termed back contact **5**. Both the front contact **4** and the back contact **5** consist of the same material, namely, of a first layer of silver, a second layer of titanium, chromium, molybdenum or tantalum, and a third layer of a precious metal, preferably in the platinum group such as palladium and platinum.

**Figure 73: Telefunken Design for Solar Cell Contacts**

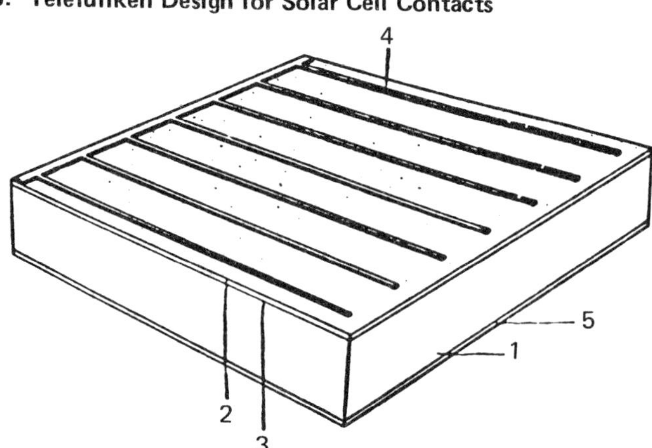


Source: U.S. Patent 3,686,036

The front contact is constructed in the form of a grid or comb, while the back contact represents a large-area electrode which, in contrast to the front contact, covers the entire back of the semiconductor body.

In order to produce the front and back contacts, a layer of titanium for example is first vapor deposited, then a layer of precious metal, for example of palladium, and finally a layer of silver on the layer of palladium. The silicon body may be maintained at a temperature of about 150°C during the vapor deposition which is preferably effected under vacuum. The thickness of the vapor-deposited titanium layer may amount to 350 A, the thickness of the palladium layer may amount to 50 to 200 A, and the thickness of the silver layer may amount to 5 μm for example.

Experiments have shown that the solar-cell contacts of the type proposed have a very satisfactory bond strength even under extreme conditions and that, apart from resistance to corrosion and temperature stability, they have a low contact resistance as well as excellent ohmic behavior with respect to N-type and P-type material.

A process developed by *H. Fischer, R. Gereth and K.-H. Kreuzer; U.S. Patent 3,736,180; May 29, 1973; assigned to Licentia Patent-Verwaltungs GmbH, Germany* for producing solar cells includes subjecting the back contacts of the cells to a separate intensified intermediate sintering prior to the application of the front contact.

In known solar cells with a semiconductor body of silicon the electrode consists, for example, of superimposed layers of titanium and silver which are applied by evaporation in a high vacuum by means of a mask. During a further process step these layers may be tin plated wholly or partially. During the application of the metal electrodes by evaporation, the side surfaces of the semiconductor body may frequently receive a vaporized deposit at the points where the metal must be applied directly up to the edge of the large area surface of the semiconductor body. This causes a short circuit of the P-N junction exposed on the surface of the side face, rendering the solar cell useless.

Although such a short circuit could be prevented by carefully covering the side faces during the evaporation, this would require a highly accurate adjustment of the solar cells relative to the masks and would, therefore, be uneconomical. Another possibility is to eliminate undesirable edge short circuits subsequently by covering the electrodes with a suitable substance, and removing the short circuiting material from the side faces by etching with a suitable solvent. However, such a method is also uneconomical because it requires a number of steps such as, e.g., covering, etching, washing, drying, removal of the cover, etc. In addition, there is the danger of etching below the cover, and this leads to strong fluctuations of the quality.

A process developed by *H.-H. Arndt and F. Belke; U.S. Patent 3,772,770; November 20, 1973; assigned to Licentia Patent-Verwaltungs GmbH, Germany* is one in which after deposition of an electrode which approaches an edge of the semiconductor body on which it is deposited, the side surface next to that edge is ground away to remove any electrode material deposited thereon.

In known photoelectric barrier layer cells based on semiconductors, the surface provided for receiving the incident light is usually covered with a thin wire metal mesh which is pressed or bonded on and which serves for collecting the electric current produced from light and to reduce the internal resistance of the cell by shortening the current paths, thereby increasing the current yield for a given starting voltage.

In the prior art, the contact resistance between the semiconductor surface on the one hand and the metal mesh on the other was reduced by gold-plating the mesh in order to prevent the formation of oxide layers. This gold-plated mesh was placed on the semiconductor surface and covered with a foil applied by means of an adhesive. Then, the whole assembly was pressed and exposed to such heat that the adhesive made a firm connection with the zones of the semiconductor surface exposed between the mesh-shaped electrode, and the electrode itself. Contact resistances between the metal mesh and the semiconductor body are comparatively large, in spite of the use of a metal electrode plated with gold.

An improved technique developed by *H. Fischer and E. Justi; U.S. Patent 3,778,684; December 11, 1973; assigned to Licentia Patent-Verwaltungs GmbH, Germany* is based on the surprising fact, supported by electron diffraction recordings, that gold is by no means a completely noble metal, but is covered by a single molecule layer of gold oxide with a very bad electrical conductivity. Systematic investigations have shown that only one of the noble metals has a good electrical conductivity of the oxide top layer, namely Pd. Furthermore, it has also been found that a less noble metal, such as rhenium, has, amongst its seven different oxides, low oxides which are excellent conductors so that the desired low transfer resistance can be achieved with a Cu mesh by covering the same, preferably electrolytically, with a Re layer of a few $\mu$m thickness. Combined oxygen is removed from this Re layer preferably subsequently either by cathodic reduction or by a later chemical reduction in a hydrogen atmosphere at a temperature of several hundred degrees.

By means of this method, the transfer resistances of all semiconductor elements which are equipped with pressed on or cemented on contacts, may be substantially reduced. Thus during tests, the power yield of a photoelement with an electrode coated with palladium could be increased by 7%, compared with the hitherto used elements with gold electrodes.

As pointed out by *J. Roger; U.S. Patent 3,793,082; February 19, 1974; assigned to Societe Anonyme de Telecommunications, France,* a number of various processes for making solar batteries contacts are already known. As these batteries are generally used on satellites and, because of this subjected over long periods, sometimes of the order of ten years, to a very large number of thermal cycles, it has been found that such assemblies stand up badly to such conditions.

The breakdowns observed are due in part to the difference in the linear coefficient of expansion of the constituents, and in part, to the degradation of the tin normally used in the manufacture of such contacts. In addition, the welding of the contacts normally used subjects the solar battery to a thermal shock which is very localized in the neighborhood of the weld, which affects the functioning of the cell adversely.

The process developed by J. Roger avoids these problems by depositing a metal comprised of silver on an electrical contact area on the first and second portions of the battery by evaporating the metal in a vacuum onto the contact areas; and fixing the metal contact pieces to the deposited metal on the contact areas by intermetallic diffusion using temperatures in the range of 500° to 700°C and a pressure of approximately 50 $kg/cm^2$.

This process enables solar batteries to be obtained fitted with their interconnection strips, and integrated during the last operation in the manufacture of a battery. In this way both the electrical efficiency and reliability of the battery are considerably improved.

High conductivity and high optical transmission are incompatable insofar as, other things being equal, the former calls for thickness and the latter for thinness of material. In general, the highest conductivity is possessed by metals which, if sufficiently thin, are also transparent. However, practical tests show that metallic films have much lower conductivities and lower optical transmission than is to be expected from the bulk values of their conductivity and optical constants. Certain semiconducting oxides, such as stannic oxide and indium oxide have also been used as transparent conductors. Unfortunately, however, these materials must be applied to hot substrates (400° to 600°C) or, subsequent to formation, heat treated at high temperatures in order to render them sufficiently conductive. Furthermore, the optical transmission and related electrical conductivity of these oxides cannot be made sufficiently high to be useful in many practical applications.

A material developed by *A.J. Nozik; U.S. Patent 3,811,953; May 21, 1974; assigned to American Cyanamid Company* is cadmium stannate ($Cd_2SnO_4$) which is shown to provide a light-transmitting electrically conducting composition in which the electrical conductivity can be varied from $10^{-7}$ $ohm^{-1}$-$cm^{-1}$ to $10^4$ $ohm^{-1}$-$cm^{-1}$ by controlling the oxygen vacancy concentration of the material. Amorphous and crystalline films of $Cd_2SnO_4$ can be disposed on cold and/or hot substrates and they exhibit high optical transparency as well as high electrical conductivity.

In an N-P type solar cell electrons will travel to the top surface where they will then be collected by a metallic grid positioned thereon. The metallic grid may typically comprise six metallic fingers separated along the top surface by a relatively large distance and connected to each other by a common bus bar. The electrons will travel either directly to the metallic fingers or approach the top surface between the fingers and then travel along the surface of the solar cell until they can be collected by one of the fingers. Holes, on the other hand, will travel to the bottom surface of the solar cell where they may be collected by a metallic sheet covering the entire bottom surface.

The six-fingered metallic grid is necessary at the top surface of the solar cells in order to enable light to enter the solar cell. However, one problem associated with the six-fingered construction relates to the relatively large separation between the fingers. Electrons which must travel along the surface to the metallic fingers encounter a high surface resistance. Therefore, due to the relatively long distance the electrons must travel before collection, and due to the problem of surface resistance, a series resistance may develop, thereby limiting the efficiency (electrical power output/solar power input) of solar cells by limiting the electrical power output.

The prior art has sought to obviate the above problem by diffusing an impurity into the surface of the solar cell in a higher order concentration, on the average of about $10^{15}$ atoms per square centimeter or higher. Higher order concentration (i.e., heavier diffusion) lowers the surface resistance but introduces other problems. Higher order concentration of impurities is obtained by a process known as "solid solubility diffusion," i.e., the solar cell is allowed to assume as many impurities as it can on the surface, e.g., approaching $10^{21}$ atoms per cubic centimeter.

However, in such a solid solubility process there is crystal lattice damage to the solar cell which propagates deep into the solar cell substrate. The efficiency of the solar cell is thereby reduced in two ways. First, the damage to the crystal lattice causes a reduction in the diffusion length or lifetime of minority carriers. This means that holes, for example, in an N-type diffused region will recombine with available electrons before they can be separated by the junction. Secondly, damage to the crystal structure affects the power output of the solar cell (which is basically a diode) by "softening" the current-voltage characteristics of the diode.

In addition, the diffusion of such higher order concentration of impurities creates a relatively deep junction of about 4000 A. This relatively deep junction means that light of relatively short wavelengths (where solar energy peaks) cannot penetrate beyond the junction, but is absorbed in the diffused region (i.e., between the top surface and the junction). Electron-hole pairs generated in the diffused region have a relatively short diffusion length (even if there were no crystal lattice damage) and therefore will largely recombine before separation by the junction.

An improved type of construction developed by *J. Lindmayer; U.S. Patent 3,811,954; May 21, 1974; assigned to Communications Satellite Corporation* utilizes a top surface current collector comprising a significantly greater number of fine metallic fingers (or other fine geometric pattern) wherein the physical separation between the fingers and the width of each finger is substantially reduced. The junction depth and/or impurity concentration is reduced in accordance with the degree of freedom provided by the use of the fine geometry cell.

The solar cell is made by first introducing impurities into, for example, a silicon slice, and then oxidizing the solar cell. Then, the fine metallic pattern is placed on the top surface of the solar cell using a photolithography technique. Finally, a plating process is used to build up the fingers of the metallic pattern to a proper thickness.

Figure 74 shows the effect of contact geometry on cell efficiency in graphic form. As can be seen, the efficiency obtainable with the fine geometry solar cell is approximately 50% greater than the efficiency obtainable with the standard six finger geometry solar cells.

**Figure 74: Effect of Contact Geometry on Cell Efficiency**

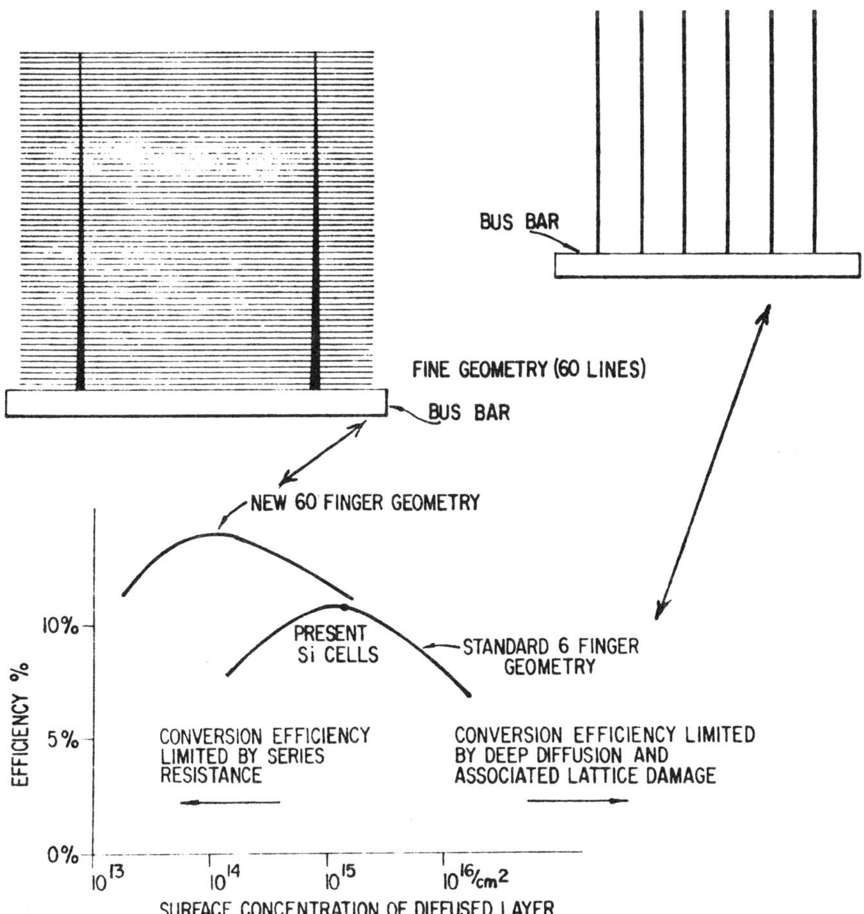


Source: U.S. Patent 3,811,954

The prior art method of assembling solar cells to create a flexible solar cell array is to interconnect the cells electrically in a series-parallel matrix by means of expandable metal interconnectors, each of which extends between adjacent columns of solar cells connecting the top edge of the cells in one column to the bottom edge of the cells in the preceeding column. The mechanical interconnections provided by the metal interconnectors are undesirable features since the interconnections are not sufficiently flexible and therefore stresses caused by thermal expansion and the vibrational environment are transmitted through the interconnectors. Failure can occur at the cell-interconnector interface.

In many solar cell assemblies the entire matrix is attached to a flexible substrate by means of an adhesive for the purpose of rolling up or folding the array. However, with the prior art metal interconnectors this puts additional strain on the interconnectors. Furthermore, whenever the cells are subjected to any thermal excursion, the plastic substrate and adhesive which connects the matrix to the substrate will stretch or shrink more than the solar cell matrix since the thermal coefficients of expansion of the plastic and adhesive are greater than that of the solar cell matrix. This creates a severe amount of stress on the solar cell interconnectors.

In a type of construction developed by *J.G. Haynos; U.S. Patent 3,819,417; June 25, 1974; assigned to Communications Satellite Corporation* a flexible solar cell assembly is provided in which the mechanical interconnection of the solar cells in a row is provided by a plurality of flexible nonconducting strips interlaced or woven among the cells in the row. At least two strips are woven in opposite manner so that if one strip goes over one cell and under the next, the other strip goes under the first cell and over the next. Serial electrical interconnection is provided by metallic patterns which are formed on one surface of the flexible strips.

A type of construction developed by *H.G. Mesch; U.S. Patent 3,833,426; September 3, 1974; assigned to TRW Inc.* involves a solar array having solar cells arranged in parallel strings at the front side of a supporting substrate with the corresponding cells in the several strings arranged in rows transverse to the strings. Each cell has a rear contact surface facing and exposed through an opening in the substrate and front contact and light sensitive surfaces. The cells in each cell string are connected in electrical series by series interconnectors, each passing between a pair of adjacent cells and through the substrate.

Each interconnector is electrically joined at its front end to the front contact surface of the adjacent cell and at its rear end to the rear contact surface of the other adjacent cell through the corresponding substrate opening. The cells in each cell row are connected in electrical parallel by parallel interconnector strips electrically jointed to the rear ends of the corresponding series interconnectors.

A scheme developed by *W.R. Baron; U.S. Patent 3,837,924; September 24, 1974; assigned to TRW, Inc.* is one in which a solar array is uniquely constructed to minimize, if not virtually eliminate, cyclic stressing of the cell interconnects and thereby prevent fatigue failure of the interconnects. This is accomplished by making each interconnect of such a length that the center spacing between its attachment points to the adjacent solar cells is related to the center spacing between the attachment points of the cells to the supporting substrate and to the coefficients of thermal expansion of the interconnect and substrate materials in a manner which results in substantial equal thermal expansion and contraction of the interconnect and substrate within the regions between the respective attachment points. Cyclic stressing and flexing of the interconnect is thus substantially reduced or eliminated. Moreover, the need for flexibility in the interconnects to accommodate thermal expansion and contraction of the substrate is eliminated.

A type of construction developed by *H.Fischer and W. Pschunder; U.S. Patent 3,887,935; June 3, 1975; assigned to Licentia Patent-Verwaltungs GmbH, Germany* utilizes a rectifying metal semiconductor contact having a metal coating split up into regions largely separated from each other but electrically connected.

A cell design developed by *A. Gauthier; U.S. Patent 3,887,995; June 10, 1975; assigned to Societe Anonyme de Telecommunications, France* consists of two electrodes, a collector grid, a copper sulfide layer and a cadmium sulfide layer. The two electrodes are constituted by a single conductive film divided into two parts separated by an insulating material, the sulfide layers being deposited on one of the parts and surmounted by the collector grid which is electrically connected to the other part of the conductive film.

Figure 75 shows a prior art electrode design (upper view) and an improved electrode design (section in middle view and plan in bottom view).

**Figure 75: Prior Art and Improved Cell Contact Designs**

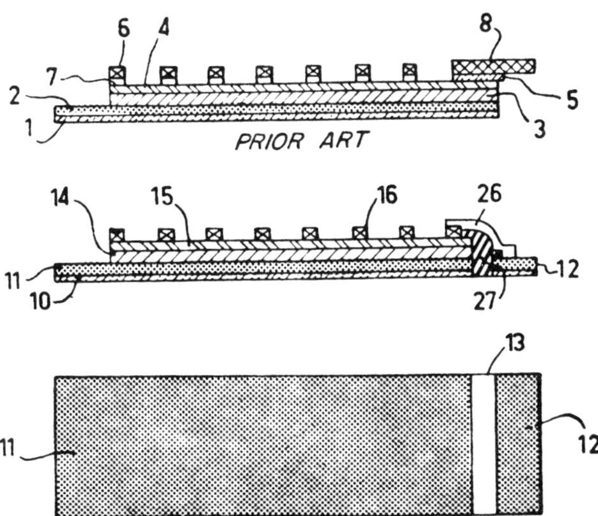


Source: U.S. Patent 3,887,995

In the upper view there are shown in succession, starting from the lower part, an insulating support **1**, constituted by a sheet of Kapton, a conductive layer **2** constituted by a silver lacquer, a layer **3** of CdS, a layer **4** of $Cu_2S$ and a grid **6** adhered to the layer **4** by a layer of conductive adhesive **7**. The extension of the grid **6** in the form of a metal plate **8** constitutes the second electrode of the solar cell, the layer **2** of silver lacquer constituting the first electrode.

In order to ensure that the metal plate **8** does not short circuit the solar cell, a piece of insulating sheet **5**, such as a sheet of Kapton, is interposed between the plate **8** and the layer **4** of $Cu_2S$. The difference in level between the two electrodes **2** and **8** is clearly shown.

In order to overcome this drawback of differences of level, the solar cell according to the process has a new arrangement which is shown in the lower views.

There is deposited on an insulating support **10**, such as a sheet of Kapton, a conductive film or layer, for example of silver lacquer deposited in two distinct plates **11** and **12** separated by a zone **13** in which the insulating sheet remains bare. Deposited on the conductive zone **11** by known means are the layers of CdS and $Cu_2S$ **14** and **15** of the solar cell. Directly fixed to the layer **15** is the collector grid **16** which is obtained by electrolytic deposition.

To this end, there is formed a photoengraving on the layer of $Cu_2S$ so as to obtain the image of the grid so that the parts of $Cu_2S$ remaining bare constitute the pattern of the desired grid, all the silver-coated zones being equally masked by the photo-resistant product with the exception of the end of the electrode **11**. Any other masking method, such as serigraphy for example, may also be employed.

Figure 76 shows the electrolytic deposition apparatus. The tank **17** contains an electrode **18** which is a gold salt solution, the cathode being constituted by the solar cell in the course of manufacture which is connected by the end **19** of its N electrode to the first terminal **20** of an ammeter **21** the second terminal **22** of which is connected to the negative terminal of a dc supply **23** whose positive terminal is connected to the anode **24** of the electrolysis device.

**Figure 76: Apparatus for Electrolytic Deposition of Cell Contacts**

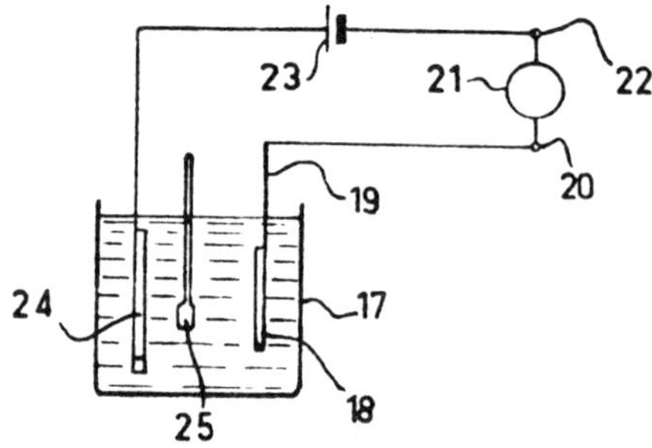


Source : U.S. Patent 3,887,995

This anode, whose surface area is at least equal to the surface area of the cell, is constituted by a platinum plate. The temperature of the electrolytic solution **18** is controlled by means of a thermometer **25** which is immersed in the solution. The pH of this solution is adjusted to between 7 and 10. The current density is so chosen as to produce an adherent metallic deposit and it must not exceed a certain value beyond which the CdS-$Cu_2S$ junction is liable to be impaired. This value is a function of the respective thickness of the layers of CdS and $Cu_2S$. Bearing in mind the usual thicknesses of these layers the current density is between 5 and 20 milliamperes per square centimeter.

The thickness of the metallic deposit constituting the grid is so determined as to reduce the resistance of the cell as far as possible. This thickness is obtained by the control of the amount of electricity employed. Good results are obtained with a thickness of 3-5 microns.

After the grid has been deposited, the masking product is removed by means of xylene or trichlorethylene. The solar cell is then rinsed with alcohol and dried in a vacuum.

The grid **16** is electrically connected to the P-type electrode **12** by means of a layer **26** of conductive lacquer which is insulated from the layers $Cu_2S$ and CdS and from the N-type electrode by a layer of insulating varnish **27** which bears on the zone **13**, which had been left bare up till now, of the insulating support **10** of the solar cell.

The solar cell of the improved design has the great advantage of having its two electrodes in the same plane. The construction of a solar panel by means of the solar cells according to the process is thus facilitated.

The manufacture of the collector grid of the solar cell by electrolytic deposition has the advantage of lowering the usual price of CdS solar cells and reducing the weight of the cells while allowing more elaborate grid conformations and avoiding any danger of deterioration of the cell by the heating of the latter in the course of polymerization of the gold-containing adhesive which serves to fix the collector grids obtained by the known processes.

An improved cell design developed by *G.J. Pack; U.S. Patent 3,903,427; Sept. 2, 1975; assigned to Hughes Aircraft Company* produces a 10% increase in useful power from a photocell area without increasing the size or weight of the cell. It also provides the capability to decrease the temperature of the cells themselves due to elimination of some of the power losses. These improvements are obtained by taking the front leads through the cell to its back surface instead of across its surface.

Such a form of construction is shown in cross-section in Figure 77. As shown there, a solar cell **10** is formed from a wafer of semiconductor material **18** of P-N or N-P construction, having a top layer **18'** of N-type silicon on P-type silicon or vice versa.

**Figure 77: Hughes Aircraft Co. Solar Cell Design Featuring Increased Output Through Improved Contact and Lead Location**

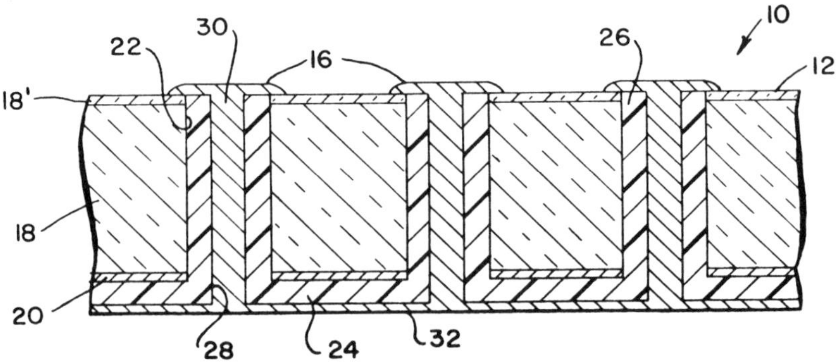

Source: U.S. Patent 3,903,427

Placed across the front surface of the cell is a plurality of one or more spaced current pick-up points **16** configurated as small, individual metallic contacts. These contacts are laid on the front surface. Attached to the back side of the semiconductor material is a first conductor **20**. Machined or otherwise formed through semiconductor material **18** and back leads **20**, such as by laser or electron beam drilling, are a plurality of holes **22** which extend from front surface **12** through the back surface and back conductor **20**. Within holes **22** and along back conductor **20** is placed a layer of electrical insulation material.

This layer includes insulation coatings **26** within each of the holes to form an insulated hole **28**. A pass through conductor **30** is electrically coupled to each contact **16** and extends through holes **28** to a metal layer **32** which is adhered to insulation layer **24**.

Although contacts **16** are shown as buttons spaced in parallel, they may take any suitable form in any suitable configuraton, whether parallel or not. Specifically, contacts **16** are so numbered and designed to minimize resistance, yet to maximize surface areas for generation of current. That is, because contacts **16** comprise the points at which current collects, there should not be so much distance between contacts as to produce an unacceptable power loss due to increased resistance. However, there should not be so many contacts that the effective current producing area is reduced to a level comparable to that of conventional cells. Thus, a balance between number and spacing of contacts vis-a-vis usable surface is attained to provide efficient use of the configuration. By passing the leads through the cell rather than across the cell, the total resistance of the front lead can be reduced to that of the back lead. To pass the front leads through the cell, holes of approximately 0.010" diameter are machined or otherwise formed through the cell, such as by laser cutting and electron beam cutting.

As a consequence of this construction, for a given area and weight of photocell arrays, an increase of approximately 8 to 10% in power can be obtained along with a reduction of temperature of the cells due to decrease in power losses at the cell itself.

In connecting one cell to another, whether in parallel or in series, all connections are made at the backside of the cells, thereby facilitating assembly operations.

In a type of construction developed by *H. Fischer and W. Pschunder; U.S. Patent 3,956,765; May 11, 1976; assigned to Licentia Patent-Verwaltungs GmbH, Germany*, the ohmic electrodes of a solar cell and the rectifying metal-semiconductor contact may be produced from the same material. The ohmic electrodes of the solar cell and the rectifying metal-semiconductor contact may comprise, for example, a metal combination. When using a semiconductor body of silicon, the metal combination may comprise, for example, two layers each, that is to say a first layer, for example of titanium or chromium and a second layer, for example of silver. In this case, for example, a solder may be additionally applied to the silver layer.

A cell design developed by *R.K. Yasui and P.A. Berman; U.S. Patent 3,966,499; June 29, 1976; assigned to U.S. National Aeronautics and Space Administration* involves providing a solar cell with a metallized grid pattern consisting of concentric geometric figures about the cell's geometric center. The concentric figures are preferably equally spaced from one another where the distance between any adjacent pair of figures is definable as L. The grid pattern is deposited on the cell so that the path length of any injected majority carrier, at any point on the cell face, is never greater than L/2, and is independent of the distance from the geometric center. The concentric figures are interconnected by one or more intersecting metallized main current-carrying connections or conductors.

A cell with such a grid pattern exhibits relatively low cell series resistance thereby exhibiting less power loss which is dissipated in the form of heat. Furthermore, the pattern is chosen to produce relatively low temperature differentials and thereby a more uniform temperature distribution across its face. Consequently, the decrease of cell power with increasing solar intensity occurs at a higher point of solar intensity than is experienced with solar cells employing conventional grid patterns.

A type of construction developed by *J. Lindmayer and J.F. Allison; U.S. Patent 3,982,964; September 28, 1976; assigned to Communications Satellite Corporation* involves an improvement in a fine geometry solar cell having a top surface contact comprising substantially more and finer metallic fingers spaced close together for collecting photocurrent. The improvement comprises a reduction in the ohmic contact area of the fine metallic fingers with the light-incident surface of the solar cell.

The improved solar cell is constructed by etching a fine line pattern into the antireflective coating of the solar cell and then depositing a fine line electrode onto the surface of the solar cell but oriented perpendicular to the etched fine line pattern in the antireflective coating. Due to the 90° rotation between the etched fine line pattern in the antireflective coating and fine line electrode, only point contacts are made between the fine line electrode and the exposed light-incident surface of the solar cell.

Figure 78 shows such a solar cell design in cross section as well as a view of the top surface and a detail of the top surface.

**Figure 78: Fine Geometry Solar Cell Contact Pattern**

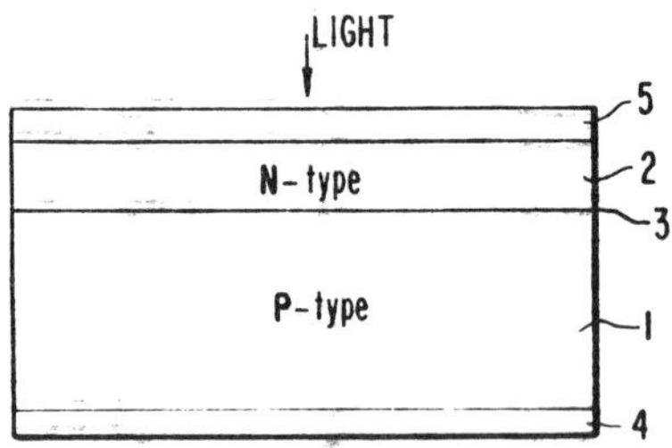


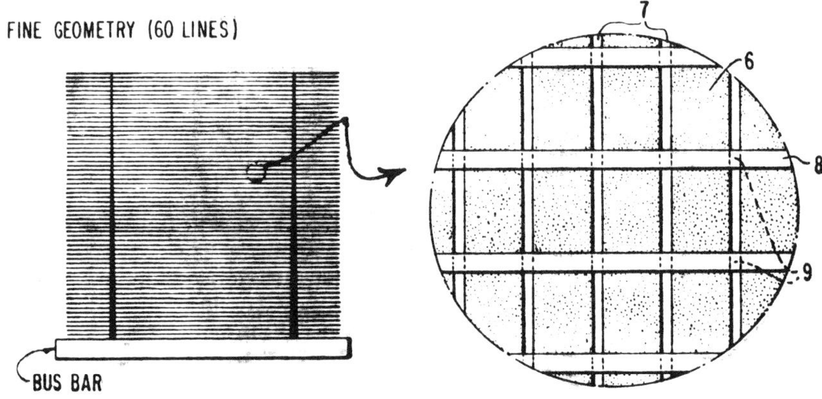


Source: U.S. Patent 3,982,964

The single crystal silicon solar cell comprises a silicon substrate **1** of P-type material and a silicon layer **2** of N-type material with an N-P junction **3** positioned a predetermined depth below the top surface of silicon layer **2**. In an N-P silicon solar cell, the junction **3** will produce an electric field directed towards the substrate **1** thereby resulting in generated electrons flowing to the top of surface **2** with holes flowing to the bottom of the substrate **1** wherein the holes may be collected by a contact **4** covering the entire bottom surface of layer **1**.

A fine geometry metallic grid as illustrated in the view at the lower left of the figure is used for a collection of electrons flowing to the light-incident surface of silicon layer **2**. A suitable antireflective coating **5** is provided over the light-incident surface of silicon layer **2**. This antireflective coating **5** has a thickness of ¼ wavelength of the desired portion of the spectrum of the light to which the solar cell is to be exposed, and the coating further has the required transparency and index of refraction. For the blue-violet portion of the spectrum corresponding to 0.3-0.5 microns, tantalum oxide has been found to be a suitable antireflective coating. A more detailed discussion of antireflective coatings is found in the next section of this volume.

As shown in the lower right-hand view in the figure, the antireflection coating **6** is etched to provide a series of parallel fine line openings **7**. Metallization fingers **8** are then deposited transverse to these openings **7** to produce a plurality of point contact areas **9**. These point contact areas **9** are the only ohmic contacts on the light-incident surface of the top layer of the solar cell.

A type of solar cell construction developed by *H. Kressel and V.L. Dalal; U.S. Patent 3,988,167; October 26, 1976; assigned to RCA Corporation* has a means for collecting electron-hole pairs with an incident surface through which solar radiation enters. The collecting means can be a P-N junction between two regions of opposite conductivity of the semiconductor body, or a partially transparent metallic film on the semiconductor body providing a metal to semiconductor material surface barrier rectifying junction.

On a surface opposite the incident surface of the collecting means is a noncontinuous oxide layer. The oxide layer is noncontinuous because of openings extending through the oxide layer to the opposite surface. The openings are distributed across the opposite surface. In the openings at the opposite surface and on the oxide layer is a reflecting contact which functions both as an electrical contact and as a reflector to solar radiation in the semiconductor body.

A problem that arises is that, during the diffusion process in solar cell manufacture, both sides of the relatively thin, substantially flat silicon wafers are exposed to the diffusant; therefore, a junction developes on both sides of the wafer. Since the junctions are the same, one of them has to be eliminated or modified while the other junction is maintained on the other side of the wafer. This may be done by protecting the one layer from contact with the diffusant.

A process developed by *J. Lindmayer; U.S. Patent 3,990,097; November 2, 1976; assigned to Solarex Corporation* involves subjecting a crystalline silicon wafer to a diffusant without protecting the wafer. In this manner junctions are formed extending inwardly from both surfaces of the wafer and films of diffusant glass coat both surfaces. A coating of aluminum is then applied to the glass on that side of the wafer whose polarity is to be changed and the so-coated wafer is again heated. On such heating it has unexpectedly been found that the aluminum has in part passed through the diffusant glass to form an aluminum-silicon junction that replaces the diffusant junction and is of a polarity opposite thereto.

Further, the glass that has been coated with aluminum now contains aluminum and has become electrically conductive. So, by failing to remove the diffusant glass before coating with aluminum, not only has an aluminum-silicon junction nevertheless been created, but the glass has been rendered conductive, thereby making optional the application of a layer of conductive metal, such as titanium-silver, titanium-gold or chromium, to that surface of the wafer.

A similar area of technology is covered by *J. Lindmayer; U.S. Patent 4,056,879; November 8, 1977; assigned to Solarex Corporation.* The steps in the process described in these patents are shown in Figure 79. The figure shows the silicon wafer, from its initial form prior to treatment until the wafer has a back contact and front and back junctions.

As illustrated, a silicon wafer of desired size, for example, a circular disc 3 inches in diameter and 10 mils in thickness, is subjected to diffusion with phosphine, oxygen and argon in an oven at a temperature of about 840°C for 12 minutes. In particular, there may be used 1% phosphine in argon at an input rate of 1,200 cc/min, oxygen at 70 cc/min and argon at 650 cc/min. A uniform exhaust from the furnace was maintained.

After diffusion, that surface referred to for convenience as the front surface **12** of the wafer **10** had been diffused with a phosphorus junction **11** that extended inwardly some 2000 A into the wafer. The front surface **12** had also been coated with a layer of phosphor-glass about 1000 A thick, which glass is an insulator and gave the underlying surface a blue appearance. In a like manner, the back surface **12a** of the wafer had a phosphor-glass coating **13a** and a phosphorus-silicon junction **11a** had been formed inwardly of the back surface.

Since this process relates to the application of a back contact to the silicon wafer, further specific reference will not be made to the front surface of the wafer.

A coating of aluminum **14a** is now applied to the phosphor-glass **13a** by evaporation, although other known techniques can be used. The aluminum coating was about 5000 A thick. Then the coated wafer was heated at 800°C, for example, well above the alloying temperature of the aluminum, for 10 minutes, and a new junction or zone **15** was formed inwardly of the back surface **12a**. Also formed was a new, electrically conductive layer **16** on the back surface. Although precise analysis has not been made, it is apparent that the back junction **15** is a combination of aluminum and silicon. The layer **16** is composed primarily of aluminum, oxygen and phosphorus. Since layer **16** is electrically conductive, it may serve as the back contact for the solar cell. For perhaps even better conductivity, a layer of metal, such as titanium-silver, titanium-gold or chromium, may be applied to the layer **16** by known methods. After front contacts have been applied, an operative silicon solar energy cell will have been formed.

An important feature of this process is the use of aluminum to form both the back junction **15** and the electrical conductive coating **16**

**Figure 79: Method of Forming Silicon Solar Energy Cell Having Improved Back Contact**

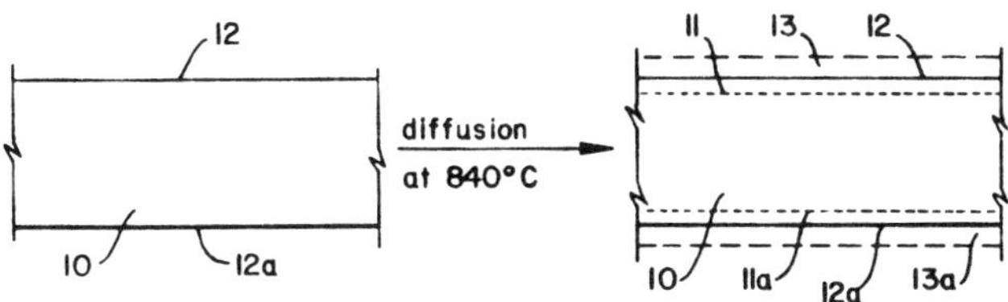


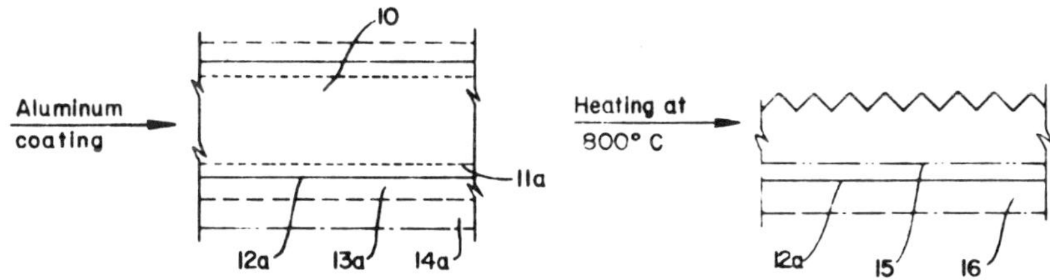


Source: U.S. Patent 3,990,097

In arrays of components such as solar cells or the like, relatively large displacements can occur between the components due to vibration and thermal stresses to which the arrays are subjected. These displacements may occur in any direction. Prior art interconnectors tend to break or tear during large amplitude displacements of the components. Even prior art interconnectors made of a mesh or having diamond shaped holes or provided with a fold or loop of extra material are unsatisfactory for large displacements. Also, the design of the interconnectors should be compatible with use in conventional automatic array assembly machines.

Some prior art interconnectors are made in intricate designs of curved conductors having many tabs that must be soldered or welded. Such intricate interconnectors are so difficult to hold and position that automatic assembly is very complex. At present, they can only be soldered or welded manually.

A type of construction developed by *G.J. Pack, Sr.; U.S. Patent 3,993,505; November 23, 1976; assigned to Hughes Aircraft Company* is one in which instead of using a vertical loop or folds of extra material which extend vertically to provide slack in the interconnector, as is done in the prior art, the extra expansion length is provided by modifying the interconnector so that some expansion modes take place in an orthogonal plane. In one embodiment the interconnector has a long, generally diagonal arm or connecting path of thin metal foil which extends from one corner to the diagonally opposite corner. When the components move apart, the ends of the arm rotate as it tends to try to extend straight across between the two components. Also, the arm lengthens by stretching because it is under tension. Due to a resulting torque, the arm goes out of plane (or buckles). To reduce the amount of this buckling out of plane, at least one slit is provided in the center of the arm. This, in effect, provides multiple parallel arms, each of which buckles slightly, but the amount of buckling is less than with a single arm. The symmetrical, generally rectangular configuration of the overall interconnector makes it compatible with being used in automatic assembly machinery.

Another feature of this type of construction is that when the interconnectors are to be soldered, rather than welded, the central portion of the interconnector is chrome plated in a manner that avoids a sharp stress transition line between the plated and unplated portions. The chrome plating is to keep the solder from filling the slits and notches, and generally covers the surface of the interconnector, but not the connecting surfaces that are soldered to the components.

To avoid having a sharp stress transition line between the plated and unplated portions, the plating terminates in a wavy line similar to a sine wave, and the plating on the reverse side terminates in a similar wavy line, except that the two wavy lines are 180° out of phase. This tends to give a gradual transition in the equivalent Young's Modulus from the plated to the unplated surface.

A type of construction developed by *W.T. Kurth; U.S. Patent 4,019,924; April 26, 1977; assigned to Mobil Tyco Solar Energy Corporation* comprises a plurality of solar cells mounted on a laminate comprising a base electrically-insulative sheet, an electrically-conductive layer disposed on the base sheet in a predetermined pattern so as to provide first and second cell-connecting sections electrically insulated from one another, and a second electrically-insulative sheet having a plurality of openings and being disposed over the conductive layer so that selected portions of the conductive layer are exposed through the openings.

The laminate is provided with a plurality of bent tabs, each of which includes an exposed portion of the first cell-connecting section so that the exposed portion can be attached to the top surface electrode of a solar cell. At least one exposed portion of a second cell-connecting section of the conductive layer is attached to the bottom surface electrode of the same cell. Various circuit patterns of the electrically-conductive layer are described for connecting the cells in a series or parallel array and for use in width-limited systems, such as solar concentrators.

A solar cell design developed by *T. Matsutani and K. Nishida; U.S. Patent 4,029,518; June 14, 1977; assigned to Sharp KK, Japan* is one in which a P-type diffusion layer is formed on an N-type silicon semiconductor wafer to establish a P-N junction in a solar cell, the diffusion layer being exposed to radiation. A pair of electrodes are formed on the surfaces of the diffusion layer and the semiconductor wafer in a desired configuration in order to provide output of electric energy generated by the solar cell.

The diffusion layer is formed in such a manner that the layer has a thickness of around 3 μm at areas where the electrode is formed and has a thickness of around or below 0.5 μm at regions on which the electrode is not formed. With such an arrangement, radiation having a wavelength of about or shorter than 400 mμm can be used for performing optoelectric generation.

A device developed by *M.G. Coleman and R.A. Pryor; U.S. Patent 4,045,245; August 30, 1977; assigned to Motorola, Inc.* is a solar cell package which includes a plurality of solar cells within a space formed by a support member and a transparent cover member. A first conductor is positioned between the support member and the solar cells and makes electrical contact to a first surface of each of the solar cells.

A second conductor is essentially coplanar with the first conductor and makes electrical contacts to a second surface of each of the solar cells. The output terminals of the solar cell package are connected to the first and second conductors. The first and second conductors are electrically isolated from each other and are also insulated from the support member and from the solar cells insulative means which is electrically insulated but thermally conductive to facilitate dissipation of thermal power dissipated in the solar cells.

A device developed by *J. Lindmayer; U.S. Patent 4,082,568; April 4, 1978* is a solar cell having a contact formed from a titanium group element in contiguous relationship with the cell, a mixture of a titanium group element and a platinum group element overlying the titanium group layer, and a layer of a platinum group element overlying that layer. A body of silver or other contact metal is adhered to the platinum group layer. The three layers may be vapor deposited on the semiconductor body, while the silver layer is more advantageously applied by plating or electroplating. Similar ground is covered by *J. Lindmayer; U.S. Patent 4,124,455; November 7, 1978.*

A scheme developed by *J.C. Evans, Jr.; U.S. Patent 4,082,569; April 4, 1978; assigned to the U.S. National Aeronautics and Space Administration* meets the need for an improved collector system for photovoltaic devices in that it has the advantages of the high conductivity of a metallic mesh, is designed to obscure only a minimum area of the surface of the photovoltaic device exposed to light with opaque material while providing good conductivity for the carriers generated by light photons to the external circuitry of the device and maintains the advantage provided by a transparent electrically conductive coating in contact with the total active diffused surface of the photovoltaic substrate.

Methods are known in the art for providing photovoltaic substrates with a transparent or at least translucent layer which functions as one of the electrodes necessary for the passage of the current generated by incident photons on the device. Thus, U.S. Patent 3,811,953 (cited earlier in this section) shows a photovoltaic device coated with a conductive, transparent layer of $Cd_2SnO_4$ which functions as a conductive electrode. U.S. Patent 2,766,144 shows a photovoltaic device coated with a translucent coating of a Group III element such as indium or Group V elements such as antimony as a conductive electrode.

U.S. Patent 2,870,338 shows a transparent conductive layer of a material such as tin oxide on a photovoltaic substrate. However, these devices have the disadvantage that such transparent, conductive surfaces as $Cd_2SnO_4$ and $SnO_2$ lack highly conductive channels within or on the conductive layer which adversely affects the ability of the device to effectively transport the current generated by the incident photons to the external circuitry of the device. Moreover, photovoltaic substrates which are coated with a layer of a Group III or V element have the disadvantage that the current generating ability of the device is diminished because of the reduced transparency of the layer.

This device is one in which a transparent, conductive collector layer containing conductive metal channels is formed as a layer on a photovoltaic substrate by coating a photovoltaic substrate with a conductive mixed metal layer. Device fabrication in this case further consists of attaching a heat sink having portions protruding from one of its surfaces which define a continuous pattern in combination with recessed regions among the protruding portions to the substrate such that the protruding portions of the heat sink are in contact with the conductive layer of the substrate; and heating the substrate while simultaneously oxidizing the portions of the conductive layer exposed to a gaseous oxidizing substance forced into the recessed regions of the heat sink, thereby creating a transparent metal oxide layer on the substrate containing a continuous pattern of highly conductive metal channels in the layer.

A similar mode of construction is described by *J.C. Evans, Jr.; U.S. Patent 4,122,214; October 24, 1978; assigned to U.S. National Aeronautics and Space Administration*. Figure 80 shows the heat sink device used in this process in combination with a photovoltaic substrate to form a transparent collector layer on the substrate containing highly conductive metal channels.

**Figure 80: NASA-Developed Solar Cell Featuring Transparent Conductive Metal Oxide Layer**

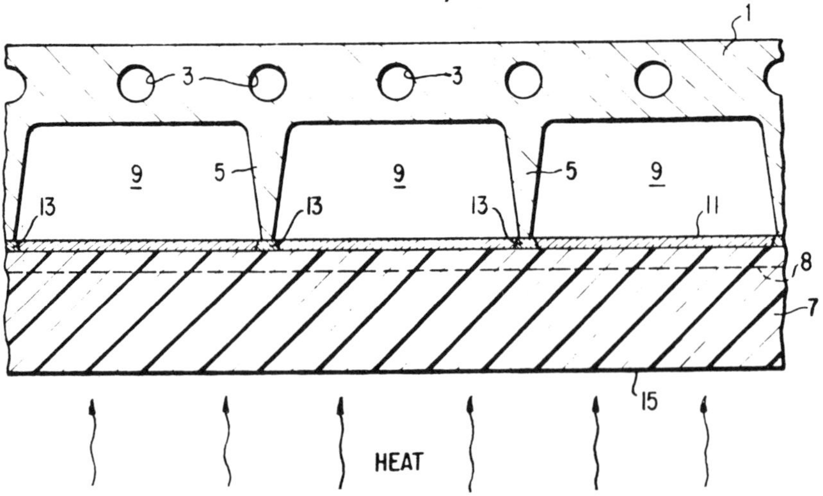

Source: U.S. Patent 4,082,569

The heat sink **1** or heat conductor is fashioned from a block of a metal such as copper, silver, gold, or platinum. The block **1** is provided with channels **3** for the flow of a coolant such as water through the heat sink, and is further characterized by raised portions **5** which protrude from the block **1** in such a fashion so as to define the pattern of the metallic grid system desired for the collector system coated on photovoltaic substrate **7**. The heat sink is of a size sufficient to define a grid system pattern on a photovoltaic substrate. Solar cells differ in sizes according to use and shape.

Typical sizes of rectangular shaped devices are 1 x 2 cm and 2 x 4 cm, while a typical square shaped device is 2 x 2 cm. Circular shaped devices typically are of 2", 3" and 4" nominal sizes. Even larger photovoltaic devices will be soon used for terrestrial use. Moreover, collector patterns employed currently on commercial solar cells vary greatly. For example, a space quality 2 $cm^2$ cell may have eight uniformly-spaced parallel wires joined to a heavier bar running at right angles to the parallel wires along an edge of the cell.

The silicon cells commonly used in satellites placed in orbit by COMSAT (Communications Satellite Corp.) of the same size may have 60 fine wires to form the collector system described above. A 3" diameter terrestrial wafer may have a dendritic or chevron configured collector. No matter what the size and shape of the individual cell, the surface area of the cell obscured by the collector grid system ranges from about 5 to 10% of the total photovoltaically active surface. In this device however, the obscured area is reduced to only about 3 to 5% while at the same time the collection efficiency of the cell is increased as a result of reduced recombination losses. Protruding portions **5** of the block **1** also define the pattern of recessed regions **9** of the block.

A photovoltaic substrate or solar cell such as a film of cadmium sulfide or a silicon substrate whose surface has been diffused with a counter dopant in a conventional manner is coated with a conductive layer **11** of indium-tin, indium-antimony or indium-arsenic by any conventional technique such as vacuum evaporation or vacuum sputtering to a thickness such that the thickness of the mixed metal oxide conductive layer formed ranges from 500 A to 10 $\mu$m. Thicker perimeter conductors may be added by known methods such as solder dipping, plating, and the like. A factor in determining the thickness of the layer **11** is that it must be thick enough so that the total collector grid system has sufficient conductivity.

Other suitable photovoltaic substrates to which this collector grid system can be applied include gallium arsenide and numerous other types of solar cells. Other suitable substrates include Schottky barrier and heterojunction devices. The P and N dopants used in the surface layers of these devices must be accounted for in the design and doping of the transparent conductive grid system. However, it has been established that properly formed transparent conductive layers are compatible with both conductivity types. Line **8** defines a surface diffused region or junction region if the photovoltaic substrate employed is an N-P or P-N photovoltaic diode or solar cell.

The block **1** is positioned and secured against the solar cell **7** such that the protruding portions **5** of the block **1** are in contact with the conductive layer **11** of the solar cell. The solar cell **7** is then subjected to heat on the bottom side **15** of the cell. The cell can be heated simply by placing the substrate in open air on a hot plate or heating block controlled by a thermocouple. Also, a flow of hot air from any suitable device can be used. However, the use of an electrically heated block probably provides better heat control.

At the same time a gaseous oxidant such as moist air or oxygen is forcefully passed through or allowed into the recessed regions **9** of the block **1** which causes the oxidation of all portions of the conductive layer **11** exposed to the oxidizing atmosphere to a transparent, mixed metal oxide layer. The heat sink **1** is positioned on the mixed metal layer **11** of the solar cell substrate such that protruding portions **5** are in contact with the metal layer **11**.

In the design of heat sink **1**, it is only necessary that recessed regions **9** be of sufficient relief or clearance above layer **11** to allow the free flow of oxidizing gas molecules in the regions as they strike and rebound from the surface and combine with the metal elements in the layer. The gas molecules will have rather large mean free paths which are dictated by the temperature of the substrate, the extent of access to more of an atmosphere and the ambient pressure of the gas used.

The depth of the recessed regions **9** can be determined empirically. Furthermore, the relative closeness of the conductive metal channels **13** in the oxide layer **11** will determine the spacings between protruding portions **5**. These factors are not likely to be critical over a great range. For example, for the honeycomb pattern and if a 3" diameter solar cell is assumed, a clearance of about 2 mm of the recessed regions **9** above the oxide layer **11** will be sufficient. If other types of collector grid patterns are used, the clearance will decrease or increase according to the desired grid geometry.

Alternatively, in some cases such as the case of honeycomb pattern, it may be desirable to provide the heat sink **1** with small diameter holes such as 1 mm centered in each of the hexagonal shaped recessed regions **9** of the heat sink **1** for the free movement of the gases of the oxidizing atmosphere. Moreover, if the recessed regions **9** are sufficiently deep in the heat sink, enough oxygen may be present in these regions such that it will not be necessary to pass additional oxidizing gas into the recessed regions **9** to complete the oxidation process.

Alternatively, the oxidation of the layer **11** may very well be accelerated by forced circulation of an atmosphere through the heat sink. It is important that the heat sink be able to effectively conduct heat away from layer **11** wherever the layer is in contact with the heat sink so that the temperature in layer **11** under protruding portions **5** never reaches the oxidation temperature of the metal alloy. The areas **13** of the conductive layer **11** under the protruding portions **5** which are protected from the oxidizing atmosphere remain unoxidized and thus define a pattern of highly conductive metal channels through conductive layer **11**. Heat is also conducted from areas **13** by the protruding portions **5** to the coolant flowing through the block.

The temperature at which the oxide layer normally forms is dependant upon several variables which include alloy constitution of the metal films, the thickness of the metal film, and the like. Thus, for the alloy systems used for the formation of the transparent metal oxide layer, any ratio of one element to the other element can be used because an oxidation temperature within the range of 400° to 700°C is sufficient to cause complete oxidation of the metal layer. Complete oxidation normally occurs in a brief time, for instance, on the order of 10 min.

An empirical determination of the conditions necessary for the oxidation of the metal surface may simply be conducted by heating a coated substrate in air and measuring the parameters to achieve a transparent mixed metal oxide layer. For example, a silicon wafer 10 to 12 mils (300 $\mu$m) thick can be surface oxidized in about 10 minutes. The oxidizing atmosphere may be ambient air which is circulated by convection or is forced through the heat sink.

Other oxidizing atmospheres can be obtained by enriching air with oxygen or by mixing oxygen with an inert gas in any amount sufficient to accomplish the desired oxidation. Also water vapor can be present in the oxidizing atmosphere up to the level of 60% humidity and even greater to increase the rate of oxidation. None of the conditions of the oxidizing gas are critical and it is only important that the temperature of oxidation be sufficient to completely oxidize the desired portions of layer **11** in the collector sheet while the metallic nature of the embedded grid structure is maintained.

The protruding portions **5** of the block **1** also prevent oxidation of the underlying areas **13** by acting as a heat sink, thereby dissipating heat from regions **13**. In an alternative method of accomplishing the oxidation of the conductive layer **11**, hot moist air can be forced through the recessed regions **9** to oxidize the exposed portions of the conductive layer while the portions **5** of the heat sink conduct heat away from areas **13** and prevent contact of the layer **11** with hot moist air.

A device developed by *J.C. Evans, Jr.; U.S. Patent 4,104,084; August 1, 1978; assigned to U.S. National Aeronautics and Space Administration* covers similar ground and further spells out the constructional details of a heterojunction or Schottky barrier photovoltaic device comprising a conductive base metal layer compatible with and coating predominately the exposed surface of the P-type substrate of the device such that a back surface field region is formed at the interface between the device and the base metal layer. A transparent, conductive mixed metal oxide layer is provided in integral contact with the N-type layer of the heterojunction or Schottky barrier device having a metal alloy grid network of the same metal elements of the oxide constituents of the mixed metal oxide layer embedded in the mixed metal oxide layer.

An insulating layer is also provided which prevents electrical contact between the conductive metal base layer and the transparent, conductive metal oxide layer, and a metal contact means is provided covering the insulating layer and in intimate contact with the metal grid network embedded in the transparent, conductive oxide layer for conducting electrons generated by the photovoltaic process from the device.

Similar ground is covered by *J.C. Evans, Jr.; U.S. Patent 4,135,290; January 23, 1979; assigned to U.S. National Aeronautics and Space Administration.*

A process developed by *E. Anagnostou and A.F. Forestieri; U.S. Patent 4,083,097; April 11, 1978; assigned to U.S. National Aeronautics and Space Administration* is one in which electrical connections to solar cells in a module are made at the same time the cells are encapsulated. The encapsulating material is embossed to facilitate the positioning of the cells during assembly.

A technique developed by *J.W. Yerkes and J.E. Avery; U.S. Patent 4,105,471; August 8, 1978; assigned to Arco Solar, Inc.* applies to a silicon solar cell having a body of boron doped P-type silicon material with a shallow P-N junction formed therein through diffusion of phosphorus into one surface thereof. A contact pattern of conductive material is formed on the surface of the solar cell in which the P-N junction is formed.

The pattern is formed by first depositing a metallic layer upon the entire surface of the body and then applying the contact pattern by printing upon the surface of the metal. The metal has a characteristic such that when heated in the presence of oxygen to an appropriate temperature to fire the conductive material, it oxidizes and forms an antireflective layer on the surface of the cell except in those areas where the printed contact pattern is disposed. In the areas of the printed contact pattern the metal forms an ohmic contact between the surface of the silicon and the printed contact pattern and provides a barrier to preclude the conductive contact pattern material from punching through the shallow P-N junction. The steps in the process are shown schematically in Figure 81.

An appropriate P-type silicon semiconductor wafer properly etched and cleaned is provided. Thereafter, the wafer is placed within an open tube system at an elevated temperature of about 1200°C. Phosphorus oxychloride is passed thereover thereby to deposit a phosphorus pentoxide glass upon the entire surface of the silicon wafer. The phosphorus contained within the phosphorus pentoxide diffuses into the silicon wafer while at the elevated temperature to thus form the desired P-N junction at the depth of approximately 0.3 micron. Thereafter the back surface of the silicon wafer is etched to remove the phosphorus pentoxide and the P-N junction therefrom.

After removal, so that there is a clean silicon surface, an aluminum paste is applied in sufficient quantities over the thusly cleaned surface to provide the desired back contact to the P-type silicon body. The back contact is perfected by heating the aluminum and silicon to an appropriate temperature of about 750°C to cause a fusing or alloying thereof to occur. Thereafter the front surface wherein the P-N junction is formed is cleaned by again etching the same for example with hydrofluoric acid to remove the residue of phosphorus pentoxide.

Figure 81: Arco Solar Process for Improved Printed Contact Production for Solar Cells

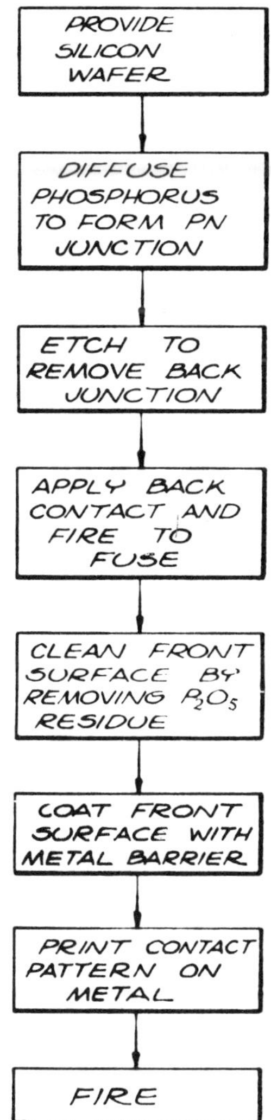


Source: U.S. Patent 4,105,471

The cleaned surface is then coated with a 450 A layer of tantalum, tungsten or niobium. Thereafter the desired contact pattern is printed on the surface of the metallic layer, for example, by placing a silver-glass frit material in a desired organic binder thereon in the desired pattern. Subsequent thereto the combination is then heated to a temperature of approximately 600°C in air thereby to convert the tantalum to tantalum pentoxide and to thus provide an antireflective coating while at the same time providing an excellent ohmic contact between the printed contact pattern and the surface of the solar cell.

It also appears that the back contact may be applied to the cleaned surface of the silicon wafer and subsequently the tantalum layer applied to the cleaned front surface thereof. Subsequently the printed contact is applied to the tantalum layer and the entire combination may then be heated to an elevated temperature sufficient to accomplish alloying of the aluminum into the back surface, the formation of the tantalum oxide antireflective layer and the formation of the front contact all at the same time thereby further enhancing the mass production of solar cells.

It has been found that the metal layer upon the surface of the silicon wafer also functions to protect the shallow junction in the surface of the cell from contaminants normally found in the production equipment, e.g., the furnace or other handling equipment where there is usually a residue of aluminum, silver, phosphorus, boron and other materials. Such protection also renders mass production techniques more feasible than heretofore possible.

A type of construction developed by *R.P. Della-Vedowa and I.M. Shahryar; U.S. Patent 4,131,123; December 26, 1978; assigned to Solec International, Inc.* is one in which a plurality of solar cell discs or wafers is first disposed between opposite faces of an electrically insulating sheet of the same thickness as the wafers, and then printed on both sides with electrical conducting material to provide a positive conducting face on one side of all wafers and integral solderless printed conductors between adjacent wafers and a negative grid on the other side of all wafers and integral solderless printed conductors between adjacent wafers.

When the wafers are series connected, apertures are provided in the sheet through which the printing material extends to series connect adjacent wafers. The wafers may be secured within the sheet by securing same within apertures therein or elongated crystals may be disposed in a mold and electrically insulating material may be cast therearound to form a block which may then be sliced into modules.

A system developed by *M.C. Keeling, W.L. Bailey, M.G. Coleman, I.A. Lesk and R.A. Pryor; U.S. Patent 4,131,755; December 26, 1978; assigned to Motorola, Inc.* provides an improved interconnection system for interconnecting a plurality of photovoltaic devices. The photovoltaic devices each have a first and a second side and the interconnect system is located on the second side of the photovoltaic devices. A sheet of dielectric material and a sheet of electrically conductive material are bonded together and positioned so that the dielectric material is next to the photovoltaic devices. A plurality of patterns is formed in the sheet of electrically conductive material. The patterns each have angled tabs punched therein so that the angled tabs are punched through both the electrically conductive material and the dielectric material. When a photovoltaic device is positioned within a group of angled tabs, the angled tabs can be brought into contact with electrical contacts on the first side of the photovoltaic device. The group of angled tabs are electrically common to a part of the pattern which has an extended portion which extends beneath an adjacent photovoltaic device. Some of the dielectric material is removed from the extended portion so that this portion then makes contact with the second side of an adjacent photovoltaic device.

The patterns can be formed in predetermined configurations to provide series or series-parallel interconnections for photovoltaic devices within an array of photovoltaic devices. The interconnect system, in turn, provides for substantially all possible series, parallel, or series-parallel interconnections of a plurality of arrays.

A type of construction developed by *S.-Y. Chiang and B.G. Carbajal; U.S. Patent 4,133,698; January 9, 1979; assigned to Texas Instruments Incorporated* provides a solar cell having first and second closely spaced, parallel P-N junctions wherein the illuminated surface is totally free of metallization, for example, the junction nearest the illuminated surface is not electrically connected, and thereby participates only indirectly in the collection of photogenerated carriers by providing a charge field to suppress the front surface recombination and to enhance collection at the back side junction. All metallization is on the back side, which preferably includes an interposed finger pattern of $N^+$ and $P^+$ zones.

Figure 82 shows such a cell in section and in back side elevation. It is seen to consist of a predominantly P-type monocrystalline silicon body **11** having a resistivity of 5 ohm-cm and a nonreflective texture **12** on its illuminated surface. A first P-N junction is formed at the textured surface by a shallow N-type region **13** having a thickness of about 0.3 $\mu$m and a resistivity of about 100 ohms per square. Texturizing is achieved by any known means, such as a nonselective etching with NaOH.

At the back side, the cell includes a second P-N junction formed by N-type region **14** having a depth of about 0.3 $\mu$m and a resistivity of 100 ohms per square. Since the total cell thickness is 50 $\mu$m, the two P-N junctions are spaced apart by slightly less than 50 $\mu$m, which is roughly equal to a diffusion length of electrons generated in the central P-type region. The two junctions are generally parallel to each other. The surface texture is somewhat exaggerated in the drawing, and does not materially alter the parallel relationship.

Although the example shown is fabricated from a P-type wafer, the process also contemplates a reversal of all conductivity types, i.e., as would result from starting with an N-type wafer.

**Figure 82: Texas Instruments Interdigitated Back Contact Cell Design**

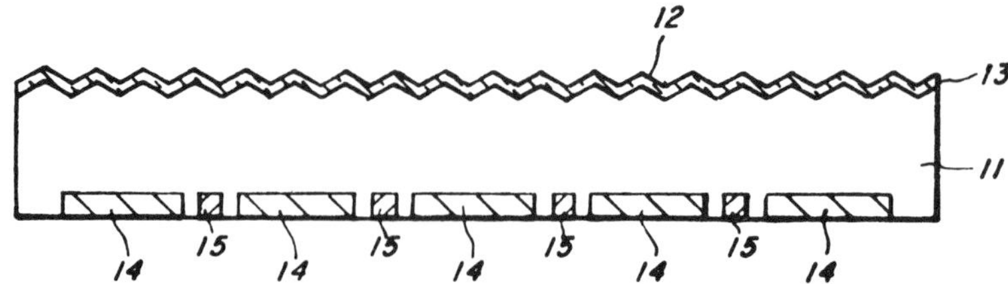


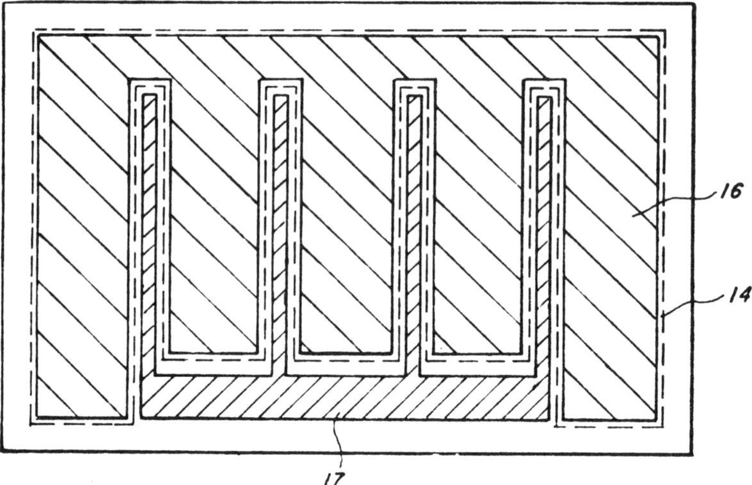


Source: U.S. Patent 4,133,698

In the lower view in Figure 82, the back side of the cell is seen to have an interdigitated geometry. This is, $N^+$ region **14** has a broad-fingered geometry to which metallization pattern **16** makes ohmic contact, while narrow-fingered $P^+$ pattern **15**, interdigitated therewith, has a corresponding metallization geometry **17**. The broader fingers for region **14** improves collection efficiency by increasing the area of the back side P-N junction.

Test results and calculations have shown that the illustrated cell has the same collection efficiency as can be achieved by electrical connection of both junctions, while retaining the cost advantage of omitting top-side metal, together with avoidance of the shadowing effect that results from top-side metal. Additional savings are realized by using less semiconductor material, compared to the thicker conventional cells.

A cell design developed by *J.I. Pankove; U.S. Patent 4,139,858; February 13, 1979; assigned to RCA Corporation* consists of a solar cell which comprises a body of silicon having a P-N junction therein with a transparent conducting N-type gallium nitride layer as an ohmic contact on the N-type side of the semiconductor exposed to solar radiation. Such a device is shown in schematic cross section in Figure 83.

**Figure 83: Solar Cell with a Gallium Nitride Electrode**

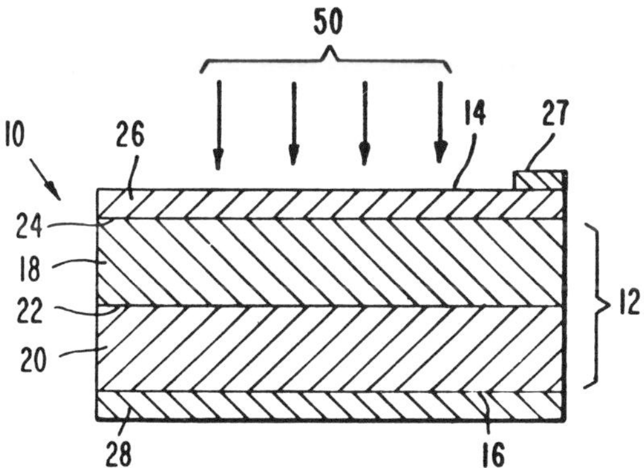


Source: U.S. Patent 4,139,858

The solar cell **10** includes a body **12** of silicon with a first layer **18** of N-type conductivity and second layer **20** of P-type conductivity. Thus, a P-N junction **22** is formed between the first and second layers **18** and **20**. Preferably, the P-N junction **22** is graded from P-type to N-type rather than an abrupt P-type N-type transition. The layers **18** and **20** are doped by methods known in the art, such as the employment of phosphorus to provide an N-type conductivity layer, and boron to provide a P-type conductivity layer, or other dopants known in the art.

The body **12** has an incident surface **14** on which solar radiation **50** impinges and an opposite surface **16**. Along the incident surface **14** is a transparent, conducting gallium nitride electrode **26**. The GaN electrode **26** forms a heterotransition **24** with the N-type first layer **18**. A heterotransition is defined as the contacting of two layers of different materials wherein both materials are of the same conductivity type. A metal contact **27** is formed on electrode **26** to facilitate attaching a wire (not shown) to withdraw the current generated by the operation of the device.

Gallium nitride (GaN) is N-type and highly conducting in an undoped state. The transparent, conducting, low resistivity gallium nitride electrode **26** forms an ohmic contact on the solar cell without the shadowing normally encountered with finger or grid electrodes.

On the opposite surface **16** is an electrode **28** which also forms an ohmic contact with semiconductor body **12** on the P-type conductivity layer **20**. The electrode **28** can be a single layer of metal such as aluminum, or multilayered, e.g., a first layer directly on the body **12** of chromium with second layer of a metal such as gold on the chromium layer. Preferably, the single layer of metal is a sintered aluminum electrode.

A process developed by *L.F. Durkee; U.S. Patent 4,144,139; March 13, 1979; assigned to Solarex Corporation* involves plating electrical contacts onto one or more surfaces of a solar cell having an electrical junction therein by immersing the cell in an electrolyte and exposing it to light so that platable ions in the electrolyte will be attracted to an oppositely charged surface of the cell. The apparatus which may be used in such a process is shown in exaggerated schematic form in Figure 84.

The process starts with a silicon wafer containing a P-N junction, perhaps subdivided from a silicon sheet. A contact base **20** is provided on the wafer (by vapor deposition for example). That base lies atop a portion of the front face segment **12a** of wafer segment **10a**. It will normally cover only a fairly small portion of the area of the front face segment **12a**, because the contact will obscure a portion of the front face and thereby diminish the amount of light that passes through the silicon and reaches the electrical junction **11a** of each wafer segment **10a**.

As so formed, the entire wafer matrix **10** is immersed in a container **21** containing a bath of electrolyte **22**, as illustrated. Such bath that has been used has been a potassium silver cyanide formulation known as Silver Sol-U-Salt (Sel-Rex Co.). The material is stated to contain 54% silver. The entire wafer matrix **10** was immersed in this bath of electrolyte. First, however, the back surface **17** of the matrix **10**, which had not been scored or grooved, was coated with a continuous layer of silver by known plating or electroplating methods, so that the entirety of that back surface **17** bore a silver coating **18** to a depth of about 10 microns.

In the form described, the wafer matrix **10** in an electrolytic bath **22** in container **21** was next passed under a 100-watt incandescent lamp. After 5 minutes the wafer matrix had been converted substantially to the form shown, i.e., the thickness of the layer of silver **18** on the back surface **17** of the matrix had been reduced in thickness by approximately one-half, i.e., to 5 microns, and a glob of silver **23** had been deposited on each contact base **20**.

The silver adhered to the base and formed a unitary contact consisting of, e.g., titanium-palladium and silver firmly adhered to the front surface **12a** of each solar cell segment **10a**. Then the matrix **10** was cut through the grooves **15** already formed therein to separate each wafer segment **10a** from the remainder of the wafer segments. The result was a plurality of small solar cells, each containing a contact properly positioned and containing a mass of silver sufficient to make the contact a good conductor of electricity.

Figure 84: Method of Plating Solar Cell Contacts Using Photovoltaic Property of Cell Itself

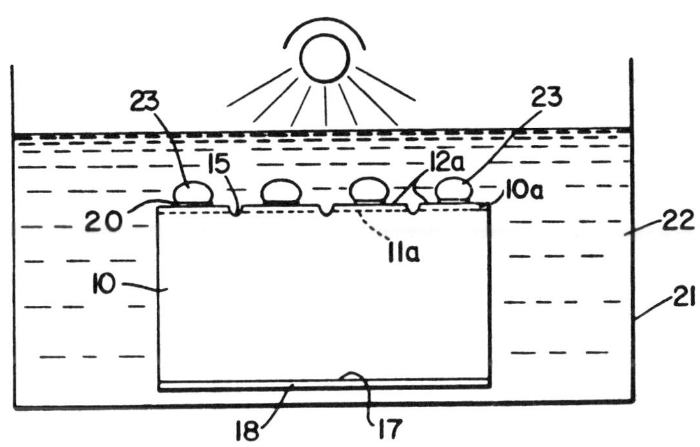


Source: U.S. Patent 4,144,139

Apparently what occurred when the matrix **10** was exposed to light in the bath of electrolyte, was that light was absorbed by the front faces **12a** of the individual wafer segments **10a** and into contact with the junction segments **11a**. Since the junction was an N-P junction, negative charges and holes or positive charges were generated, with the negative charges collecting at the front surfaces **12a** of the matrix segments **10a**, while positive charges collected at the back surface **17**.

Since the contact bases **20** were the most conductive portions of the front surfaces, the negative charges tended to congregate at such bases. A migration of silver ions then occurred from the electrolyte to the contact bases **20**, with the silver ions in the electrolyte immediately thereafter being replaced by silver from the layer **18** that had been plated on the back surface **17** of the wafer matrix **10**. This process continued so long as the front surface **12** of the matrix was exposed to light, in this case, incandescent light.

The bath was then removed from beneath the light source, whereupon the matrix ceased to be activated and silver plating ceased. Thereafter the cell matrix was washed, subdivided and a highly corrosion resistant contact had been formed on each wafer segment **10a**. As explained, the use of the solar cell to generate its own electricity in lieu of the conventional outside source of energy utilized in electroplating made possible a simple and convenient method for placing a contact on a solar cell. It is particularly convenient here where the contact area is so located that it is not readily subject to electroplating because it is extremely difficult, if not impossible, to connect each contact base **20** with an electrical lead.

## ANTIREFLECTION COATINGS

As pointed out by *J. Lindmayer and J.F. Allison; U.S. Patent 3,922,774; December 2, 1975; assigned to Communications Satellite Corp.,* the efficiency (i.e., power output-power input) of a solar cell is directly related to the amount of useful light, i.e., carrier generating light, which is absorbed by the solar cell. The efficiency of the solar cell is limited, however, by a known optical phenomenon whereby some of the light (both useful and nonuseful) striking the top surface of the solar cell is partially reflected from the solar cell. To reduce the problem of light reflection prior art solar cells employ an antireflective coating positioned on the surface of the solar cell through which light enters.

The particular environment in which a solar cell will be used will determine the specific mechanical, chemical and optical properties that its antireflective coating must have.

In space applications, for example, where reliability in a hostile environment and over-extended time period is required, the optimum determination of these properties is essential to a successful space mission.

As pointed out by *R. Gonsiorawski; U.S. Patent 4,078,945; March 14, 1978; assigned to Mobil Tyco Solar Energy Corp.,* solar cells generally are sensitive only to incident radiation lying within a relatively narrow portion of the solar radiation spectrum. For example, in the case of conventional silicon cells the sensitive band is considered to fall in the range of between 0.4 and 1.1 microns in wavelength, with the upper limit being determined for silicon by the fact that its energy gap causes it to be transparent to wavelengths substantially greater than about 1.1 microns. Actually, most silicon solar cells intended for terrestrial use are generally designed to operate at a principal peak of about 0.55 micron; cells intended for use in space generally are designed to operate at a principal peak of about 0.4 micron. However, silicon solar cells are relatively reflective to solar radiation in this wavelength range. For example, reflectivity from a substantially flat uncoated silicon solar cell of normally incident light of 0.55 micron wavelength is about 30 percent. Such reflection obviously lowers the conversion efficiency of the solar cell.

To function properly the antireflective coatings thus must possess, among other things, certain optical properties. With respect to one of its optical properties, the antireflective coating should reduce reflection of the useful light. More specifically, in space applications, for example, wherein a quartz cover slide is usually placed over the antireflective coating to prevent harmful radiation such as protons from damaging the solar cell, the index of refraction of the antireflective coating should be between that of the quartz cover slide and the underlying solar cell, as is known. In connection with one other optical property, i.e., its absorption property, the antireflective coating should not absorb the useful light, but should enable the passage of such light to the underlying solar cell. The use of a particular antireflective material is, therefore, dependent upon the refractive index of the underlying solar cell and the cover slide, as well as the wavelength response of that solar cell.

As pointed out by *J. Lindmayer and J.F. Allison; U.S. Patent 3,949,463; April 13, 1976; assigned to Communications Satellite Corp. (COMSAT),* the antireflective coating also must satisfy certain mechanical and chemical criteria. In addition to environmental and life-time considerations, these criteria would be determined by the physical characteristics of the solar cell. A solar cell responsive to light in the short wavelength region would require an antireflective coating to satisfy certain specific criteria. For example, in one short wavelength responsive solar cell the N-P junction is only about 1000 to 2000 A from the top surface of the solar cell. Under these conditions the antireflective coating would damage the shallow junction if the coating penetrated into the solar cell. Also, any mechanical stress produced at the antireflective coating-semiconductor interface must be small so that such stress will not damage the junction.

In addition, the antireflective coating should not degrade upon exposure to ultraviolet light in a vacuum. The effect of such degradation could be a change in the index of refraction of the antireflective coating and the absorption of light at short wavelengths. Also, with respect to silicon solar cells, there is a phenomenon known as dispersion whereby the index of refraction of the silicon becomes greater at the shorter wavelengths. Therefore, the antireflective coating should have a relation to wavelength that matches the variable refractive index of silicon.

Still other criteria of an antireflective coating relate to its stability, adhesion qualities and hardness. The antireflective material should be chemically stable in that it should not change composition during processing where it may be exposed to temperature, chemicals and moisture or during shelf storage in order to assure constant optical properties. The adhesion of the antireflective coating to the solar cell should be excellent so as to ensure that delamination would not occur during processing or exposure to moisture or temperature cycling. Finally, the antireflective material should be hard enough so that it would not be damaged during manufacture or use, particularly during cover slide attachment.

To minimize this undesired reflection, the art has proposed various antireflective coatings for covering the radiation-receiving surfaces of the cell. Such coating generally comprises a material which is substantially transparent to solar radiation incident upon the cell, but is capable of reflecting back towards the radiation-receiving surfaces of the cell, much of the light energy reflected from the cell. Known antireflective coatings generally comprise a thin layer of a material such as silicon monoxide, tantalum oxide, titanium oxide, silicon dioxide, or magnesium fluoride.

While such antireflective coatings are fairly effective, they are relatively expensive to produce. The coatings are usually formed in situ on the cell using classical deposition techniques such as vapor deposition or sputtering, which are relatively costly techniques and also may involve substantial capital investment. Furthermore, some of the known antireflective coatings are relatively heat sensitive. For example, silicon monoxide coatings degrade when exposed to temperatures above about 50°C for 200 hours. Thus, prior art solar cells generally also require cover slides or interference filters which are substantially reflective to incident light in the near and infrared wavelength region. Such reflected light is not available for conversion to electrical energy, so that the solar cell must operate on only that portion of the total solar energy lying within the pass band of the filter. Further, known solar cells may also require means for cooling the solar cell, i.e., to protect the antireflective coating, according to R. Gonsiorawski in U.S. Patent 4,078,945.

A composition developed by *K.S. Tarneja and W.R. Harding; U.S. Patent 3,533,850; October 13, 1970; assigned to Westinghouse Electric Corporation* is an antireflective coating consisting of a material selected from the group consisting of titanium dioxide, tantalum oxide, cerium oxide, zinc sulfide, and tin oxide which is deposited on the surface of the shallow region of semiconductivity of a solar cell. A quartz cover is cemented to the antireflective coating and the electrical contact of the shallow region. The combination of a quartz cover and the antireflective coating provides for a more efficient solar cell.

As noted by *S. Harrison; U.S. Patent 3,539,883; November 10, 1970; assigned to Ion Physics Corporation*, the reflectivity of the silicon surface in solar cells is such that about 35% of the photons incident on the cell surface suffer reflection and are lost. To eliminate this loss an 800 A thick outer reflection layer of silicon monoxide may be deposited on the cell surface. This layer reduces the reflection of photons in the peak response area and the average cell efficiency increases from 8.5 to 10.5%.

High energy, incident, charged particles such as electrons and protons, rapidly degrade these cells and it is necessary to protect the cells by mounting quartz cover slips over the monoxide layer. These cover slips may range in thickness from 0.006 to 0.020 inch and are secured to the cell surface by a several-mil thick coating of a room temperature vulcanizing silicone cement. The application of this cover slip-cement combination disrupts the antireflective effectiveness of the monoxide coating such that the overall cell efficiency drops by about 1%. This decrease in cell efficiency is considered necessary by the prior art in order to obtain the increase in cell lifetime.

Harrison, however, discovered that the introduction of this cover slip actually creates significant loss in the efficiency of the cell by increasing losses in the peak response area but because it simultaneously decreases the losses at the ends of the spectrum by an amount which approximately equals the increase in loss in the center of the spectrum, the overall loss appears very small.

Apparently workers in the prior art observed only the overall decrease and this misled them to believe the efficiency of the cell was only slightly affected by the introduction of the cement-cover slip combination.

Further, as discovered by Harrison, solar cells can be made to absorb and utilize more of the solar spectrum by applying a coating of cerium oxide between the cell and its cover slip.

A process developed by *A.G. Revesz and R.J. Dendall; U.S. Patent 3,904,453; September 9, 1975; assigned to Communications Satellite Corp. (COMSAT)* utilizes an antireflective coating formed of an oxide of niobium, zirconium, hafnium, or tantalum. The oxide is formed by oxidizing a layer of metal deposited on the surface of the semiconductor portion of the solar cell. A pattern for the top electrode of the solar cell is etched into the oxide layer by a technique which includes the formation of a metallic and a photoresist mask and also includes lift-off photolithography.

A process developed by *J. Lindmayer and J.F. Allison; U.S. Patent 3,922,774; December 2, 1975; assigned to Communications Satellite Corporation* is a process for making a solar cell, responsive to light in the short wavelength region, including a noncrystalline tantalum pentoxide antireflective coating. The method includes placing the antireflective coating and a metallic current collector on the top surface of the solar cell using the technique of lift-off photolithography and the oxidation of elemental tantalum which is evaporated onto the solar cell. Figure 85 shows a perspective view of such a cell design.

**Figure 85: Solar Cell with Tantalum Pentoxide Antireflective Coating**

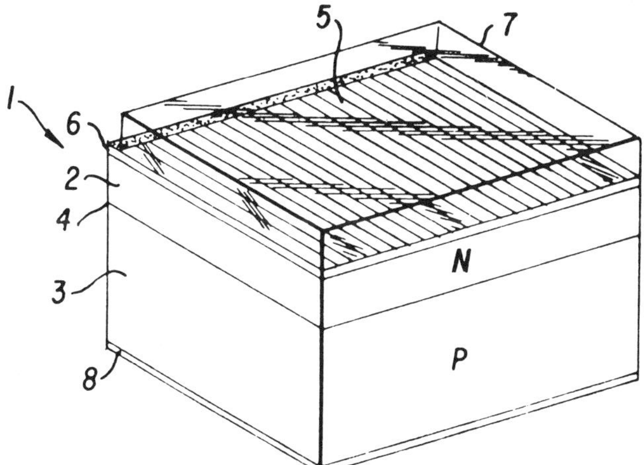


Source: U.S. Patent 3,922,774

There is shown a solar cell **1** having a layer **2** of first-type conductivity separated from a layer **3** of second-type conductivity by a junction **4**. Of course, it should be realized that such dimensions as the size of the solar cell and relative thickness of the several layers shown are not representative of an actual solar cell, but are shown, as is, merely for purposes of illustration. The silicon solar cell has an N-type layer **2** separated from a P-type layer **3** by a shallow P–N junction **4** which is about 1000 A from the top surface of layer **2**. On the top surface of the N-type layer is a metallic grid **5** and an antireflective coating **6**. The antireflective coating may occupy those areas of the top surface of layer **2** not occupied by the metallic grid. The metallic grid may be of a fine geometry type comprising about 60 metallic current-collecting fingers. The antireflective coating comprises amorphous tantalum pentoxide. A quartz cover slide **7** of a type known in the art, covers the antireflective coating and metallic grid. On the bottom of the P-type layer is a back metallic contact **8**, also of any type known in the art, which may fully cover the entire back surface of the layer. Not shown are interconnectors which may interconnect the metallic grid of one solar cell to the metallic contact of another cell for purposes of forming a series-parallel solar array, as is well known.

According to *R.J. Dendall; U.S. Patent 3,943,003; March 9, 1976; assigned to Communications Satellite Corporation*, in a solar cell a pad of material may be placed between the cell semiconductor surface and the portion of the top electrode which is bonded to an interconnector. The pad prevents destruction of the cell P-N junction during bonding. The pad may be the same as an antireflective coating and may be formed simultaneously with the formation of the antireflective coating.

A process described by *J. Lindmayer and J.F. Allison; U.S. Patent 3,949,463; April 13, 1976; assigned to Communications Satellite Corporation (COMSAT)* involves the preparation of an antireflective coating for a semiconductor wafer.

Figure 86 shows several illustrations of a side view of the wafer during the successive steps of the preferred method for coating a solar cell with a metal oxide.

The preparation of this wafer does not form a part of the present process but a typical method will be described briefly for background purposes. The starting point is a slice of silicon which is cut into a predetermined size suitable for use as a solar cell. Then, the silicon slice is subjected to a diffusion process by which an impurity (typically phosphorus) will be diffused into one surface of the silicon to provide the N-P junction **4** that is approximately 1000 A from the top surface of layer **2**, as is shown in illustration **A**.

Referring to illustration **A** there is shown a silicon slice having layer **2** of N-type conductivity separated from layer **3** of P-type conductivity by a shallow P-N junction **4**. A layer **9** of elemental niobium is shown. The layer **9** may consist of other metals whose oxides are useful as antireflective coatings such as titanium, tantalum, yttrium, erbium, manganese, thorium and tellurium. The choice of these metals would depend on the type of properties the oxide was required to have.

In the first step of the present process, the slice of silicon having the required N-P junction **4** will have a layer **9** of elemental niobium evaporated over the entire top surface **2** of the solar cell. In a preferred embodiment of the process, the evaporation of niobium may take place by means of the standard electron beam evaporation process. The niobium layer is approximately 250 A thick; however, the layer may be 100 A or greater, depending upon the desired thickness of the resultant oxide that will be the antireflective coating. The oxide thickness should be equal to one-quarter of the wavelength of light in the center of the spectrum of useful light for a particular solar cell. A 250-A layer of niobium may be oxidized to a 550-A layer of niobium pentoxide. For other metals, the resultant oxide and its characteristics may be found in standard reference books such as the *American Institute of Physics Handbook,* 2nd Edition, (1963).

Although the electron beam evaporation technique is well known in the art, certain considerations should be recognized. In order to assure proper deposition, a high-purity niobium source, small enough to prevent undue thermal radiation from the hot niobium metal, should be bombarded by the electron beam in a high vacuum. In addition, the P–N silicon slice should be shielded from any electron damage which may result from the beam of electrons focused on the niobium. A shield, comprising a positively charged metal electrode screen, which will attract any stray electrons, may be used. The niobium metal itself should be free of any impurities which, if deposited on layer **2**, may diffuse into layer **2** during this or subsequent steps of the coating process and thereby damage the P–N junction of the silicon solar cell.

In the second step, illustration **B**, a layer of photoresist material **10** is placed on the entire top surface of the niobium layer **9**. The photoresist may be any commercially available photoresist used in the photolithographic process of manufacturing microcircuits . In order to prepare the desired pattern of developed photoresist, a photomask having a pattern which will pass light, identical to the pattern desired for the top metallic grid (for a positive photoresist; a negative photoresist would have a photonegative pattern), is placed over the photoresist. The photomask would be sized to cover the silicon wafer surface and may be made by techniques well known in the art.

Next, the top surface covered by the photoresist is exposed to ultraviolet light through the photomask. Then, the photomask is removed and the layer of photoresist is developed. The solar cell is rinsed in water or other solvent, thereby removing the soluble photoresist which was exposed to light and leaving a pattern of unexposed but developed photoresist material **10a** on the top surface of layer **9** as shown in illustration **C**. At this point, the pattern of photoresist material **10a** is the photonegative of the metallic grid pattern to be eventually placed on the top surface of layer **2**.

In the fourth step, as shown in illustration **D**, a metal contact layer **11**, such as a mixture of chrome and gold approximately 2000-A thick, is deposited by vacuum evaporation over the entire top surface of the solar cell, including photoresist **10a** and the pattern of niobium **9**. Since chromium adheres well to niobium, initially, pure chromium is evaporated to a layer of about 200 A with gold gradually being mixed into the metal deposit. Toward the end of the evaporation process, the chromium is phased out and only the pure gold is deposited to complete the 2000-A contact. The conditions for this two-metal evaporation process may be found in standard text books and are well known in the art.

The fifth step (**E**) in the process is to remove the developed photoresist material from the surface of elemental niobium. Removal of the developed photoresist is accomplished by the lift-off technique in which the developed photoresist to be lifted off is dipped in acetone or some other recommended chemical. The acetone is contained in an ultrasonic bath to facilitate removal of the developed photoresist. The result of the lift-off process is to remove not only the photoresist, but also the metallized layer **11** above the photoresist. This lift-off process will leave exposed areas of elemental niobium and a pattern of chrome and gold on the top surface of the cell. The maximum thickness of the chrome and gold should be about 2000 A to enable the underlying developed photoresist to be lifted off easily. The resultant structure is shown in illustration **E**.

In the sixth step (illustration **F**), a standard electroplating technique may be used to deposit a layer of silver **13** over the chrome and gold layer to build up the thickness of the metallic grid to approximately 5 microns. At the same time, the silver back contact **8** having a thickness of 5 microns will be plated on the back layer **3**. It has been found that the silver will not plate-out onto the elemental niobium during this process, thereby leaving a silver-coated finger pattern over a niobium layer on one side of the cell and a silver back contact on the other side.

In the seventh and final step (illustration **G**), the silicon solar cell with the silvered metallic contact pattern and elemental niobium surface is exposed to oxygen in either a thermal or anodic oxidation process. In thermal oxidation, the silicon slice is heated in air at a temperature of about 400°C for 3 minutes. The parameters may be in the range of 350° to 450°C for 3 to 15 minutes and should be selected to provide a uniform oxide without grain boundaries (noncrystalline). In addition, it should be recognized that higher temperatures will induce undesired alloying of the chrome, gold and silver, due to the low eutectic temperatures of these mixtures, while oxidation at lower temperature and longer time will allow diffusion of the chrome or gold atoms into the semiconductor junction area. Under proper conditions, niobium pentoxide is formed having a resulting index of refraction 2.4, while the remaining elements of the cell are unaffected. The silver layer does not oxidize at the temperature and time required for niobium and will remain essentially in its elemental state.

In the alternate oxidation technique, known as anodic oxidation, a platinum electrode as the cathode and silicon slice as the anode would be used. Both electrodes are immersed in an electrolyte and a current allowed to pass therethrough. As one example, the anodic oxidation process may last for approximately 20 minutes commencing with an initial current of 1 milliampere. The use of organic, nonaqueous electrolyte, such as tetrahydrofurfuryl alcohol is preferred since this will result in a uniform niobium pentoxide coating which firmly adheres to the top surface of layer **2**.

The result of both thermal or anodic oxidation is a layer **12** of niobium pentoxide on the top surface of layer **2** as shown in illustration **G**. The layer of niobium pentoxide would be approximately 550 A thick in order to produce a one-quarter wave match at 0.5 micron.

The niobium layer between the silicon surface and the metallic layer provides an excellent contact for the collection of carriers by the metallic grid.

**Figure 86: Steps (A–G) in Method of Applying an Antireflective Coating to a Solar Cell**

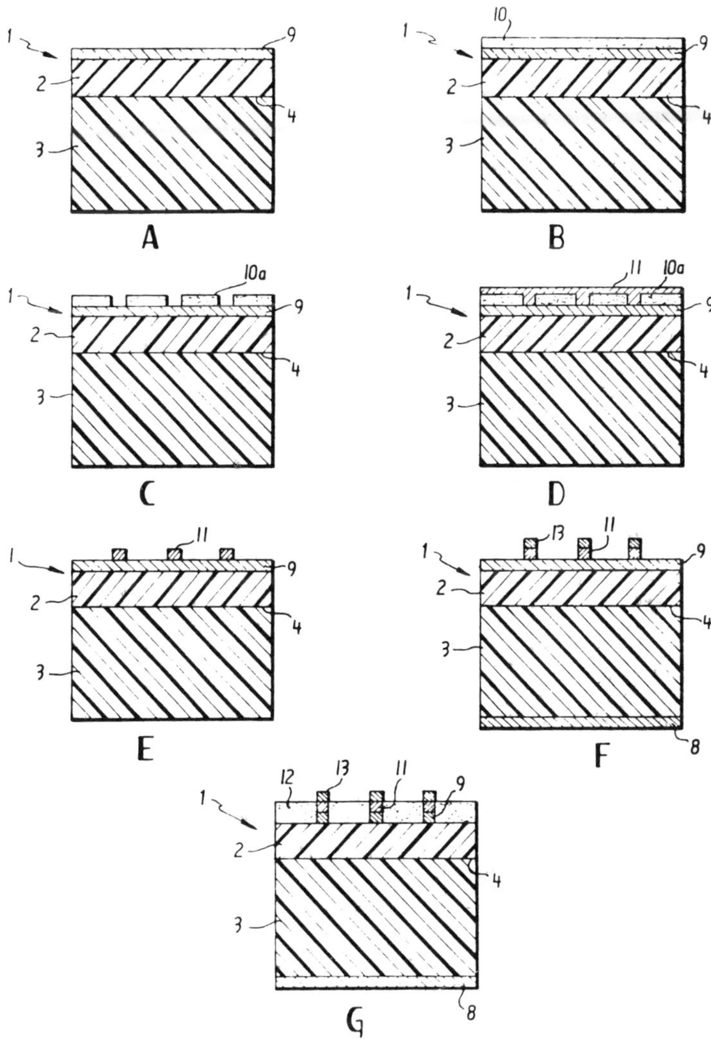


Source: U.S. Patent 3,949,463

Very similar ground is covered by *A.G. Revesz and J. Lindmayer; U.S. Patent 3,977,905; August 31, 1976; assigned to Communications Satellite Corporation (COMSAT)* with specific attention to niobium pentoxide antireflective coating for use on a solar cell responsive to light in the short wavelength region.

A process developed by *F.K. Crosher; U.S. Patent 4,055,442; October 25, 1977; assigned to Optical Coating Laboratory, Inc.* yields first and second layers which are formed on the surface of the solar cell and serve to provide an antireflection coating which is effective within the spectral range of 400 to 1,200 nanometers. A glass solar cell cover is provided which is secured to the body having the first and second layers fixed thereon by a cement. The first layer is, counting from the body, formed of material that has an index of refraction which is less than that of the body and which is greater than that of the glass cover. The second layer is formed of a material that has an index of refraction which is greater than that of the glass cover but which is less than that of the first layer.

By way of example, the first layer is formed of a titanium oxide which has a higher index of refraction of approximately 2.35 to 2.4. It can be formed of a gas-reacted titanium oxide. Also, if desired, it can be formed from the evaporation of titanium sesquioxide or titanium dioxide (rutile). The titanium oxide layer is deposited to a thickness of one-quarter wavelength at a 600 nanometer design wavelength. The second layer is then deposited. The material which is utilized for the second layer has a layer index of refraction of approximately 1.6 to 1.7, which is less than that of the material for the first layer and which is greater than that of the cement which is approximately 1.4. It has been found that aluminum oxide forms a satisfactory second layer. It is deposited to the thickness of one-quarter wavelength at the 600 nanometer design wavelength and has an index of refraction in the vicinity of 1.65.

A solar cell design developed by *M.G. Coleman and F. Restrepo; U.S. Patent 4,070,689; January 24, 1978; assigned to Motorola Inc.* is one in which the dielectric antireflective coating is a composite of silicon dioxide and silicon nitride layers. The steps in forming such a cell are shown in Figure 87, on the following page.

Referring to Step A, a semiconductor wafer or substrate **10** is used as the starting point for fabricating the solar energy device. The semiconductor wafer or substrate can be formed by various techniques, including crystal pulling and subsequent wafer cutting and polishing or other techniques such as dendritic growth or ribbon growth. Preferably, the starting substrate for the wafer is of P-type conductivity, which means that it is a semiconductor substrate such as silicon doped with an impurity which would exhibit P-type electrical conductivity characteristics. Examples of such dopants that would provide a P-type substrate are boron or aluminum. If desired, the starting wafer can be of N-type conductivity and the subsequently doped regions would be of opposite type conductivity to that described below. The P-type substrate preferably has a thickness of about 10 mils, but can be made with a thickness in the range of about 1 to 30 mils, and a resistivity in the range of about 0.1 to 20 ohm-centimeters.

**Figure 87: Steps in Forming a Solar Cell Incorporating Antireflective Coating**

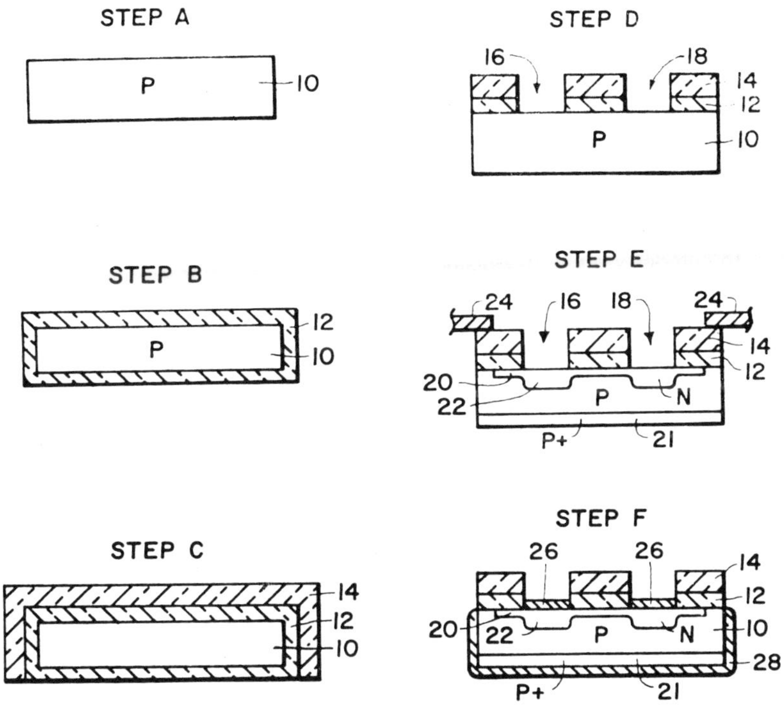


Source: U.S. Patent 4,070,689

Referring to Step B, a thin silicon dioxide layer **12** is preferably grown on the starting P-type substrate and forms a coating around the substrate. The thin silicon dioxide layer preferably has a thickness of between 50 and 300 A, and most desirably between 75 and 100 A. This thin silicon dioxide layer is necessary for achieving minimum surface state density and low recombination velocity.

Referring to Step C, a thicker silicon nitride layer **14** is deposited or formed on the top and incidentally on the side portion of the thin silicon dioxide layer. The silicon nitride layer is deposited by various well known silicon nitride chemical vapor deposition techniques and is preferably deposited at a temperature of 750°C or less to preserve etchability in HF solutions. The silicon nitride layer has a thickness in the range of 600 to 1300 A (preferably 900 A) and the silicon dioxide layer together with the silicon nitride layer form the antireflective coating that is needed for the solar energy device.

The function of the antireflective coating or layer is to enhance the absorption of photons of light striking the antireflective layer into the semiconductor substrate or wafer. Accordingly, the thickness of the silicon nitride layer and the silicon dioxide layer, together with the index of refraction of both the silicon nitride material and the silicon dioxide material serve to provide an excellent antireflective layer or coating.

Referring to Step D, holes **16** and **18** are opened up through the silicon nitride layer and the silicon dioxide layer by using conventional photolithographic masking and etching techniques using a HF solution. The holes are part of a continuous opening through the antireflective coating. The purpose of this step in the process of fabricating the solar energy device is to define the geometry of subsequently formed metal ohmic contacts that will provide an electrical contact to the underlying semiconductor region. However, before the ohmic contacts are formed or deposited in the openings, these openings are used to provide a differential P-N junction depth, as shown in Step E.

With regard to Step E of the figure, doped regions **20** and **22** are formed preferably using ion-implantation techniques, but conventional diffusion techniques can be combined with the ion-implantation doping techniques.

A doped region **20** which is of N-type conductivity and having an impurity concentration in the range of about $10^{14}$ to $10^{16}$ atoms per cubic centimeter is formed by an ion-implantation technique using a substrate or wafer holder **24** which is preferably made of metal and serves to hold the entire structure in position for the ion bombardment of the N-type impurities into the substrate. As can be seen with reference to the ion-implanted region **20**, a region beneath the metal holder in the P-type substrate **10** is not converted to N-type conductivity due to the shielding effects of the metal holder. Thus, it can be readily seen that the P-N junction formed between region **20** and substrate **10** does not extend out to the side of the substrate, thereby eliminating any chance of shorting when the ohmic contact is made to the backside portion of the semiconductor device structure.

As can be seen with reference to Step E, the ion-implantation process that is carried out causes a doped region to form beneath the antireflective layer or coating. This is created beneath the antireflection layer or coating by the penetration of ions therethrough.

A $P^+$ region **21** located on the back or bottom portion of the substrate is preferably formed by means of an ion-implantation technique. The $P^+$ region has a surface impurity concentration of about $10^{21}$ atoms per cubic centimeter and serves to provide a semiconductor region that will permit a good ohmic contact to be made to the backside portion of the device and thereby provide an excellent electrical contact to the semiconductor substrate. Additionally, the $P^+$ region provides an electric field to aid in collecting minority carriers generated in the substrate.

Still referring to Step E, a deeper doped region **22** is shown in the region beneath the openings **16** and **18**. This occurs because the implanted ions in the semiconductor material beneath the openings did not have to go through the antireflective coating. The portion of the doped region **20** that is underneath the antireflective coating has an impurity concentration profile which preferably provides the maximum concentration of impurities at the silicon-silicon dioxide interface and the impurity concentration is graded and decreases in concentration with increasing depth from the silicon-silicon dioxide interface. This is accomplished by selecting an ion implantation energy which will insure that the maximum concentration is preferably at the silicon-silicon dioxide interface or slightly above this interface and in the antireflective coating. This portion of the doped region **20** that is underneath the antireflective coating will have a P-N junction depth of about 0.3 micron or less. The purpose of the deeper doped regions **22** is to provide deeper doped regions beneath the subsequent location of metal ohmic contacts and thereby serves to protect the device from shorting when the ohmic contact is sintered to provide a good electrical and mechanical contact to the semiconductor device.

In Step F, an electroless metal deposition operation is carried out wherein metal contacts **26** are formed in the openings located in the antireflective coating. These electroless metal coatings are formed in an electroless-plating operation and create ohmic contacts to the N-type doped region **20**. Simultaneous with the formation of the electroless metal ohmic contacts is the formation of a thin metal ohmic contact **28** which is also an electroless contact and is located on the backside of the substrate and is specifically in contact with the doped region **21**. The electroless-plating process to provide these ohmic contacts utilizes a plating solution which deposits a metal such as nickel into the exposed bare silicon regions. Thus, there is no need for separate photoresist, alignment or etching operations. If desired, contacts can be made to the front and backside of the semiconductor solar silicon energy device shown in Step F by using standard metal evaporation or sputtering techniques using a metal evaporation mask that has been aligned to the preohmic pattern. Also, electroplating can be used. If desired, sintering or alloying steps can be carried out subsequent to the deposition of the metal contact material to provide a good mechanical bond and electrical ohmic contact to the substrate and the doped regions thereof.

A related development by *I.A. Lesk and R.A. Pryor; U.S. Patent 4,131,488; December 26, 1978; assigned to Motorola, Inc.* involves the use of a rough or textured pyramid shaped silicon surface beneath the antireflective coating to increase solar cell efficiency.

The utilization and desirability of a rough solar cell surface, consisting of a uniform distribution of minute pyramids, to increase solar cell efficiency has been demonstrated. This rough or textured surface causes all the light that is reflected from the first impingement on the rough solar cell surface to strike the solar cell at least a second time (assuming initial normal incidence). This second impingement increases the amount of light absorbed in the solar cell, improving cell efficiency.

There are other advantageous features of such a textured surface on a solar cell. The major effect is that, since light is refracted into the silicon at an angle to the normal of the overall solar cell plane, more light is absorbed within a given thickness of silicon than would occur with normally incident sunlight on a smooth-surfaced solar cell.

A process developed by *R. Gonsiorawski; U.S. Patent 4,078,945; March 14, 1978; assigned to Mobil Tyco Solar Energy Corporation* is one in which an antireflective coating is formed integrally with a silicon solar cell by treating the cell with an acidic solution comprising a mixture of HF and $H_2O_2$. The following is one example of the utilization of this process.

An acidic treating solution is prepared by mixing three parts of 49% HF (bottled concentration) and one part of 30% $H_2O_2$ (stabilized), all parts by volume. The two components are miscible. The solution is maintained at room temperature.

A silicon solar cell of the type which has electrodes attached thereto is immersed in the treating mixture for 3 seconds. The solar cell is then removed from the treating solution, is immediately washed with deionized water, and dried under nitrogen. A dark blue stain coating is formed on the areas of silicon which were exposed to the treating solution. The electrodes appear substantially unchanged. The coating is approximately 0.13 micron thick. The P-N junction depth appears to be approximately 0.07 micron thinner immediately below the stained areas of the cell. The resulting solar cell has a reflectivity of about 5% when exposed to a sunlamp. Reflectivity of the same cell without the antireflection coating is about 30%. The short circuit current density of the coated cell is measured to be about 30 $mA/cm^2$, which is about 30% increase as compared to the same cell without the antireflection coating. 83% of the observed increase of current density is believed attributable to the corresponding 25% reduction in reflectivity. The remaining 17% gain is believed attributable to improved cell efficiency resulting from thinning of the junction.

The coated cell is tested for thermal stability by heating the cell to a surface temperature of 150°C for 1,000 hours. No change is observed in either the physical appearance of the coating or the short circuit current density.

A scheme developed by *A.E. Bell, B.F. Williams and D.E. Carlson; U.S. Patent 4,126,150; November 21, 1978; assigned to RCA Corporation* provides a solar cell body which includes a thin film active region and a layer substantially transparent to solar radiation. The thickness of the transparent layer is such that a first antireflection condition is present for solar radiation at a wavelength which is relatively highly absorbed by the material of the active region. Furthermore, the combined thickness of the transparent layer and active region are such that a second antireflection condition is present for solar radiation at a wavelength which is poorly absorbed by the material of the active region. As a result of the prevailing antireflection conditions the solar radiation absorption efficiency of the photovoltaic device is increased.

Such a cell construction is shown on the following page in Figure 88.

**Figure 88: RCA Design Photovoltaic Device Having Increased Absorption Efficiency**

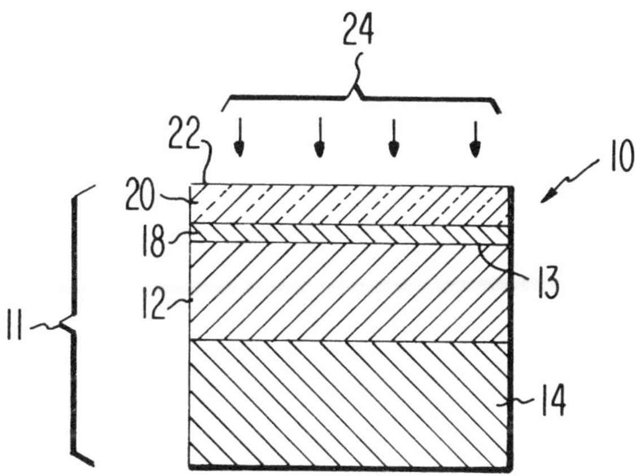


Source: U.S. Patent 4,126,150

A metallic film **18** is contiguous to the active region **12** at a surface **13** of the active region opposite the substrate **14**. The metallic film is a continuous film of a metal which is capable of forming a Schottky barrier, i.e., a surface barrier junction, with the amorphous silicon active region at the surface. Typically, undoped amorphous silicon is slightly N-type, thus the metallic film should be of a high work function metal, i.e., of a work function of 4.5 eV or greater, in order to form a Schottky barrier. Suitable high work function metals are platinum, indium, rhodium and palladium. In addition, the metallic film should be thin so that it is at least semitransparent to solar radiation, and should be of a relatively low sheet resistance. The transparency and low sheet resistance required of the metallic film, together with the fact that it is a continuous film, are factors which determine its thickness, as is known to those skilled in the art.

On the metallic film opposite the active region is a transparent layer **20** which has an incident surface **22** on which solar radiation **24** is capable of impinging the body. The transparent layer is of a material substantially transparent to solar radiation such as dielectric materials including titanium dioxide, zirconium oxide or silicon nitride.

The active region of amorphous silicon fabricated by a glow discharge in silane has an absorption profile over the solar spectrum such that solar radiation in the range of approximately 3500 to 5000 A is highly absorbed by the active region **12** (hereinafter referred to as the highly absorbing region). However, absorption by the amorphous silicon active region **12** decreases for solar radiation greater than 5000 A in wavelength (hereinafter referred to as the poorly absorbing region).

One function of the transparent layer is to ensure that solar radiation in the highly absorbing region of the amorphous silicon absorption profile is allowed to impinge the incident surface **22** substantially unreflected. Thus, the thickness of the transparent layer should be such that the incident surface is nonreflective to solar radiation at a first predetermined wavelength in the highly absorbing region. The first predetermined wavelength is chosen so that the maximum measure of solar radiation of the highly absorbing region is absorbed in the active region as a result of this first antireflection condition being present. The thickness of the metallic film to a small degree will influence the thickness of the transparent layer in meeting the first antireflection condition.

Most beneficially in the present device, the combined thickness of the transparent layer and the active region is tuned to a second predetermined wavelength in the poorly absorbing region, such that the absorption capabilities of the device **10** are increased for this poorly absorbing region. The second predetermined wavelength is chosen so as to maximize the measure of solar radiation of the poorly absorbing region being absorbed in the active region. This increased absorption is a result of this second antireflection condition being present at the incident surface for solar radiation in the poorly absorbing region.

## PROTECTION FROM ENVIRONMENTAL FACTORS

Cover slides have been used with solar cells, in order to increase the emissivity of the solar cell and thereby provide a lower operating temperature, consequently raising the electric power output of the solar cell; to protect the solar cell from radiation damage by attenuating electrons and protons in the space environment; and as a substrate for blue or ultraviolet and red or infrared reflection filters, to decrease the operating temperature, and to ensure higher solar cell efficiency.

Solar cells used in space missions generally carry transparent cover slides, typically ranging in thickness from 6 mils of glass to 60 mils of quartz, to provide a reasonable degree of thermal emissivity and/or to protect the cells from damage from charged particles or micrometeorites. These cover slides generally are held onto the cells by a thin adhesive layer. Not only does the adhesive layer sometimes have a poor heat transmissivity characteristic, but also usually requires that expensive ultraviolet rejection filters be evaporated onto the cover slide to prevent darkening of the adhesive layer. Additionally, thin cover slides are difficult to manufacture and apply to the solar cell and, accordingly, the typical cover slides used, as well as the adhesive, add significant weight to the cell.

A type of construction developed by *P.A. Iles; U.S. Patent 3,653,970; April 4, 1972; assigned to U.S. National Aeronautics and Space Administration* is one in which a lightweight protective glass coating over the radiation-receiving surfaces of a solar cell is formed integrally with the cell by depositing a layer of glass particles to such surfaces and heating the cell and glass particles to an elevated temperature sufficient to fuse the glass particles and to form the coating. In another embodiment, a conventional protective glass slide is applied to the glass particles prior to heating, and the cell, particles and slide are heated to fuse the glass and form a fused glass bond between the cell and the slide.

There have been a number of problems associated with cover slides for solar cells, as listed by *B.S. Marks and P.J. Chiesa, Jr.; U.S. Patent 3,714,119; January 30, 1973; assigned to Lockheed Aircraft Corporation.* These cover slides have given rise to problems such as high material costs, high labor costs, reduced emissivity, high weight factors, adhesive degradation, and unsuitability to new types of solar cells.

The high material costs arose because the cover glass or slide has generally been microsheet glass or fused silica with a dual coating system; namely, an antireflective inorganic salt coating on one side, and an ultraviolet reflective diffraction coating for the protection of the adhesive on the opposite side. The latter coating has been particularly expensive, being a multicoated system, with exacting thickness requirements laid down by vacuum deposition techniques.

High labor costs arose because the job of placing the slides and adhering them individually to each solar cell has required great care and precision. Furthermore, frequent breakage, particularly of the thinner cover glasses, added to the cost.

The emissivity of a cover glass coated with the antireflective inorganic salt is usually about 0.76, which, though high compared to an uncovered solar cell, is lower than is desirable.

Thick cover slides have been required for radiation protection for many missions. For others, thinner ones were sufficient. Due to the fragility of the slides, a thickness of 6 mils has usually been chosen as the minimum, yet for other requirements the thickness could be less than half this amount. So the cover slides have weighed twice or more than twice, as much as is desirable.

An adhesive was required to fasten the cover glass to the solar cell; however, the adhesive used had to be protected from ultraviolet radiation to prevent its degradation, with resulting discoloration and some loss in transmission of light.

Cover glasses may not be readily usable for new solar cell types. In particular, the use of cover slides is not easily adaptable to thin-film or dendritic types of solar cells.

Many of these problems are said to have been overcome by the use of an integral protective coating in place of the complex slide-and-adhesive approach. The integral protective coating composition proposed by Marks and Chiesa is a zinc-phenylsiloxane polymer made by reacting a zinc salt such as anhydrous zinc acetate with a difunctional diphenylsiloxane monomer such as diphenyldialkoxysilane or with a trifunctional or polyfunctional silane polymer such as the anhydrous hydrolysis and alcoholysis product of phenyltrichlorosilane, using heat sufficient to insure that reaction occurs (about 200°C) which can be noted by the distillation of alkyl acetate, a by-product of the reaction.

The zincphenylsiloxane polymer may thereafter be blended with a nonmigrating plasticizer and coreactant silicone monomer and cured. Such a coating is claimed to have the following advantages:

1. It is inexpensive materialwise in comparison to a cover glass assembly, for it can be prepared and formulated as a solution ready for application.
2. It is inexpensive laborwise, since the coating can simply be sprayed on. Furthermore, a solar cell panel can be sprayed after it is assembled, obviating special handling.
3. The coating materials of this process have higher emissivity than cover slides (on the order of 0.84).
4. Coatings weigh very little, and the weight factor for a solar cell coating can be decided definitively. A selected coating thickness can be spray-applied to give exactly what is desired for a particular mission. Furthermore, areas which are not to be sprayed can be taped or masked during spraying.
5. New solar cell types, in particular the thin-film dendritic growth and the wrap-around types of solar cells, are admirably suited for coating protection.

Materials useful in solar cell fabrication, according to *D.A. Gorski; U.S. Patent 3,969,743; July 13, 1976; assigned to Aeronutronic Ford Corporation*, are optically transparent coatings of barium fluoride, calcium fluoride, or strontium fluoride. These coating materials are insulators, chemically and electrically inert to the semiconductors, easily applied, and mechanically compatible thus making excellent protective coatings. In addition the coatings can act as diffusion masks in device fabrication thus enhancing planar device processing.

A process developed by *H. Gochermann and D. Rüsch; U.S. Patent 4,009,054; February 22, 1977; assigned to Licentia Patent-verwaltungs-GmbH, Germany* is one by which a terrestrial solar cell generator that is able to withstand thermomechanical stresses without damage can be produced economically, employing a radiation-resistant material, i.e., a material which will not become cloudy or dull when exposed to ultraviolet radiation, for the housing.

This objective is achieved by forming the housing for the solar cell generator entirely of the same material which encases the solar cells and their connecting elements or leads on all sides. Suitable materials for the housing are, for example, modified polyester resin, acrylic resin, and epoxy resin with or without glass fiber reinforcement. It is also possible to provide a thermoplastic material from among the polycarbonates which are known by various trade names or to use glass with the appropriate properties.

A solar cell encapsulation technique described by *J.S. Walker and W.C. Kittler; U.S. Patent 4,067,764; January 10, 1978; assigned to Sierracin Corporation* is one in which a solar cell panel is produced which consists of an outer rigid transparent faceply of glass or plastic material to which are applied at least two layers of plastic such as polyvinyl butyral between which are positioned a plurality of solar cell wafers. A thin flexible film of polyethylene terephthalate forms the other outer surface of the panel. The panel is manufactured by laminating the materials together and allowing the margins of the plastic film to extend beyond the polyvinyl butyral layers so that the film can be brought into direct contact and sealed to a rigid base plate, forming a fully encapsulating structure. The assembled structure is then evacuated to withdraw air and to squeeze the layers together to promote adhesion. The evacuated laminated structure is then placed in an oven for applying heat and pressure to the laminated structure for permanent bonding. After cooling, excess film is trimmed from around the edges of the rigid face plate.

A technique developed by *A.I. Mlavsky; U.S. Patent 4,078,944; March 14, 1978; assigned to Mobil Tyco Solar Energy Corporation* yields an encapsulating and solar cell supporting structure which is relatively inexpensive to produce, is stable under most normally expected environmental conditions and particularly those encountered in terrestrial and space environments, and, will remain substantially impervious to moisture over relatively long periods of time.

Such an assembly consists of at least one solar cell mounted and hermetically sealed in a substantially rigid, elongated, tubular envelope which is transmissive to actinic radiation to which the photovoltaic solar cell is sensitive. The assembly can include an antireflection coating for reducing the reflection of solar radiation from the envelope as well as reflective coatings for uniformly distributing solar radiation over the light gathering surfaces of the cell.

As illustrated on the following page, Figure 89 shows the components of such an assembly (upper view) as well as the assembled structure (lower view).

There is shown an elongated, flat solar cell structure **8** which comprises a plurality of photovoltaic cells **10** of conventional construction which are preferably supported on a flat, rigid support member **12** so that the cells are spaced from one another to provide an elongated panel-like assembly. Although not shown in detail, it is to be understood that each cell is constructed generally as shown in U.S. Patents 3,686,036 and 3,811,954 and thus is a silicon N-P solar cell with one electrode attached to and covering most of the bottom surface thereof (i.e., the surface that faces the support member **12**) and an electrode in the form of a grid attached to the opposite or upper surface thereof. The latter surface forms the radiation-receiving portion of the cell. The support member is made of an electrically-insulating material and the bottom electrode of each cell is connected to the top grid electrode of the next adjacent cell by a wire **13**.

**Figure 89: Solar Cell Assembly Encapsulated in Fluorescent Lamp Tube**

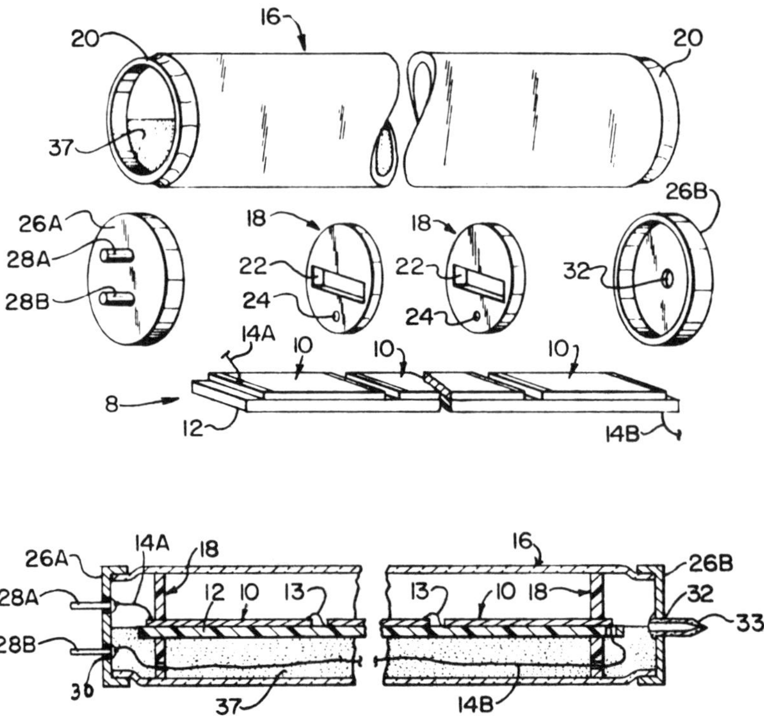


Source: U.S. Patent 4,078,944

Two terminal wires **14A** and **14B** are respectively connected to the top electrode of the cell at one end of the array and the bottom electrode of the cell at the other end of the cell array. Although only three cells are shown, it will be understood the assembly may include any desired number of cells and the manner in which they are interconnected can vary without departing from the scope of the present device. Thus, for example, a plurality of cells can be mounted in parallel (i.e., the top electrodes of the cells are all connected together and to one of the terminal wires while the bottom electrodes of the cells are connected together and to the other terminal wire) or in a combined series and parallel matrix. Further, the support member **12** may include electrically-conductive material to interconnect the electrodes of the cells. Accordingly, the support member may be in the form of a printed circuit board.

The solar cell structure is supported in an elongated tubular envelope **16** by at least two mounting elements **18**. The envelope is made of a material which is substantially transmissive to radiation to which the solar cells are sensitive. Preferably, the envelope is made of a clear vitreous or ceramic material which is transmissive to visible and ultraviolet radiation. The envelope comprises a cylindrical tube having a uniform wall thickness for structural integrity. The ends of the envelope are preferably reduced in diameter so as to facilitate attachment of end caps, but the minimum inner diameter of the envelope is greater than the width of the support member, to permit the structure to be inserted into the envelope. Such envelopes are commercially available and commonly used in certain gaseous-vapor discharge devices such as fluorescent lamps of the type employed for lighting living and work areas in buildings.

Preferably, each mounting element **18** is in the form of a disc which is made of a resilient, electrically-insulative material and is sized so that it can be inserted into the envelope through the ends and yet make a tight press fit with the interior surface of the tube when it is positioned so that its plane extends at a right angle to the longitudinal axis of the envelope. The resiliency of the disc-like mounting element is advantageous where the assembly is subjected to relatively large temperature changes such as those encountered in space applications, since the mounting elements can yield to prevent buildup of stresses in the unit. Each member includes an elongated centrally located aperture **22** which is sized and adapted to receive an end of the solar cell structure and hold it centered on the longitudinal axis of the envelope.

The mounting elements preferably include holes **24** to permit the terminal leads **14** to be brought out to one end of the envelope. The opposite ends of the envelope are fitted with end closures or caps **26A** and **26B**. These caps can be made of any suitable material, but preferably they are made of metal and are like the end caps which are used to close off the tubular envelopes of conventional fluorescent lamps. Cap member **26A** is provided with two terminal posts **28A** and **28B** which are electrically connected by soldering or the like to the respective terminal leads **14A** and **14B**. Where the cap **26A** is made of an electrically-conductive material, posts **28** are insulated from the cap by any suitable electrically-insulative material **30** as well known in the art, e.g., by glass seals. Caps **26A** and **26B** fit on the ends of the envelope and are hermetically sealed in place by means of a suitable cement or by a glass or ceramic bonding agent.

Preferably, the sealed envelope is either evacuated to create a vacuum or is filled with an inert gas such as nitrogen. To this end the cap member **26B** may be provided with a hollow fitting **32** which as installed is open at both ends so that it can be used for evacuating and filling purposes, after which its outer end is pinched off as shown at **33**. Other techniques well known in the art may be employed for evacuating or filling the sealed envelope. For example, one of the terminal posts can be made as a hollow member and its outer end pinched off after the evacuation or filling of the envelope has been completed.

When assembling the device shown in the figure, the structure is positioned in the envelope by support elements, the terminal wires are connected to the terminal posts of the cap member and then the cap members are sealed in place. Thereafter, the envelope is evacuated (or evacuated and filled with an inert gas) and the fitting is pinched off to fully seal the envelope.

As an alternative measure, the support members need not be resilient but instead may be stiff and may even be made of a ceramic material or glass, in which event they are sized so as to make a close but sliding fit with the envelope and are secured to the solar cell support member so that they cannot move relative to the solar cell structure. Also, in such event, the support member is made long enough to engage the end caps, whereby it is restrained against lengthwise movement, and at least one end of the tube is not reduced in diameter so as to permit the members to be inserted therein.

A solar cell design developed by *T. Hirano; U.S. Patent 4,104,083; August 1, 1978; assigned to the Japanese Government, Japan* consists of a solar battery package having at least one solar cell substantially embedded in a block of fiber-reinforced thermosetting resin with a transparent flexible covering. The block has a portion, overlaying a light-receiving area of the solar cell, which is transparent and has another portion which is either transparent or substantially opaque. A protective coating of fluorine-containing compound may be applied to the block to improve the weatherability of the solar battery package.

A device developed by *H. Gochermann; U.S. Patent 4,147,560; April 3, 1979; assigned to Licentia Patent-Verwaltungs-GmbH, Germany* is a solar cell generator for producing electrical energy for terrestrial use including a plurality of interconnected solar cells encased on all sides by a radiation resistant plastic material and a foil of a weather resistant material which adheres to the plastic material covering the outer surfaces of the casing.

### Protection from Heat

During their exposure to light, which is ordinarily sunlight, the cells and the panels in which they are housed often experience an increase in temperature, such rise in temperature being due to the heat generated by the light impinging on the cell. Under normal operation, solar cells have a lower efficiency at higher temperatures. Consequently, a decrease in efficiency is often effected by the increased temperatures. Thereby, despite the strong light concentration, the output is lower than expected upon direct exposure of cells to high light levels.

While a slight decrease in efficiency of operation of solar energy cells exposed to high light intensities may be tolerable in some instances, it is nevertheless desirable to cool the cells by some means. This is especially true if the solar cells are to be exposed to light intensities greater than one sun. This is the case when light concentrators are utilized. Such concentrators may be lenses, mirrors, or other means used to focus light from the sun. When the light rays are concentrated and are directed or reflected on the cells, an increased flow of electrons results.

The use of concentrators to focus the light on solar energy cells results in a magnification of the problem that might, under circumstances where no concentrators are utilized, be tolerable. Thus, if a concentrator is capable of increasing the light intensity from the sun on a solar energy cell by a multiple of three, then the electricity that the cell generates from such light will be substantially increased, optimally by the same factor of three. Under such conditions, however, the temperature of the cell will rise considerably and there will be a corresponding decrease in cell efficiency. As a result, particularly when concentrators are used but also when optimum efficiency is necessary for solar cells exposed only to ambient light, it is highly desirable to cool the cells, and thereby maintain maximum cell efficiency, according to *P.F. Varadi; U.S. Patent 4,056,405; November 1, 1977; assigned to Solarex Corporation*.

Since the output of cells varies as a function of temperature it is necessary to compute cell temperatures in order to obtain an accurate estimate of the output of cells installed in various types of collectors. The temperature of the cells depends on the effective heat-transfer coefficient of the system used to provide cooling. In simple flat-plate systems, the heat loss from the front (and possibly also the back) of the arrays provides adequate cooling. In concentrating devices, it is usually necessary to provide more sophisticated cooling systems. These systems can be passive (i.e., large, finned radiating surfaces attached to the cell), or active cooling can be provided by pumping liquids across the back of the cell (3).

A type of construction developed by *R.R. Scott; U.S. Patent 3,532,551; October 6, 1970; assigned to U.S. National Aeronautics and Space Administration* is one in which a fused silica cover plate is bonded over a solar cell with portions of the plate overhanging the edges of the cell and a reflective coating is formed on the second or lower surface of overhanging portions to reduce the amount of heat absorbed by the cell and hence reduce the overall temperature.

A device developed by *J.V. Goldsmith and G.P. Rolik; U.S. Patent 3,615,853; October 26, 1971; assigned to U.S. National Aeronautics and Space Administration* is a cover plate for a solar cell panel, which defines a plurality of recesses having inclined sidewalls. The cover plate is made of a material such as glass through which light directed to the solar cells can pass. The sidewalls of the recesses or apertures are coated with a light-reflecting material so that the sidewalls act as light reflectors. The shapes and dimensions of the apertures in the plate are chosen so that when the vehicle, on which the solar cell panel is mounted, is near the earth, the light-sensitive surface of the panel is substantially perpendicular to the light direction, enabling substantially all the light to reach the solar cells. This enables maximum power to be provided at the start of the mission near the earth. However, as the vehicle moves toward the sun, by controllably inclining the panel's light-sensitive surface with respect to the light direction, a substantial portion of the light is reflected back into space by the light-reflecting sides of the apertures. By preventing a controlled part of the light from reaching the solar cells their temperature of operation is controlled to be within safe limits.

A type of solar cell construction developed by *J. Lindmayer, D.J. Curtin, J.G. Haynos and A. Meulenberg, Jr.; U.S. Patent 3,888,698; June 10, 1975; assigned to Communications Satellite Corporation* is one in which the problem of decreased power due to increased thermal temperature of solar arrays is alleviated by providing a cell having back electrodes which cover less than 10% of the back surface of the cell and preferably less than 5% of the back surface of the cell. Photolithographic techniques are used to place an electrode comprising extremely fine lines of metal separated by short distances onto the back surface. The electrode allows deep infrared to pass through the nonelectroded surface areas and out into space. The amount of deep infrared absorbed by the electrode is substantially reduced and thus the thermal equilibrium temperature of the cell in a deployed array is not appreciably increased, when compared with prior art cells on the body-mounted array.

The semiconductor material adjacent the bottom surface may be more heavily doped than the bulk of the bottom semiconductor layer to increase the lateral surface conductivity, but this added feature is not necessary because the relatively large bulk of the bottom semiconductor layer is sufficient to provide good lateral conductivity.

At least the exposed areas of the bottom surface are covered with an insulating layer, e.g., $SiO_x$, $TiO_x$, which is optimized to achieve maximum emission of the deep infrared toward space.

A design for a solar cell developed by *H.W. Brandhorst, Jr.; U.S. Patent 3,989,541, November 2, 1976; assigned to U.S. National Aeronautics and Space Administration* is an improved solar cell assembly for use under high intensity illumination conditions where heat is a problem.

The solar cell assembly includes a solar cell having an overlay of a semitransparent coating of a metal such as aluminum or silver, which covers the entire surface thereof. The purpose of the coating is to lower the amount of incident radiation on the cell and thereby lower cell temperature. The use of the semitransparent coating over the entire cell surface uniformly limits incident radiation and hence reduces cell heat without any temperature gradients. The coating also lowers series cell resistance. The coating may be directly deposited on the cell surface or on the undersurface of a cover plate bonded to the cell.

This type of solar cell thermal protection is particularly applicable to cells used in close-solar-approach space missions.

A mode of construction developed by *P.F. Varadi; U.S. Patent 4,056,405; November 1, 1977; assigned to Solarex Corporation* provides a panel for mounting solar energy cells, and particularly those cells upon which light is to be concentrated, which includes an enclosure for holding the cells and has at least one wall formed from a good conductor of heat. The cells are mounted within the enclosure on a resinous cushion that is a relatively good conductor of heat and a poor conductor of electricity, so that when heat is generated by impingement of light on the cells, it will be carried by the cushion to the enclosure wall and dissipated therefrom.

*W.E. Horne; U.S. Patent 4,097,309; June 27, 1978; assigned to The Boeing Company* has developed a type of solar cell construction which contains thermal isolation shields consisting of two glass slides separated by insulating standoffs which are positioned upon the front radiation-receiving surface of a solar cell and/or upon the back surface of the solar cell. One of the two glass plates is made from material selected to absorb and radiate electromagnetic wave energy with a wavelength above 5 microns to prevent overheating of the cell. The space between the two cover plates forms a thermal gap that is, if desired, bridged by a bimetallic strip. The strip is adhered to one of the plates and has a reverse bend to extend along the face surface of the opposed cover plate. The strip distorts under the increased temperature to break the bridge between the two plates and thereby isolates the solar cell from the thermal shield formed by the outer cover plate until there is a sufficient reduction in temperature at which the bimetallic strip reestablishes conductive contact between the cover plates.

Each solar cell assembly in an array is adhered to a substrate by first bonding glass pads to the substrate and then heating the cell and glass pads to a temperature of about 400°C while applying an electrical potential of the order of 400 volts with the glass being negative potential with respect to the cell. A hermetic bond is achieved without the use of adhesives.

Such a cell construction is shown in sectional perspective in Figure 90.

**Figure 90: Boeing Co. Thermally Isolated Solar Cell Construction**

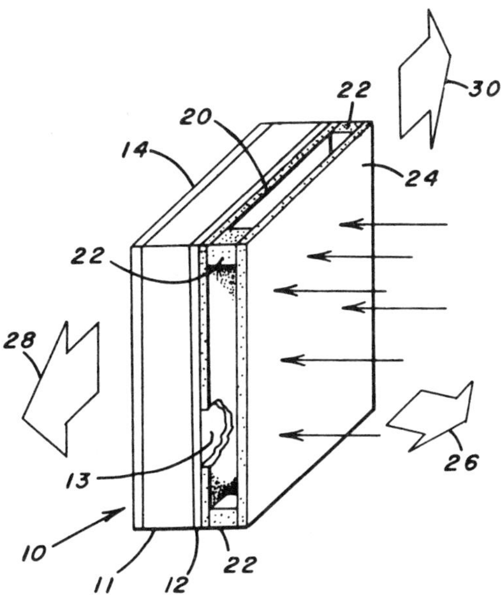


Source: U.S. Patent 4,097,309

Figure 90 (illustrated on the previous page) shows the thermal isolation shield wherein a solar cell **10** is constructed in a manner which is well known in the art and includes a substrate **11** of semiconductive material of one type conductivity, typically silicon, forming a P–N junction with a layer of semiconductive material **12** of the opposite type conductivity. The outer face surface **13** of the cell is adapted to receive incident electromagnetic wave energy for conversion to electrical power by the cell. A soldered interconnection usually extends between the layer of semiconductive material and power distribution network. A second electrode formed, for example, by a deposited layer **14** of aluminum material upon the back-face surface of the substrate is also connected by a soldered interconnection to the power distribution system.

The thermal isolation shield for the solar cell includes a composite of two cover slides separated by thermally insulated standoffs. It is to be understood that the composite of two cover slides may be arranged as shown upon the radiation-receiving surface **13** of the solar cell and/or upon the layer **14** at the back surface of the solar cell. The construction of the thermal isolation shield is the same in either event.

More specifically, an inner cover slide **20** is adhered to the face surface of the solar cell by the method of electrostatic bonding. Thermally insulated standoffs **22** are arranged at the four corners of the cover slide to carry an outer cover slide **24** in a spaced-apart relation with cover slide **20**. The material used to form the standoffs preferably is the same as the material which is used to form the cover slides. The material selected may be any suitable well known form of glass including quartz. In this regard, it will be noted that quartz has a transmission cutoff of incident solar radiation at about 5 microns. Thus, quartz is the particularly suitable material to form at least one of the cover slides since such a cover slide functions effectively as a thermal isolation shield by absorbing long wavelength heat energy (i.e., above 5 microns) and irradiates it preferably back to space (i.e., away from the cell). However, the thermal isolation shield formed by the composite of cover slides **20** and **24** must be transparent to light in the visible spectrum.

Figure 90 further illustrates the energy transfer mechanism provided by the thermal isolation shield. In this regard, the electromagnetic energy from the sun has an intensity of about 0.140 W/cm$^2$ which is incident upon the cover slide **24**. The radiated energy from this cover slide is indicated by arrow **26** by the expression $\epsilon AT^4$. In a similar manner, arrow **28** represents radiated energy from the substrate given by the same expression. Arrow **30** represents the output of electrical energy at a conversion efficiency rate of about 10%.

Thus, it can be seen that the energy absorbed by an array of such solar cells from the sun must be either converted to electrical energy or irradiated back into space via the front and/or back surface of each cell. The spaced-apart arrangement of the cover slides provides a secondary shield for the cell whereby, under normal operating conditions, an increased operating temperature of the cell may be anticipated.

Similar ground is covered by *W.E. Horne; U.S. Patent 4,108,704; August 22, 1978; assigned to The Boeing Company.*

A device developed by *R.M. Graven, A.J. Gorski, W.W. Schertz and J.E.A. Graae; U.S. Patent 4,118,249; October 3, 1978; assigned to the U.S. Department of Energy* consists of a modular assembly of a solar energy concentrator having a photovoltaic energy receiver with passive cooling. Solar cell means are fixedly coupled to a radiant energy concentrator. Tension means bias a large area heat sink against the cell thereby allowing the cell to expand or contract with respect to the heat sink due to differential heat expansion.

## Protection from Radiation

Natural radiation sources of penetrating particles are generally from random directions in space, such as cosmic, solar flare, particle and Van Allen radiation. Electromagnetic radiation, commonly called solar radiation, is predominantly unidirectional from the sun, reflectable, and is of interest here primarily for its visible radiation.

Solar cell radiation shielding is conventionally provided to extend the life of solar cells such as N–P silicon solar cells, which are exposed to the penetrating particle radiation to a degree sufficient to deteriorate the solar cell. Such radiation shielding is customarily provided by coverings of glass or quartz, usually as films or layers deposited on the solar cells. The protective layer provided by glass or quartz on the surface of solar cells allows electromagnetic radiation to pass through somewhat attenuated by the film material, and absorbs the penetrating particle radiation to the limit of the energy which it can contain in the thickness of protective shielding used. Such films pass a high proportion of the visible light, or the energetic electromagnetic radiation, but they do attenuate the desired radiation and they fail to take any advantage of directionality or reflectivity of electromagnetic radiation which is normally utilized as an energy source for solar cells.

It is well known that solar cell materials are subject to substantial degradation under penetrating particle radiation but not under electromagnetic radiation. For this reason, glass or quartz shields, which may be 60 mils in thickness or more, may be bonded to the surface of a 10 to 12 mil thick solar cell to reduce particle radiation damage. Often the bonding materials, such as epoxy resins, are themselves sensitive to radiation in certain wavelengths; and, to protect the bonding materials, additional filter or shield material is often placed upon the glass or quartz shielding to filter out the radiation to which the bonding material is most sensitive, thus further reducing the transmission to the solar cell itself of the desired electromagnetic radiation.

A type of radiation-resistant construction developed by *J.H. Myer; U.S. Patent 3,490,950; January 20, 1970; assigned to Hughes Aircraft Company* takes advantage of the fact that penetrating particle radiation does not reflect substantially from polished or mirror surfaces, and the desired electromagnetic radiation may be so reflected with efficiencies up to about 90%. Although the space environment will reduce reflection efficiency in time, reflection efficiencies in excess of 45% can be maintained over extended periods while subject to penetrating particle radiation damage.

The preferred mode of construction for a radiation-resistant solar cell thus involves a solar cell array having support wall structure arranged to shield cells from direct nuclear radiation and to reflect solar radiation onto the cell.

A process developed by *S. Kaye, L. Garasi and G.P. Rolik; U.S. Patent 3,513,040; May 19, 1970; assigned to Xerox Corporation* is one for fabricating a radiation-resistant solar cell by alloying a graded base region to a low resistivity substrate of semiconductor material of the same conductivity type, the graded base region being substantially thinner than the substrate and increasing in resistivity away from the substrate, and providing a thin region of the opposite conductivity type atop the graded base region.

A solar cell array design developed by *J.F. Wise; U.S. Patent 3,620,847; November 16, 1971; assigned to the U.S. Secretary of the Air Force* is one having the substrate, the cell-supporting grid structure, the electrical connecting leads, the cell contacts, and the terminal connections all fabricated from low atomic number material such as aluminum, beryllium, or magnesium. The cell also incorporates glass, silicon oxide and anodized aluminum insulation, silicone adhesives, and ultrasonic welded electrical connections which provide a hardened silicon solar cell array that is relatively impervious to damage from space nuclear blast radiation.

A cell design developed by *J.D. Broder; U.S. Patent 3,912,540; October 14, 1975; assigned to U.S. National Aeronautics and Space Administration* is one using a transparent plastic film of fluorinated ethylene-propylene copolymer for a binding agent to attach a cover glass. Such a binding material does not degrade in its bonding characteristics under particulate radiation bombardment.

A technique developed by *R.L. Russell; U.S. Patent 3,925,103; December 9, 1975; assigned to TRW Inc.* is one whereby a power-generating solar cell for a spacecraft solar array is hardened against transient response to nuclear radiation while permitting normal operation of a cell in a solar radiation environment by shunting the cell with a second solar cell whose contacts are reversed relative to the power cell to form a cell module. The mode of operation consists of exposing the power cell only to the solar radiation in a solar radiation environment to produce an electrical output at the module terminals, and exposing both cells to the nuclear radiation in a nuclear radiation environment so that the radiation-induced currents generated by the cells suppress one another.

A type of construction developed by *J.D. Broder; U.S. Patent 3,996,067; December 7, 1976; assigned to U.S. National Aeronautics and Space Administration* is one in which a nonoxide antireflective coating (such as silicon nitride) is used with a transparent plastic cover of fluorinated ethylene-propylene copolymer on a silicon solar cell to increase the resistance to damage caused by electron bombardment.

A process developed by *A. Meulenberg, Jr. and J.H. Reynolds; U.S. Patent 4,072,541; February 7, 1978; assigned to Communications Satellite Corporation* is one by which a solar cell is constructed such that the losses resulting from radiation damage are reduced without sacrificing efficiency in the unirradiated cell. This is done by using a first conductivity type junction on the front of a high resistivity cell and a second, opposite conductivity type junction, on the back of the cell. The two junctions are separated by an intrinsic region, and the cell potential, which is normally reduced by using high resistivity substrate material, is instead raised by using the second junction.

The cell includes a reflective back surface and a nonreflective front surface to further enhance light absorption within the active cell volume. The reflective back surface is formed by polishing the substrate. The nonreflective front surface is a multipyramidal or V-grooved surface formed by preferential etching of the substrate material. The front electrode is a fine geometry grid deposited over the front surface of the cell. An antireflection coating may additionally be provided over the front surface of the cell to further promote the nonreflective characteristics of that surface. Further, a cover slide may be cemented to the front surface of the cell.

Normally, solar cell assemblies for extraterrestrial use comprise semiconductor wafers which are covered on their light-incident surface with transparent cover slides to protect the cell from radiation which can damage the semiconductor material. To protect the edges of the wafer from particulate radiation incident at angles other than 90°, prior art cover slides generally are oversize, by approximately 5 mils on a side for a 2 x 2-centimeter (798 x 798-mil) cell. When several cell assemblies are grouped into an array for use on spacecraft, spaces remain between the individual cells which are required for the electrical interconnection mechanisms and are designed to accommodate thermal stresses on the array. Typically, these spaces would be covered substantially by the oversize portion of the cover slide. Thus, a significant amount of light falling on the array falls perpendicularly on the inactive spaces between the cells, thereby resulting in a decrease of the array's overall ability to convert illumination into electrical energy.

An improved type of construction developed by *A. Meulenberg, Jr.; U.S. Patent 4,133,699; January 9, 1979; assigned to Communications Satellite Corporation* is one in which a solar cell cover slide, oversize to protect the cell against damage from radiation, is modified to have an edge portion configured to bend light incident on the edge portion inwardly toward an underlying solar cell. An adhesive fillet may be formed between the undersurface of the overhanging portion of the cover slide and the top edge of the solar cell to facilitate further the transmission of light striking the configured edge portion into the solar cell.

A process developed by *E.S. Rittner; U.S. Patent 4,135,950; January 23, 1979; assigned to Communications Satellite Corporation* is one in which a radiation-hardened junction solar cell is prepared by etching V-grooves in the surface of a semiconductor substrate. The overall thickness of the substrate and the depth of the V-grooves are chosen as functions of the diffusion length of the minority carriers in the substrate material at beginning of life and at end of life for a given design application. An opposite conductivity type layer is then formed over the V-grooved surface of the semiconductor substrate. The planar back surface of the solar cell is highly doped to convert it into a minority carrier-reflecting region.

### Protection from Pollutants and Corrosives

In the past, solar cell panels have been generally satisfactory in clean environments such as in outer space, desert areas, and at high elevations. However, in other environments, and in particular for marine uses, solar panels have had an unexpectedly short life when they have encountered typical degrading conditions, including humidity, salt water, atmospheric pollutants, chemical agents, muds, sulfur compounds, bird droppings, degrading sunlight (both ultraviolet and infrared), and extreme temperature changes.

While the partial use of glass has been suggested, the more modern tendency has been to utilize various plastic materials to enclose the solar cells. However, the various types of solar packages in use have been generally unsuccessful when subjected to adverse conditions, such as marine use, because of various modes of degradation. The most common mode of failure is the separation occurring between the various layers of the solar panel resulting in excessive light reflection with a consequent reduction of electrical output and eventual failure of the electrical connections between photovoltaic cells. Failures have been most often due to failure of the panel enclosure rather than failure of the cells, and is caused by the abovementioned degrading agents encountered in hostile environments.

A type of protective construction developed by *W.R. Klein, C.L. Kotila and I.L. Krams; U.S. Patent 4,097,308; June 27, 1978; assigned to Tideland Signal Corporation* consists of a solar cell panel having a top and bottom molded glass plate. The glass plates are advantageous in that: (1) they are relatively impermeable to degrading agents; (2) they do not undergo UV degradation; (3) the glass top cover reduces transmission of degrading ultraviolet radiations which attack potting compounds; (4) glass and the commonly used solar photovoltaic cell materials have a similar thermal expansion property thereby minimizing mechanical stresses; (5) the bottom glass plate permits transmission of unused infrared radiation through the panel resulting in a cooler, more efficient panel; (6) the chemically inert properties of glass gives stability in the presence of corrosive organic and inorganic agents; (7) the surfaces of the glass are easily washed by rain; and, (8) undesirable coatings are easily removed from the glass plates.

A particular feature of the present device is the provision of recesses molded in either the bottom of the top plate or the top of the bottom plate for receiving solar cells whereby the thickness of the compartment is minimized. Solar cells are positioned in each of the recesses and a potting compound having an index of refraction similar to the index of refraction of glass fills the compartment. The provision of the recesses provides the following advantages: (1) uniform thickness of the embedding compound, resulting in more uniform stresses; (2) reduction of potting compound volume with consequent cost savings; (3) reduction in the thickness of the potting compound thereby overcoming the problem of thermal expansion of the potting compound; and, (4) floating of the solar cells in the potting compound reducing point contact stresses. However, the thickness of the compartment is made sufficient to allow the potting compound to withstand temperature changes.

## MASS PRODUCTION TECHNIQUES

The quantities of solar cells required for commercial-scale power generation will be huge, as illustrated by the following calculation:

> The solar energy flux in space is 0.14 W/cm$^2$. On the ground, there is much variation due to the earth's rotation, the atmospheric absorption, and weather conditions. On a clear day at noon, this value is approximately 0.10 W/cm$^2$. The efficiency of current silicon solar cells to convert solar energy into electrical energy is from 8 to 13%. For an efficiency value of 10%, therefore, on the ground solar cells can produce $1.0 \times 10^{-2}$ W/cm$^2$, or equivalently, 260 MW/mi$^2$.

This huge area in turn requires minimizing costs per unit area which can only be done by application of mass production techniques.

Photon Power, Inc., of El Paso, Texas, is developing a chemical spray process which forms cadmium sulfide solar cell panels directly on hot float-glass as it comes out of the glass factory (3).

Photon Power has completed a pilot plant in El Paso for testing the technique. If the process works as well in large-scale production as it has in laboratory tests, the facility will be expanded into a small manufacturing plant which, it is hoped, will be able to produce arrays which can be sold for $2 to $5 per watt by 1980. Low costs are possible because all of the processes in cell manufacturing (cell growth, junction formation, applying contacts, and encapsulation) involve spraying a series of chemical layers onto hot glass moving through the plant on a continuous conveyor. Laboratory devices produced with a similar process have yielded 5.6% efficiencies, and cells with 8 to 10% efficiencies appear possible. Even higher efficiencies may result if the promising results of experiments mixing zinc with cadmium can be integrated into the process.

Photon Power is working with Libby-Owens-Ford (a part-owner of the company) to design a process which can be attached to a float-glass plant. A preliminary calculation indicated that a plant costing about $140 million would be able to produce 2 GWe/yr for $0.05 to $0.15/W. The major technical challenge in increasing output will be finding a way to increase the speed of the spray-application process from the 2 cm/min in the pilot plant by about an order of magnitude to match the rate at which glass is produced from a flat-glass facility. It will also be necessary to determine how much of the cadmium not initially captured on the glass can be recycled (this is important since cadmium is expensive and toxic) and to determine how well the hydrogen chloride, sulfur, and sulfurous acids can be recycled.

The French National Petroleum Company, Compagnie Francaise des Petroles, owns controlling interest in Photon Power. Libby-Owens-Ford and D.H. Baldwin Co. own minority interests (3).

Patcentre International of Hartfordshire, England is working on a spray process very similar to Photon Power's. They report 5.5% efficiency for their laboratory devices, and hope to market a commercial product in 1980.

The process developed by *J.F. Jordan and C. Lampkin; U.S. Patent 3,880,633, April 29, 1975; assigned to D.H. Baldwin Company* is one for making low-cost solar cells on a large-scale basis by means of a continuous process of fabricating float glass and coating the float glass, in sequence, with tin oxide, cadmium sulfide, and copper sulfide, while the glass floats atop tanks of molten material in a furnace of proper temperature for each step of the process.

The application of the coatings, in a preferred embodiment, is accomplished by depositing materials which form the coatings on contact with heated surfaces at such slow rates and, in the case of spray application, via drops of such uniformity that the float glass may (1) remain at uniform temperatures by virtue of the superior thermal conductivity of the molten material and retain those temperatures despite the abstraction of heat from the glass by evaporation of liquids and/or formation of crystalline layers; and, (2) be substantially free of temperature gradients along the surface of the sheet glass.

Figure 91 (on the following page) shows a perspective view of the construction of such a cell. As shown there, **10** is a plate of double-strength window glass, one-eighth inch thick, which serves as a substrate for the present large-scale mass-produced solar cells. On the window glass substrate has been deposited a layer **14** of $SnO_x$, constituting a negative electrode **12** common to the entire substrate, and the x indicating that the precise composition is not known. On the layer **14** is deposited a layer **15** of crystalline CdS, which is about 2-micron thick. Superposed on layer **15** is a layer **16** of $Cu_2S$, which, after suitable heat treatment, serves to form a heterojunction with the CdS. On the layer of $Cu_2S$ is superposed a layer **17** of copper, which serves as a positive terminal. The layers **15, 16** and **17** may be etched through to the $SnO_x$ layer **14** at intervals to provide channels in which may be deposited buses of Inconel or chrome **18**, and superposed aluminum **19**, which provide negative terminals having multiple areas of contact with the $SnO_x$.

**Figure 91: Cell of Type Continuously Produced on Glass Sheet**

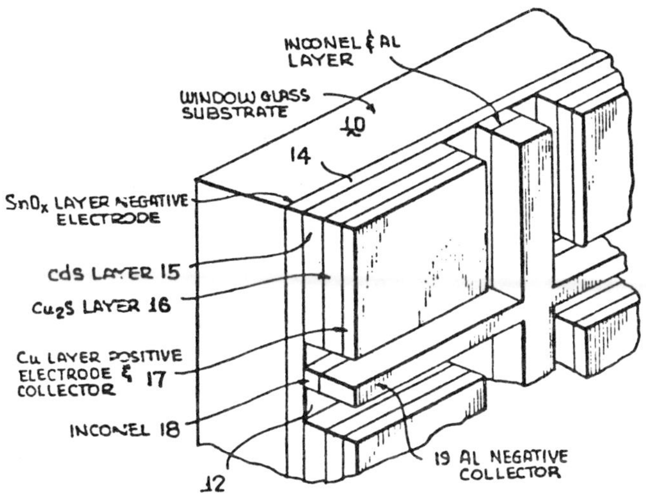


Source: U.S. Patent 3,880,633

The float-glass process itself is shown in Figure 92, on the following page. The raw materials **20** are continuously fed to a furnace **21** and melted. The resulting liquid is poured into a bath of molten tin **22**, where, by controlling the feed rate of materials and the velocity of the floating-glass ribbon, its thickness can be controlled to ±0.002 inch. The glass is one-eighth inch thick (so-called double-strength window glass). Glass is fed as a continuous ribbon 10 feet wide from the liquid tin bath to a cooling or annealing lehr **23**, where it is cooled gradually, before being automatically cut to size. A typical annual output for such a plant would be 200 million square feet per year.

Figure 93 shows the adaptation of the basic float-glass plant to solar cell manufacture by the present process. As shown, there are interposed three or more chambers. The first of these, **25**, will be used for spraying a solution of $SnCl_2$ and reactants and dopants to provide a low-resistance transparent conducting $SnO_x$ layer. In the chamber it is necessary to maintain the top glass surface temperature equal to the tin temperature, though the impingement of the spray causes a momentary temperature decrease in the glass. The second chamber **26** will be used to spray a CdS film, and the third **27** in the cooling line at a much lower temperature for spraying a $Cu_2S$ layer.

After further cooling, the glass will be cut automatically to panel size by cutters **28**. After cutting the glass into panels, these will then be treated with resist through a suitable screen, leaving exposed the areas required for channels. These areas are then etched out at station **31**, then washed at station **32**. Inconel or chromium is evaporated onto the exposed glass in the required pattern, followed

Figure 92: A Float-Glass Plant

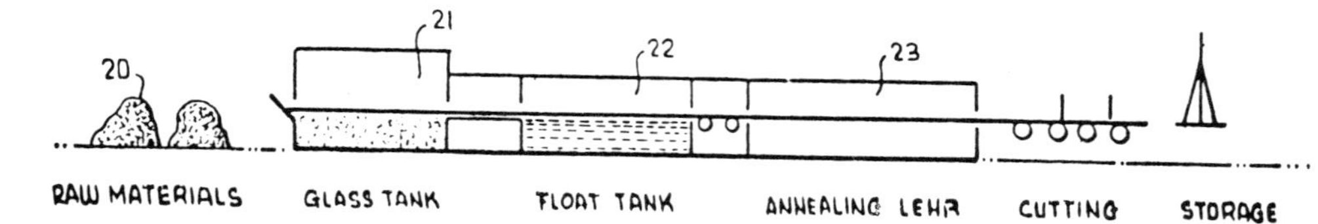


Figure 93: Adaptation of Float-Glass Plant to Continuous Solar Cell Manufacture

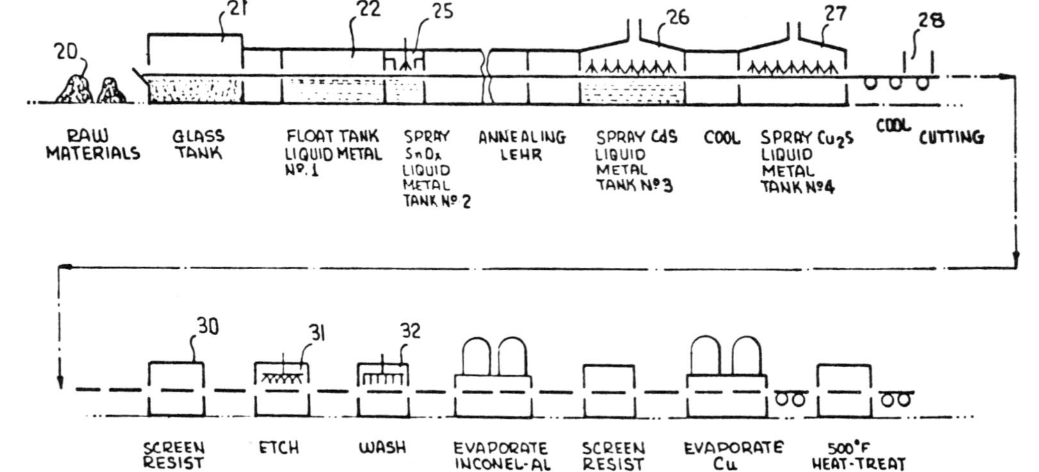


Source: U.S. Patent 3,880,633

by an evaporation of Al to form the negative collector. Cu is then evaporated over the $Cu_2S$ layer not etched at station **31** to form the positive collector and the entire panel slowly heated to 500°C and then slowly cooled. Upon cooling the panel is ready for installation in a macrosystem involving square miles of such panels interconnected among themselves and to suitable power transmission equipment.

Figure 94 shows a pilot plant for the application of this solar cell manufacturing process to prefabricated panels of window glass. In the figure, **40** represents storage of panels of glass, which may not be of the desired dimensions. The panels provided may be precut to size at station **41**, and cleaned at station **42**, after which they are fed to a first elongated furnace **43**, which raises the temperature of the panels from an assumed temperature of 70°F, induced in the cleaning station to 700°F, and thereafter in a second furnace **44**, which raises the temperature of the panels to approximately 950°F. Following furnace **44** is a float station **45** for spraying the panel with $SnO_X$.

The coated glass panel now proceeds through an annealing lehr **46**, in which the temperature of the panel is slowly reduced to 700°F, so that the glass is annealed when it arrives at a float chamber **47**. The float chamber contains a vat of liquid tin, in which the panel floats, and the liquid tin is maintained at 620°F, so that the panel is maintained uniformly over its area at this temperature, while a water solution capable of providing CdS and other elements is slowly deposited by spraying. The spraying process is carried out by spraying in a moving pattern, covering a small area at a time, so that heat will not be abstracted at a substantial rate from the panel in response to evaporation of water or formation of CdS crystals, whereby the temperature of the panel can be maintained uniform.

As the panel proceeds along the float chamber **47**, successive small areas extending transversely of the panel are coated by moving the nozzles transiently of the panel as it moves, until the entire panel is completely coated to the required thickness. Many spray nozzles may be employed, and the surfaces continue to move during spraying of the glass until the required thickness of CdS crystals have been uniformly deposited. The length of the float chamber must be adequate to provide adequate spray times, giving consideration to the number of nozzles employed and the speed of travel of the glass.

While the CdS is coated on the panel, the float material for the latter is maintained at 620°F. Coating of a layer of $Cu_2S$ on the CdS layer must be carried out at 300°F. Accordingly, the panel is cooled to nearly 300°F in chamber **48** and from the chamber the panel is introduced into a metal float chamber which maintains the panel at 300°F while it is being sprayed with a solution which produces $Cu_2S$, in a slow-scanning process essentially like that described as taking place in chamber **47**.

After the panel has been coated with $Cu_2S$ to the requisite thickness, the panel is slowly cooled to 70°F, in chamber **50**, and from that chamber it proceeds to

Figure 94: Pilot Plant for Continuous Solar Cell Manufacture from Prefabricated Panels of Window Glass

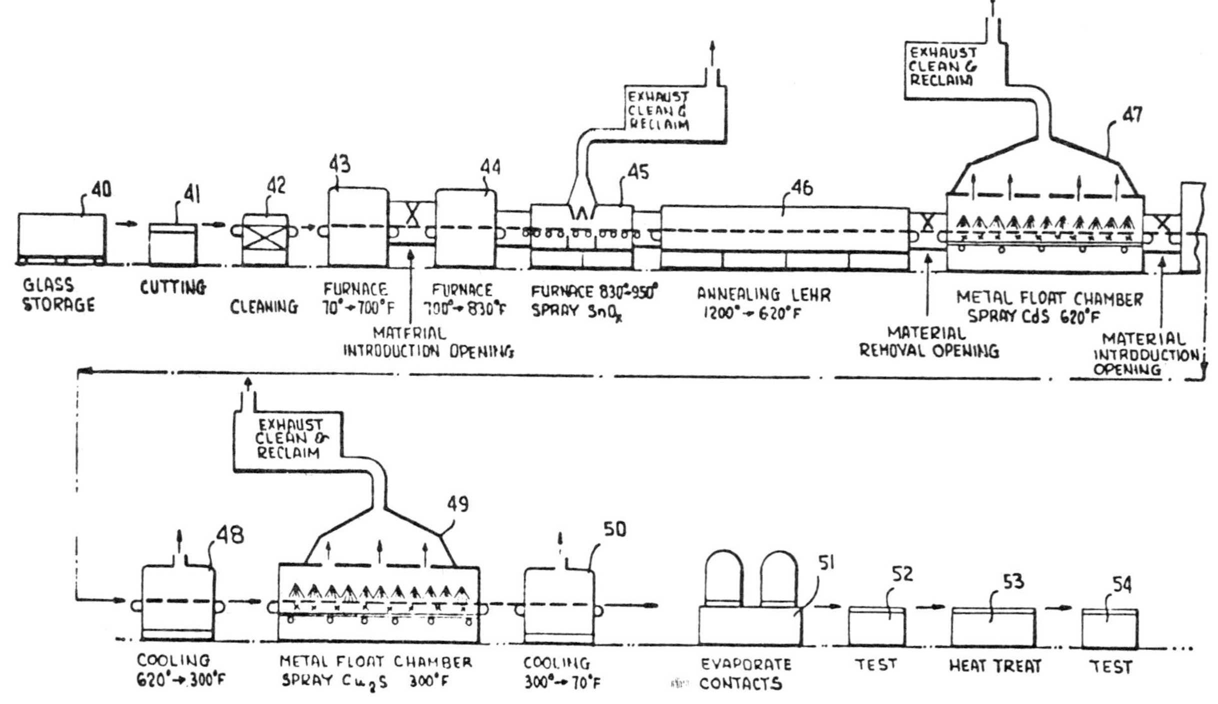


Source: U.S. Patent 3,880,633

station **51**, in which the requisite positive and negative contact areas are deposited, to station **52**, where the completed panel is tested, to station **53** where it is slowly heated to 500°F in a curing process, then slowly recooled, and thence to station **54**, where it is finally tested prior to delivery.

A later coverage of this process is presented by *J.F. Jordan and C.M. Lampkin; U.S. Patent 4,086,101; April 25, 1978; assigned to Photon Power, Inc.*

It describes in detail a method of making low-cost photovoltaic cells on a large-scale basis by means of a continuous process of coating sheet glass while the sheet glass moves in and has its under surface immersed in a tank of molten material, forming the film of CdS microcrystals on the glass sheet which has previously been coated with transparent $SnO_x$ to a thickness of about 0.3 to 0.6 micron. A water solution of a cadmium salt, a sulfur compound, and an aluminum-containing soluble compound is intermittently sprayed on the glass while its exposed surface is maintained at a constant temperature in the range of 500° to 1100°F and while irradiating the surface with intense ultraviolet light so as to form a film of CdS microcrystals. This CdS film has Al impregnated within the stratum of the film adjacent to the $SnO_x$, but only optionally has Al impregnated in the stratum of the CdS film adjacent to the exposed surface of the CdS film.

After the spray process is completed, the temperature of the sheet of glass is brought to the range of 450° to 550°C, 525°C being optimum. After heating, the glass is cooled to approximately room temperature and the exposed surface of the CdS is converted to $Cu_xS$, the x being as close to 2 as possible, by dipping the coated glass in a solution comprising: a solvent, which may be water; a weak acid, such as tartaric acid, citric acid, or lactic acid; a copper-containing compound; and, optionally, a quantity of $H_2Ce(SO_4)_4$ and NaCl, or some other chloride; or by electroplating to form a film of $Cu_xS$; or, by a combination of dipping and electroplating. The $Cu_xS$-forming process proceeds by ion exchange, i.e., S from CdS combines with Cu to form $Cu_xS$. Cu is then applied over the $Cu_xS$, and the cell is cured at a temperature in the range of 400° to 500°F.

A process developed by *P.-H. Fang; U.S. Patent 3,914,856; October 28, 1975* is one for economically and rapidly producing solar cells, made of either silicon or of germanium. Complete solar cells are deposited in a carrier substrate by processes geared to large-scale production at high production rates not obtainable from the present state of the art.

A basic component of the system is a carrier substrate to support the fragile layer of silicon film. There are two stringent requirements for this carrier substrate: a matching of the thermal expansion coefficient with that of silicon; and, economy in cost.

One possibility involves the use of a nickel-iron alloy, commonly referred to as a glass-sealing alloy, with the major composition of 42% nickel and 58% iron. An alternate substrate is polyimide with a trade name Kapton. This material is more stable, but the thermal expansion match is poor, and the cost is considerably more.

The basic steps and configurations to produce the solar cell will be outlined in a format of eight adjacent stations:

(1) Start with a roll of flexible substrate which is unrolled with a conveyor belt arrangement and fed into station (i) where on the substrate will be evaporated a layer of silver and, on top, a layer of titanium, or by a single aluminum layer to form an electrode with an atomic contact to silicon to be deposited.

(2) Continue on to station (ii), a silicon evaporator source with proper impurity (either N- or P-type) and concentration. This source will have one of the following forms: a chemical solution as used in the epitaxial evaporation; a solid for electron-beam evaporation; or, an electrode plate for ion sputtering. In this station, a silicon layer in the thickness of about 10 microns is grown.

(3) Next, the silicon film with the substrate will be fed in a continuous fashion from station (ii) through (iii) into (iv), where either a high voltage ion accelerator will implant, or a thermal process will diffuse, ions of opposite type of impurity from that of original silicon, so that a P–N junction would be formed.

(4) At the end of steps (3) and (4), temperature stations (iii) and (v) should be inserted to perform annealing. The temperature of these stations will be kept at about 500°C and annealing time will be about 10 minutes.

(5) Station (vi) is to evaporate grid lines as in the ordinary solar cell for the front electrodes.

(6) Station (vii) is to perform a thermal-compression binding or an electrode-forming process to attach electrical lead wires.

(7) Station (viii) is $SiO_x$ or another type of coating which performs dual purposes: antireflection coating; and a protection of grid from environmental corrosion.

(8) In addition, one can spray on a layer of plastic coating such as clean Teflon. This is a supplement to $SiO_x$ for protection on the front and back of the cell, if this is necessary.

(9) The continuous roll of substrate fed in station (i) will come out of station (viii) as a roll of completed solar cell sheets. These sheets have intermediate connection leads which could be designed for either a series or a parallel connection to produce the desired voltage and current.

These steps are by no means exclusive nor inclusive. Some steps could be combined into a single step, e.g., the thermal annealing steps can be combined into a single step, or, several additional steps could be considered:

(a) Introducing a silicon oxide layer to the substrate before the metallic electrode evaporation. This oxide layer serves three purposes: as an electrical insulation from the substrate; to reduce the diffusion between the substrate material and the semiconductor layer; and, as a better-matched substrate for growing silicon films.

(b) When the titanium-silver alloy is used as a base electrode, if sufficiently high temperature is applied, titanium could act as a gettering agent to remove the harmful impurities from the semiconductor layer.

In resume, the system consists of a large deposition and annealing system with locking facility between different stations to prevent contamination. The external contamination and mechanical strain and stress is kept to a minimum because the product, until completed, will not be exposed to an environment outside of the closed system.

A process developed by *K.-C. Chang, A. Heller and B. Miller; U.S. Patent 4,082,602; April 4, 1978; assigned to Bell Telephone Laboratories, Inc.* is applicable to production line quality control in solar cell manufacture.

This quality control technique is based on the observation that trapping centers contribute to inefficient operation of junction devices. Devices are irradiated by a first radiation source of intensity sufficient to populate traps and a second radiation source of varying wavelength. Trapping centers are detected by a deviation from the expected photovoltaic output-incident wavelength relationship.

The measurements can be translated into changes in the processing parameters, such as etching solution or time, that compensate for changes in the etchants and for other and unforeseen circumstances that affect the quality of the junction devices produced. The translation of the monitoring measurements into changes in the processing parameters may be either manual or automatic.

A mode of construction, developed by *W.J. King; U.S. Patent 4,086,102, April 25, 1978* is one in which a single protective layer (e.g., of alumina) acts as an antireflection coating and an encapsulation. In cases where the junction is formed by ion-implantation techniques, the same layer also serves as the implantation oxide. In addition, this multipurpose layer may also serve as a mass analyzer, allowing the desired species of ions to reach the surface of the semiconductor but blocking the heavier undesired species. The necessary contacts may be formed prior to implantation, and the use of alloyed aluminum contacts with aluminum oxide passivation permits a simplified contacting procedure. A fully automatic contacting method is described by King.

A process developed by *P.L. Shah and C.R. Fuller; U.S. Patent 4,101,351; July 18, 1978; assigned to Texas Instruments Incorporated* is based on an alternate technique of junction formation using arsenic, for example, as dopant. The process is uniquely different in the fact that it simplifies the number of process steps by using the doped oxide for junction formation, then as a metallization mask, and then as an antireflection surface layer. The process has some advantages over the conventional process.

The deposited dopant layer, either vapor-deposited or spun-on polymer doped oxide, acts as dopant source for junction formation.

The technique can be used to form $N^+$ and $P^+$ layers simultaneously by deposition or spin-on application of N and P doped oxides on each side of the cell. This will result in a $N^+$-$PP^+$, or $P^+$-$NN^+$, structure depending on N or P type substrate used. These structures are more desirable for the solar cell performance since they allow simpler metal systems and yield higher efficiencies through improved open-circuit voltages.

The doped glasses or polymer oxide layers lend themselves to the use of arsenic as a dopant in addition to boron or phosphorus. Arsenic is known to have better lattice match to silicon as compared to phosphorus, as is obvious from the atomic radii of silicon, arsenic and phosphorus, 1.176, 1.18 and 1.07 A respectively. Arsenic is also mostly substitutional and diffuses more controllably as compared to phosphorus which is known to diffuse with an anomalous diffusion constant especially at lower temperatures forming tails and kinks in the impurity distributions. Phosphorus top layer deposited at high temperature is known to yield in a "dead" layer due to high concentration of electrically inactive phosphorus. This dead layer causes drastic photoresponse degradation. Use of arsenic alleviates all these difficulties. Doped oxides and polymer-based dopants allow such arsenic junction formations without need for using toxic arsenic compounds.

The doped glass or oxide can be selectively patterned to expose the silicon surface for subsequent metallization step. The oxide or the glass over the rest of the active area will protect the silicon surface during contact formation, preventing degradation from metal removal or metal-removal etching.

Since the doped oxides have an index of refraction intermediate to that of air and silicon, the thickness of the oxide can be adjusted to form a single layer antireflection coating. This antireflection coating will be in situ and is obtained without use of expensive evaporation techniques.

The top layer is deposited and formed on the clean starting silicon in the beginning of the process. This layer protects the surface throughout the subsequent operations, preserving high lifetimes and surface recombination velocity—the two most important solar cell parameters required for efficient collection of photoexcited carriers.

This process eliminates or minimizes the expensive photolithographic and vacuum process steps. With a protective patterned oxide in front the metallization can be achieved by simple, high-throughput-nonvacuum process, such as electroless metal deposition technique. The protective oxide can alternately be formed by anodic oxidation which can be used as antireflection coating and protection mask for metallization. This can be used on a diffused junction formed either by doped oxide or conventional vapor-phase technique. This technique can then be used in conjunction with phosphorus for vacuumless metallization and in situ antireflection coating. Figure 95 shows the application of the improved simple patterning technique along with electroless metallization to give an inexpensive high-volume process.

A process developed by *R.L. Mueller and R.K. Yasui; U.S. Patent 4,133,697; January 9, 1979; assigned to U.S. National Aeronautics and Space Administration* is designed to produce a flexible solar array strip by a method which readily lends itself to automated production. Such a strip may be economically provided for general usage in converting solar flux to electrical energy.

The process consists of: providing printed circuitry in a sandwiched relation with a pair of flexible layers of nonconductive material; depositing solder pads on the printed circuitry and storing the resulting substrate on a drum; withdrawing the substrate from the drum and incrementally advancing it along a linear path; serially transporting solderless solar cells into engagement with the pads and thereafter heating the pads for thus attaching the cells to the circuitry; cleaning excess flux from the solar cells; encapsulating the cells in a protective coating; and, thereafter spirally winding the resulting array on a drum.

A device developed by *E.N. Costogue, R.G. Downing, O. Middleton, R.L. Mueller, R.K. Yasui, F.J. Cairo and J.K. Person; U.S. Patent 4,149,665; April 17, 1979; assigned to U.S. National Aeronautics and Space Administration* is a machine for attaching solar cells to a flexible substrate having printed circuitry deposited thereon.

The strip is fed through a first station in which solar cells are elevated into engagement with solder pads for the printed circuitry and, thereafter, heated by an infrared lamp; a second station at which flux and solder residue is removed; a third station at which electrical performance of the soldered cells is determined; a fourth station at which an encapsulating resin is deposited on the cells; a fifth station at which the encapsulated solar cells are examined for electrical performance; and, a final station where the resulting array is wound on a takeup drum.

A process developed by *C. Deminet, W.E. Horne and R.E. Oettel; U.S. Patent 4,152,535; May 1, 1979; assigned to The Boeing Company* is a continuous process for fabricating solar cells.

The process consists of: forming a substrate for the solar cell; forming a first electrode which is supported by the substrate; depositing a layer of small-grain semiconductor material, which is originally doped N or P, in such a manner

**Figure 95: Texas Instruments Process for Fabricating Inexpensive High Performance Solar Cells Using Doped Oxide Junction and In Situ Antireflection Coatings**

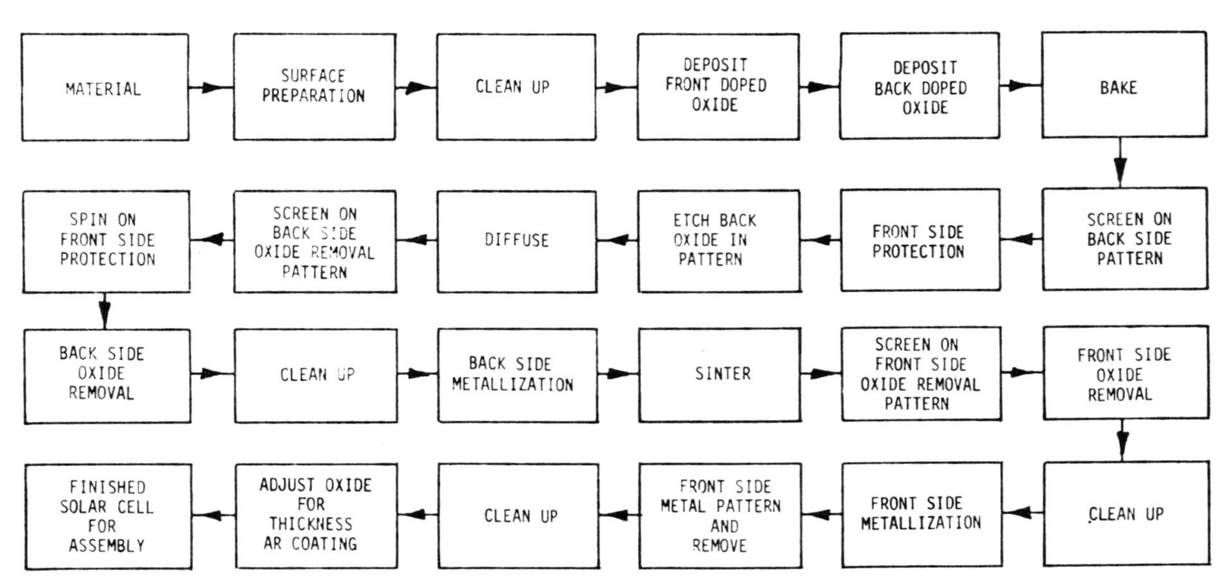


Note: AR = antireflection

Source: U.S. Patent 4,101,351

that a first surface of the semiconductor layer is in electrical contact with the first electrode; recrystallizing the semiconductor material to increase its grain size to the point where the semiconductor layer may be used as the active element in a solar cell, while simultaneously removing substantially all electrically conducting impurities from the semiconductor material; doping the upper portion of the semiconductor layer opposite to the original doping, so as to form a P-N junction in the semiconductor layer; and, forming a second electrode in such a manner that it is in electrical contact with a second surface of the recrystallized semiconductor layer, the solar cell produced thereby being capable of providing a current output in response to light impinging on the recrystallized semiconductor layer.

# Solar Cell Testing

Solar cell arrays are usually assembled first in groups of cells in parallel. These groups are then connected in series to obtain the required output voltage of a string. The detection of faulty cells or connections in an assembled string or larger submodule or array has been difficult. Measuring the electrical resistance of each group requires repeated probing and is not a positive test for the open circuit failure mode because the combined resistance of the remaining cells in parallel may be approximately the same as that of a satisfactory group.

Stimulation of the entire module and measurement of electrical output, then proceeding to group-by-group trouble-shooting, requires skillful investigative techniques which are even more necessary in determining which cell within a group is discrepant. If there is no visible evidence of damage, careful thermal techniques or trial and error methods become necessary to locate the faulty cell.

A technique developed by *C.W. Cable; U.S. Patent 3,630,627; December 23, 1971; assigned to U.S. National Aeronautics and Space Administration* includes a means for illuminating an electrically interconnected group of solar cells, means for applying a differential illumination to selected cells in the group and electrical measuring means connected to the group for determining an electrical output parameter thereof. The differential illumination may be conducted by shadowing or masking at least one selected cell or by applying a local higher intensity illumination to the selected cells.

The test procedure is conducted during illuminating individual or groups of cells at an intensity level sufficient to provide a significant electrical output from the cells. An electrical parameter of the output of the group is measured as selected cells are differentially illuminated. Using this technique, a single open circuit cell or a shorted group may be readily identified without detailed probing or trouble-shooting. Usually only one electrical hookup is necessary.

There is a need to qualify P-N junction photovoltaic solar cells as to their ability to convert solar energy into electricity in a manner that may be automated. It has been shown that if a solar cell is subjected alternately to high frequency (blue) and low frequency (red) light pulses, the minority carrier lifetimes of the diffused N layer and of the bulk P material can be measured. Furthermore, the bulk diffusion length can also be ascertained. Since these parameters determine the efficiency of a solar cell, an object of a testing technique is to provide a method and apparatus for quick and easy quality control of solar cell production.

A method has been developed by *O.H. von Roos; U.S. Patent 4,122,383; October 24, 1978; assigned to U.S. National Aeronautics and Space Administration* in which carrier lifetimes and bulk diffusion length are qualitatively measured as a means for qualification of a P-N junction photovoltaic solar cell by alternately applying high frequency (blue) monochromatic light pulses and low frequency (red) monochromatic light pulses to the cell while it is irradiated by light from a solar simulator, and synchronously displaying the derivative of the output voltage of the cell on an oscilloscope.

This output voltage is a measure of the lifetimes of the minority carriers (holes) in the diffused N layer and majority carriers (electrons) in the bulk P material, and of the diffusion length of the bulk silicon. By connecting a reference cell in this manner with a test cell to be tested in reverse parallel, the display of a test cell that matches the reference cell will be a substantially zero output.

# Solar Cell Enhancement

The contribution of photovoltaic cell costs to the overall cost of an installed photovoltaic system can be greatly reduced if an optical system is used to concentrate sunlight on the cell, even though cells designed for use in concentrators may cost more per unit area of cell surface than flat-plate cells. If such systems are used, problems of reducing cell fabrication costs are replaced with problems of mechanical engineering. In most cases, the energy required to manufacture a concentrator array is many times lower than the energy required to manufacture a flat-plate cell array with a similar area (3).

The thermophotovoltaic cells shown in Figure 96 may be able to achieve efficiencies as high as 30 to 50% by making an end-run around the fundamental limits on cell performance. This is accomplished by shifting the spectrum of light reaching the cell to a range where most of the photons are close to the minimum excitation threshold for silicon cells. The sun's energy is used to heat a thermal mass to 1800°C (the effective black body temperature of the sun is approximately 5700°C). A large fraction of the surface area of this mass radiates energy to a silicon photovoltaic device. (Reradiation to the environment can occur only through the small aperture where the sunlight enters.) A highly reflective surface behind the photovoltaic cell reflects unabsorbed photons back to the radiating mass and their energy is thus preserved in the system.

A device developed by *A.D. Adler; U.S. Patent 3,935,031; January 27, 1976; assigned to New England Institute, Inc.* is based on the discovery that when an exposed semiconductor surface of a photovoltaic cell is coated with a layer, ranging in thickness from a monomolecular layer to a thickness of up to 1 micron, of any one of a vast number of porphyrins, the photovoltage and the photocurrent of the cell are markedly enhanced.

**Figure 96: Thermophotovoltaic Converter**

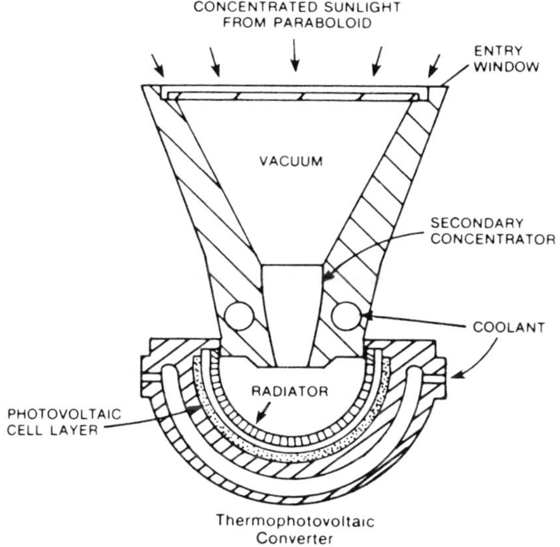


Source: Reference (3)

In the past it was not the practice in making solar cells of materials like silicon to use thin layers of active material because it resulted in a severe loss in solar cell light absorption.

The loss of light absorption is due to the fact that for the active material, typically silicon, the longer the wavelength of the incident light, the weaker the active material's light absorption. If absorption is weak, the incident light will penetrate through the active material and not be absorbed. To assure absorption of that portion of the solar spectrum consisting of long wavelength radiation, a thick layer of active material was needed. Once the light has been absorbed in the active material, an electron-hole pair is generated. To assure that the lifetime of the generated minority carrier, be it a hole or an electron, was sufficient for the generation of current, the active material had to be of a high quality. In other words, the lifetime of the generated carrier is a dominant factor in determining the diffusion length of carriers generated in the active material. Diffusion length is the average distance a carrier can travel before it recombines. Since a thick layer of active material was required for absorption, a diffusion length at least as great as the active layer thickness was needed to assure current generation by the carriers formed deep in the active layer. Therefore, in the past, a solar cell consisted of a thick layer of highly pure active material.

If a thinner layer of active material could be used, a lower-cost solar cell would result for two reasons. First, a thin layer of active material would greatly reduce the required amount of active material, which is costly. Second, a thin layer of active material would require a proportionally shorter diffusion length for optically generated electrons and holes. The ability to use shorter diffusion lengths would allow the use of active material of lower quality, an additional cost saving. Therefore, a much desired need is a solar cell with a thin layer of moderate purity active material with little loss in solar light absorption.

*D. Redfield; U.S. Patent 3,973,994; August 10, 1976; assigned to RCA Corporation* provides such a solution, including a transparent substrate having a pair of opposed surfaces. An active layer of semiconductor material having a surface of incidence is on one surface of the substrate. Grooves are in the other surface of the substrate and a reflective layer is on the surface of the grooves. Figure 97 shows such a device in schematic form.

**Figure 97: RCA Solar Cell with Grooved Surface to Improve Energy Absorption**

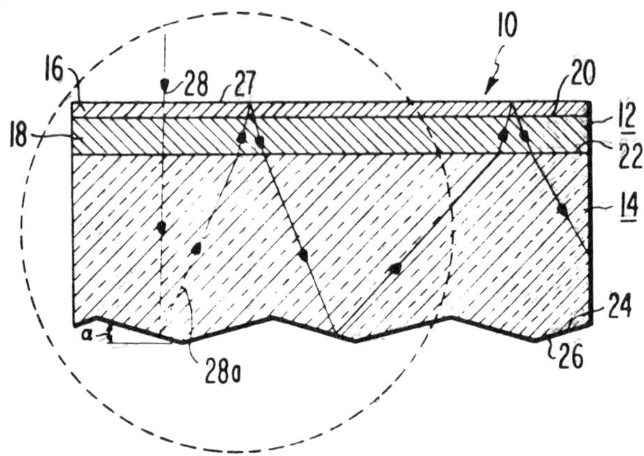


Source: U.S. Patent 3,973,994

Referring to the figure, the solar cell **10** includes a thin layer of active material **12** on a substrate **14**. The active layer is a semiconductor material, such as silicon, with a light incidence surface **27**, an N-type region **16**, a P-type region **18**, and a P-N junction **20** therebetween. Region **16** can as well be of P-type and region **18** can be of N-type with no essential change in results.

The substrate **14** is of any material which is transparent to light. For example, the substrate can be of sapphire, spinel, high temperature glass or quartz. If the substrate is of sapphire or spinel, it has the advantage of permitting the epitaxial growth of the active layer **12** thereon.

The substrate has two opposing surfaces, which are the flat surface **22** and the grooved surface **24**. The active layer is on the flat surface. The grooved surface is in the form of a polished sawtooth pattern with the surface of each sawtooth at an angle $\alpha$. On the grooved surface is a reflective layer **26**. The reflective layer is typically of a metallic material with good light reflection characteristics.

The solar light ray **28**, falling on the active layer at incident surface **27**, optically generates electrons and holes in the active layer at the points where the light ray is absorbed. A characteristic of silicon, which is typically of what the active layer comprises, is a weak absorption for low-frequency radiation. The spectrum of solar light comprises colors of varying frequencies, many of which are of a sufficiently low frequency that their radiation will penetrate deeply into an active layer of such a solar cell before it will be absorbed. A thick active layer is needed to assure total absorption of solar light.

A typical solar cell has an active layer thickness on the order of 250 microns, but in the present device the active layer is 2.5 microns or less in thickness. The ability to use this reduction in active layer thickness is the result of the grooved surface and reflective layer.

In the present device, portions of the light ray having low frequencies will tend to penetrate the active layer without being absorbed. However, they will not penetrate the reflective layer and are reflected back into the active layer of the solar cell. By the proper choice of the angle $\alpha$, the reflected ray **28a** can be made to undergo total internal reflection upon reaching the incident surface. This retains the reflected ray within the solar cell, giving it multiple opportunities to be absorbed by the active layer.

A type of cell construction developed by *C.M. Lampkin; U.S. Patent 3,971,672; July 27, 1976; assigned to D.H. Baldwin Company* is one which addresses the problem of nonabsorption of solar energy in areas where electrodes are located. These electrodes may be located in channels on the back side of the cells. Light which falls on the channels is wasted, and this reduces output per square of glass, everything else being equal. About 15% of light flux is wasted in a typical prior-art design. The latter waste can be in major part avoided according to the present process by providing a diffusing surface superposed over the channels, and such surfaces can readily be formed by etching or sand-blasting the glass surface where diffusion is required. The diffusing surface refracts about 75% of the sunlight, which would otherwise fall on the inactive areas of the cell, to the active areas.

Such a construction is shown in Figure 98, on the following page. As shown there, **10** is a sheet of glass 3 mm thick, having on its underside arrays of photovoltaic cells **11**. The entire underside is coated with a thin transparent layer **12** of tin oxide, $SnO_x$, where x indicates that the precise relation of oxygen to tin is unknown, but where x is about 1.9. Coated on the tin oxide are elongated strips **13** including cadmium sulfide incorporating aluminum, according to the teaching of U.S. Patent 3,880,633. The widths of the strips **13** are **A**, the strips being separated by spaces of width **G**, where **G** is about 1 mm, and where (**G** + **A**)/**A** = 0.85, as exemplary values.

On the cadmium sulfide, which forms microcrystals about 2 microns thick perpendicular to the glass, is a thin layer of cuprous sulfide **14**, which, with the cadmium sulfide, forms a photovoltaic heterojunction. The $SnO_x$ now provides a negative electrode. Coated over the cuprous sulfide is a layer of copper **16**, and over the latter is coated a layer of lead **17**, to protect the copper from oxidation. The lead is inert and does not oxidize, it forms a perfectly adherent seal to the copper, and leads can be readily soldered to it. Such copper leads are **18**. In order to make contact to the tin oxide layer, a strip of Inconel **19** is laid down between the photovoltaic strips, on the tin, and to the Inconel are soldered copper wires **20**.

**Figure 98: Light Diffuser for Photovoltaic Cell**

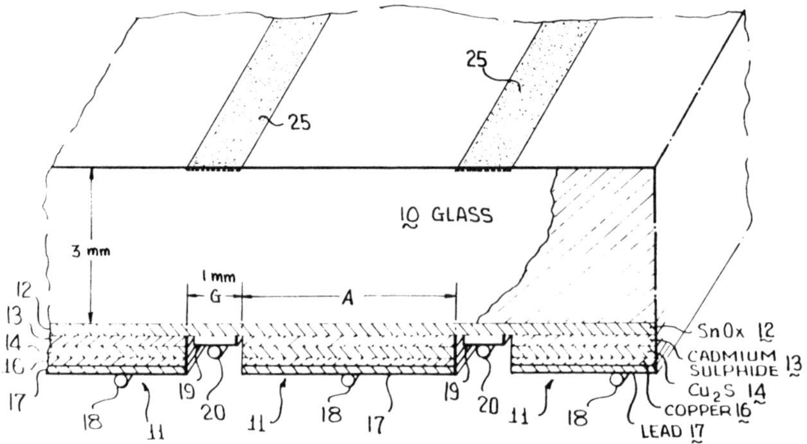


Source: U.S. Patent 3,971,672

Still referring to the above figure, the sun appears to the photovoltaic cells as a point of radiation, and the cells are oriented parallel to the daily path of the sun.

It is now true that about 15% of the area of the glass is not occupied by heterojunctions, since that much area is allocated to negative electrodes. This reduces the possible efficiency of an array of cells, in terms of conversion efficiency of sunlight to electricity, by 15%.

In accordance with the present construction the upper surface of the glass is provided with radiation-diffusing strips **25**, by etching strips of the glass, or by sandblasting, the strips being of width **G**, and being directly superposed over the nonproduction areas of the under surface of the glass. The glass is 3 mm thick and the width of the diffusing strip is 1 mm, and it can be shown that about 75% of the light falling on the diffusing strip will be diverted to the active areas of the cells and about 25% will fall on the inactive spaces despite the diffusers.

If area efficiency is 85% in absence of the diffusers, which is the case for the exemplary dimensions, the inclusion of a diffuser increases area efficiency to about 95%.

Similarly, in solar cell panel constructions, round cells are normally used and this leaves open interstices between the cells. Some of the light that hits these open interstices passes straight on out of the photovoltaic array and is lost for conversion to electricity. Some of the light which hits these interstices is reflected back towards the side of the light-transmitting member through which it initially entered. Such back reflection is at essentially the same angle as initial entry and the back-reflected light therefore escapes out the first side rather than being reflected back into the light-transmitting member. If the interstices are darkened they absorb light and are thereby heated, which can be detrimental to the efficiency of the photovoltaic cells.

This problem has now been addressed in a mode of solar cell construction proposed by *J.W. Yerkes and J.E. Avery; U.S. Patent 4,116,718; September 26, 1978; assigned to Atlantic Richfield Company*. It provides a diffusive member which fills the area between the cells. An array having such a diffusive member will have an efficiency which is higher than an array with the same close packing of cells but no diffuser. An array of this type could employ a smaller number of cells spaced apart over the same area light-transmitting member and achieve greater efficiency per solar cell or per dollar of cost for the total system than a prior-art array using close packing but no diffuser.

Diffusive members can be made from any material which efficiently diffuses or otherwise breaks up visible light into subrays. This includes most any type of rough-surfaced material. For example, etched glass or frosted glass or crinkled, or otherwise uneven-surfaced metal foil such as silver foil, tin foil, etc., can be used. Paint produces a satisfactory diffusive surface. Roughened or uneven surfaces made from anodized aluminum or mirrors, etc., can be used.

## LUMINESCENT AND FLUORESCENT MEANS

It is known that a photovoltaic semiconductor P-N junction can convert to electricity only that portion of the incident photon energy spectrum, typically solar radiation, which creates hole-electron pairs within a given semiconductor material. For example, in a silicon photovoltaic cell only that portion of the solar spectrum with energy in the vicinity of the 1.1 electron volts per photon and which exceeds the band-gap energy of silicon is converted into electricity. Photons of lesser energy do not generate electricity. More energetic photons are strongly absorbed, but much of the energy is lost in heating the cell, which heat can degrade the cell's energy conversion efficiency. To maximize the efficiency of a given photovoltaic cell, it is advantageous to convert as much of the available light as possible into an energy range to which such cell can respond in the generation of electricity before the light strikes the cell's surface.

One technique for achieving such conversion takes advantage of the fact that light falling upon a luminescent agent is characteristically reradiated or emitted in a narrow band of wavelengths of known energy content. Also, light absorbed by such an agent in one direction is reradiated in random directions. Such agents include, for example, pigments such as metal oxides and organic dyes which are used in scintillation counters, lasers, and the like. The term "luminescent agent" includes all types of luminescent agents exhibiting all species of luminescence, including, but not limited to, fluorescence and phosphorescence.

It has been shown that the dispersal of a luminescent agent within an internally reflective sheet of transparent glass or plastic, one of whose major surfaces is exposed to light, concentrates and focuses a flux of light of known energy level toward one or more of the thin upstanding edge faces of the sheet. If a photovoltaic cell responsive to light at that energy level is optically coupled to such edge face, the energy conversion efficiency of the cell increases several times. In this technique a light-transmissive member of such construction and properties is termed a "luminescent member" and a photovoltaic solar collector employing such a member is termed a "luminescent solar collector."

There are, however, problems in finding a luminescent agent or agents which have all of the desired properties for a solar collector application. The optimum luminescent agent absorbs over a very broad range of visible light and emits at a wavelength at or just slightly shorter than the band gap of the photovoltaic cell employed. It is important to the efficiency of the solar device that the luminescent agent not absorb the emitted light at all, or at least very little. It is difficult to find a luminescent agent which covers a broad part of the light spectrum, so in practice a plurality of luminescent agents have been employed. It is also difficult to find a luminescent agent that does not absorb, at least to some extent, in the same light region where that agent also emits light. That is to say, the agent reabsorbs light emitted by another particle of the same agent, and this results in scattering of the emitted radiation more than necessary or desired and the loss of some of this scattered radiation out of the collector or conversion into heat.

When a plurality of luminescent agents are employed in the same collector, the situation becomes even more complicated because it then becomes even more difficult to select groups of agents which interact in just the desired way. Further, beyond the optical requirements, all luminescent agents employed have to be compatible with one another and the materials employed in the member of the collector which contains those agents. For example, the light-transmitting member which contains the luminescent agent or agents must not degrade in sunlight so as to interfere with the operation of the device, such as by developing cracks or crazing which will divert the incident light, render the device less transparent, form peroxides or other decomposition materials which could be harmful to the luminescent agents themselves, and the like.

Accordingly, it is exceedingly difficult to find an overall combination of a number of materials which are optimum for solar collector requirements and still stand up to exposure to sunlight and other weather elements for a matter of years, even decades, without undue degradation. Even if such a combination of materials could be found, they most likely would be very expensive and might even have other objectionable features not yet known.

However, there are known luminescent agents whose properties and peculiarities are sufficiently understood that they can be employed successfully in a solar collector, but these agents may not be as stable as desired and, therefore, do not maintain their efficiency over the large period of years desirable for solar collector installation. Generally, the known luminescent agents will diminish totally or at least in part in their ability to function in the manner desired for solar collectors. Some dyes are known to have a half-life in sunlight of only a matter of weeks, while others may last longer, e.g., one or more years, but they all lose some efficiency or even approach cessation of operation over long periods of time, such as ten years or more, and this causes a gradual and continuing decrease in the efficiency of the solar collector in its ability to generate electricity. Thus, it is very desirable to have a means by which this decreased efficiency is corrected.

Replacement of the entire solar collector is possible, but this is quite expensive and troublesome since such installations can be quite large. It is also wasteful because the photovoltaic cells themselves and other parts of the solar collector assembly have much longer useful lives than the luminescent agents themselves.

A type of solar cell construction developed by *P.E.L.A. Gravisse and M. Prevot; U.S. Patent 3,912,931; October 14, 1975* contains a series of thin luminescent layers of different compositions, which are laid over the surface of the photovoltaic cell, the order of succession and the composition of these layers being selected in such manner that the light energy, in a spectrum zone, falling upon the outermost thin layer is transferred in cascade, through the intermediary of the interposed individual layers, to the spectral sensitivity zone of the photovoltaic cell itself.

Preferably, also, the thin layers will be selected with a sufficient transparency for the usual spectral zone of the photovoltaic cell so that radiation in this zone may reach this cell, with the consequence that the electric output current will be further increased.

A device developed by *A. Goetzberger and W. Greubel; U.S. Patent 4,110,123; August 29, 1978; assigned to Fraunhofer-Gesellschaft zur Forderung der Angewandten Forschung eV, Germany* utilizes fluorescent centers (light concentrators) which consist of thin layers of transparent solid or liquid materials with embedded fluorescent centers and which, in conjunction with solar cells, serve to convert solar energy into electrical energy.

A type of solar cell developed by *R.R. Chambers; U.S. Patent 4,127,425; November 28, 1978; assigned to Atlantic Richfield Company* is one in which the upper and lower surfaces of a luminescent sheet are flared apart or otherwise contoured in the vicinity of at least one of its upstanding edge faces so as to substantially widen the edge face and increase its surface area. At least one photovoltaic cell is carried on the widened edge face for receiving the light flux concentrated thereon.

Figure 99a shows a prior art luminescent solar collector and Figure 99b shows the improved present design.

**Figure 99: Luminescent Solar Collector Designs of Prior Art and Improved Present Design**

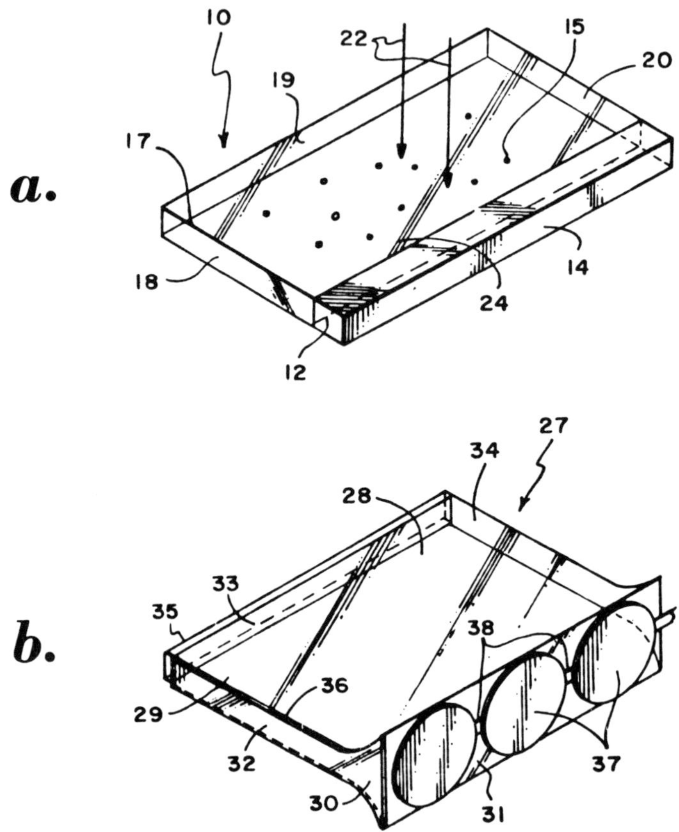


Source: U.S. Patent 4,127,425

Figure 99a shows a prior art luminescent solar collector consisting of a planar luminescent sheet **10**, one upstanding edge face **12** which is substantially covered by solar cell **14**. The sheet is composed of a matrix material impregnated with one or more luminescent agents represented diagrammatically by dots **15**. The bottom surface **17** and the remaining three upstanding edge faces **18, 19** and **20** may be mirrored for high reflectance.

In accordance with the prior art, a ray of sunlight represented by any of arrows **22** (Figure 99a) impinges on the transparent, polished upper surface **24** and into the interior of the luminescent sheet. When the ray **22** impacts a luminescent agent **15**, it is partially absorbed and reemitted in a narrow wavelength band efficiently usable by the photovoltaic cell. This reemitted light is scattered in many directions, which increases the probability of internal reflection. With successive reflections of these reemitted rays, it has been established that an intense flux of light at the desired wavelength is funneled to the edge **12** where it impinges on the photocell and stimulates the desired energy conversion.

The view in Figure 99b then shows the nature of the improved design of solar cell. Upper and lower surfaces **28** and **29** are spread apart adjacent one end of sheet **27** to form an upwardly and downwardly projecting flange portion **30** which defines the significantly widened, upstanding edge face **31**. Edge faces **32, 33** and **34** and lower surface **29** are made suitably reflectant so that all the light incident on upper surface **28** is concentrated and focused in the direction of face **31**. One or more of edge faces **32, 33** and **34** can be covered with a mirror, e.g., mirror **35** on face **33**, or other reflecting medium or with a diffusive medium, e.g., white plastic coating **36** on face **32**, to break up the incident light to maximize internal reflection. Positioned in adherent, contiguous contact with edge face **31** are one or more photocells **37** which, for purposes of illustration, are shown to be of circular configuration. It will be understood, however, that the cells **37** may be shaped by well-known means to further enlarge the percentage of the available area of the edge face **31** covered thereby.

Light in a desired energy band generated by luminescent sheet **27** will now distribute itself over the enlarged surface area of the edge face **31** and the correspondingly enlarged surface area of cells **37**. The cells are thereupon stimulated to provide an electrical output through conventional conductors **38** as shown. In this manner, one may couple the edge-directed light energy of a thin, luminescent sheet into solar cells of significantly greater size than taught by prior-art practice.

For purposes of illustration, the size of flange portion **30** in relation to sheet **27** has been exaggerated. As an example, the sheet is two feet square and one-eighth inch thick. The flange begins one-fourth inch from edge face **31**, or approximately one-hundredth of the length of the edge face. Further in this example, the surface **31** is widened to four times the overall thickness of sheet **27**, i.e., one-half inch, so that its area is approximately four times that of the area of the surface of the edge faces **32, 33** and **34**. Thus, the total area of the cell may also be increased by a factor of four. The enlarged cells will, therefore, tend to operate at a lower temperature, without any loss in power output. Further, conductive elements of any given size and capacity (not shown) on the surface of the cells contiguous with face **31** will clearly occupy a much smaller percentage of such area than would be the case if such cells were placed on an unaltered edge face such as face **33**.

The size and shape of flange **30** and the thickness of enlarged edge face **31** may be varied in accordance with factors such as cell characteristics, cell cost per unit area, expected average light intensity, external cell cooling means, power requirements, physical installation limitations, and the like. Those skilled in this art will have no difficulty in assigning appropriate weight to such factors in designing such a luminescent solar collector.

A cell design developed by *H.R. Blieden; U.S. Patent 4,130,445; December 19, 1978; assigned to Atlantic Richfield Company* is a luminescent photovoltaic device or array which carries in heat-exchange association therewith at least one conduit means adapted to have a cooling fluid passed therethrough to cool the luminescent member.

Even though a larger amount of energy from light is converted into electricity by combining a luminescent member with a photovoltaic means, there is still a substantial amount of light energy which is not converted to electricity and manifests itself essentially as heat in the luminescent member. Unduly elevated temperatures can have a deleterious effect on the photovoltaic cells by reducing the efficiency of the photovoltaic means for the conversion of light energy to electricity. Since the photovoltaic cells are normally in close physical association with the luminescent member that is heated under normal operating conditions, it is highly desirable from the standpoint of long life as well as efficiency of operation of the cells to cool the luminescent member.

Figure 100a, on the following page, shows a luminescent photovoltaic device **1** having an upper planar luminescent member **2** with a plurality of sides **3** through **6**. A plurality of photovoltaic cells **7** is carried on the bottom side **6**, the cells being wired to one another by wires **8** so that electricity can be withdrawn from the device. The bottom side **6** has integrally fixed thereto conduit means **9** which is composed of upstanding sides **10** and **11** and bottom side **12**, the upper side being formed by the luminescent member **2**.

It should be understood that the conduit means can, if desired, have a separate side closing the top, which separate side is in intimate contact, such as by gluing, etc., with the bottom side of the luminescent member, the primary requirement being that the conduit means be in good heat-exchange contact with side **6**. This way, when a cooling fluid such as air, water, glycol, and the like, including mixtures of gas and liquid, is passed into the conduit, as shown by arrow **15**, the fluid will come in contact with large surface area **6** either directly or indirectly through the upper side **18** (Figure 100b) of the conduit means as shown by arrow **16**, thereby picking up heat from the luminescent member and removing that heat from the member, as shown by arrow **17**. If, as shown, the photovoltaic cells are also exposed to the cooling medium of the conduit means, they will be directly cooled by that medium, thereby maintaining them essentially at, or more closely to, the temperature of the cooling fluid itself.

This is better illustrated in Figure 100b, which shows a cross-section of device **1** wherein the photovoltaic cells extend down into the conduit, thereby exposing them to direct cooling. If an upper side is employed for the conduit to physically isolate the cells from the cooling medium or for any other reason, the upper surface can be made to conform to and intimately contact the cells for good indirect cooling of the cells, as shown by dotted line **18**.

The conduit means can be formed from many materials so long as it will channel the desired gaseous and/or liquid cooling medium and is compatible with the material from which the luminescent member and photovoltaic cells are made. The conduit means could be formed from the same material as the luminescent member or a different material, e.g., polymeric, glass, metal, or anything else desired and obvious to one skilled in the art. Of course, thermal insulation could be applied to the conduit means.

**Figure 100: Luminescent Photovoltaic Device with Cooling Means**

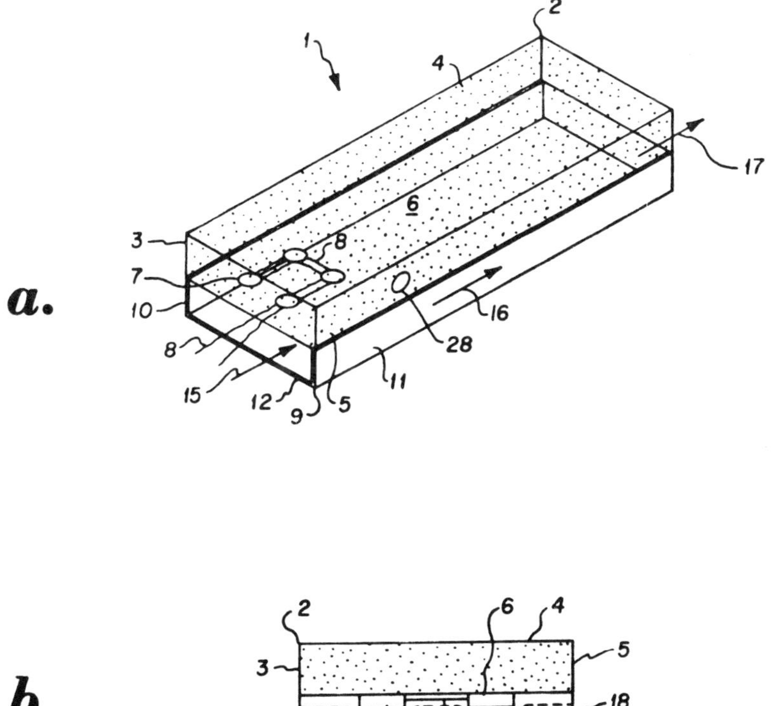


Source: U.S. Patent 4,130,445

A device developed by *R.C. Sill; U.S. Patent 4,140,544; February 20, 1979; assigned to Atlantic Richfield Company* has a luminescent collector member with divergent upper and lower surfaces. These surfaces are divergent in at least that part or area of the collector where solar radiation enters the collector. The direction of the divergence is selected to direct the internal surface angle of reflection of collected light energy toward one or more photovoltaic cells mounted on or coupled to one or more edge surfaces of the collector. The angle of divergence is such that the maximum thickness of the collector in the area of divergence is at least 1% thicker than the thickness at the point of start of divergence. At this angle and direction, the divergently sloping lower surface or lower and upper surfaces tend to reduce the number of internal light reflections and the path travelled by the light in reaching the photovoltaic cell.

Normally, photovoltaic cells will be mounted on only one edge surface and only one direction of divergence will be used, but cells may be mounted on two or more edge surfaces, for example, on opposite edge surfaces. In this case, the divergence will start near the center of the collector and the divergence will be at the appropriate angle in two directions, one toward one edge surface and one toward the opposite edge surface. By the same token, the divergence could be in all directions from the center of the collector when cells are mounted around the entire collector.

In some embodiments, the open edge surfaces not covered by cells and the lower surfaces are covered with a suitable reflective material, preferably a reflective diffusive material.

A method for increasing the efficiency of a solar cell proposed by *R.R. Chambers and P.G. Wohlmut; U.S. Patent 4,144,097; March 13, 1979; assigned to Atlantic Richfield Company* involves the use of luminescent materials in a particular way.

The present device consists of a light transmitting member which carries at least one photovoltaic means, a luminescent member removably attached to the light transmitting member, and interface material between the two members assuring the transmission of light between the members. A method for increasing the efficiency of a luminescent photovoltaic device which contains a luminescent member of reduced efficiency involves attaching to the low efficiency luminescent member a higher efficiency luminescent member with an interface material, the interface material assuring transmission of light from one luminescent member to the other.

A device developed by *P.B. Mauer and G.D. Turechek; U.S. Patent 4,149,902; April 17, 1979; assigned to Eastman Kodak Company* is a solar energy concentrator and converter comprising:

(1) a slab of optically transparent material having major opposed face surfaces defining an edge surface;

(2) one or more removable layers of a first fluorescent dye disposed in optical contact with one or more of the face surfaces;

(3) one or more removable layers of a second fluorescent dye disposed in optical contact with at least one of the first fluorescent dye layers;

(4) the first fluorescent dye being capable of absorbing a shorter wavelength of light than the second fluorescent dye;

(5) the refractive indices of the slab and the layers being substantially equal; and,

(6) radiation conversion means positioned along the edge surface of the slab.

## OPTICAL CONCENTRATORS

A concentrator system presents a solar cell with focused sunlight. The advantages of reducing the required photovoltaic (PV) cell area must be balanced against the cost for devices that focus and track the sunlight. The problem of designing a concentrator system involves a large number of choices such as cell material, support structure, and type of optics (reflecting or refracting). While many of these options interact strongly in determining the capital investment required for a given generating capacity, the variety of exploitable possibilities also presents attractive opportunities for cost reduction.

An especially promising approach to competitive photovoltaic is the use of novel concentrators and high efficiency cells. Recent concentrator designs that make use of inexpensive materials to focus sunlight show promise for substantial reductions in structural cost. Since the PV cell itself can be small, the permitted cost per unit area of cell may be quite high. This combination of inexpensive durable materials for focusing the sunlight, together with sophisticated electrical devices capable of efficient energy conversion, appears advantageous relative to the flat plate approach. At the same time, concentrator systems are in a far more primitive state of development and there is little operating experience. Sufficient information concerning the installed costs and reliability of complete concentrator systems is not yet available (1).

Thus, the business of using optical concentrators involves a complicated choice of overall system design, with particular attention to the choice of photovoltaic material which best integrates with the optical concentration system.

Because of the potential of concentrators, the recent increased emphasis in the DOE budget on addressing the technical problems of concentrator structures is most appropriate.

The combination of single-crystal silicon cells and low-cost concentrator support structures may be a particularly promising candidate for a near-term photovoltaic technology.

The efficiency of a solar cell is fundamentally limited by the ability of a single semiconductor to use optimally only a small fraction of the broad solar spectrum and by the probability associated with the Carnot efficiency that a portion of the absorbed light will be reradiated. For example, the ideal efficiency of a GaAs-GaAlAs cell operating at 1,000-fold concentration is 36%. Currently available cells have additional losses that reduce this efficiency to about 24%. Further development activities may improve concentrator cell performance substantially. Indeed, there are a number of possible ways of making cells with efficiencies of 25% or more. Such cells, although too expensive for flat plate systems, would be highly desirable for concentrators. Promising routes include GaAs-GaAlAs cells and the development of integrated multicolor cells (1).

A device developed by *R.H. Dean, L.S. Napoli and S.-G. Liu; U.S. Patent 3,999,283; December 28, 1976; assigned to RCA Corporation* is one in which solar radiation is concentrated on a solar cell of a photovoltaic device in the range of 500 to 1,600 suns. The photovoltaic device includes a plurality of solar cells on a flat surface of a heat sink, and means for concentrating solar radiation on the solar cells. The solar cells have a surface on which the solar light is incident. This high concentration of solar energy on the solar cell will increase the solar cell operating temperature. The dimensions of the solar cells and the center-to-center spacing between solar cells is such that good thermal dissipation is maintained in the photovoltaic device.

Figure 101 shows the concentrator and finned cooler (top view), a transverse section of the device (center) and a perspective view of the mounting (at bottom). The photovoltaic device having solar radiation concentration is designated as **10**. The device comprises a heat sink **12** having a flat surface **14** and a surface **16** opposite the flat surface. Projecting from the opposite surface and extending substantially perpendicular to that surface are a plurality of fins **18**. The fins are spaced from each other and are substantially parallel to each other. The spacing between fins is designated by **G**. The heat sink is of a thickness from the flat surface to the opposite surface, designated as **w**. The fins are of a length **H**, extending from the opposite surface. The heat sink and the fins are of a material having good thermal conductivity such as copper or aluminum.

On the flat surface are a plurality of solar cells **20**. Each of the cells is cylindrical in shape, having a first region **22** of one conductivity type in intimate contact with a second region **24** of the opposite conductivity type and a P-N junction **26** therebetween. The cells are mounted in a spaced relationship on the heat sink with the second region contacting the flat surface of the heat sink. Each of the solar cells has an incident surface **31**, on which impinges solar radiation. The incident surface is circular in shape and faces away from the flat surface of the heat sink.

Figure 101: RCA Design for Solar Concentration Combined with Solar Cell Cooling

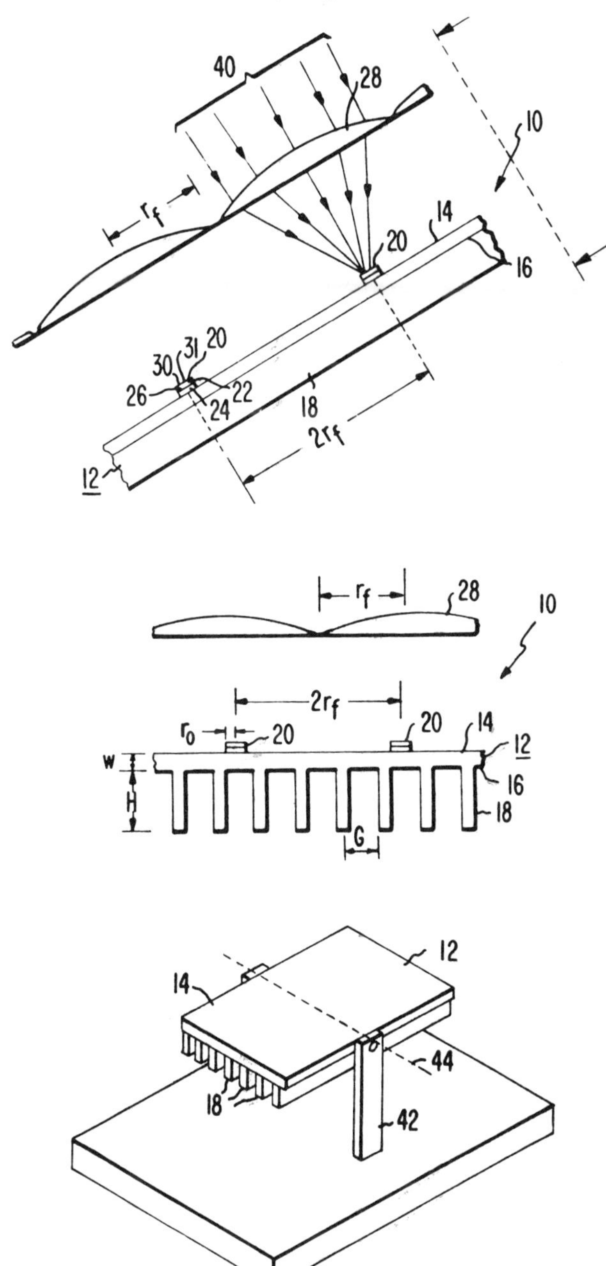


Source: U.S. Patent 3,999,283

For purposes of this discussion, the photovoltaic device has eight solar cells and the first region is of P-type conductivity, while the second region is of N-type conductivity. Typically, the solar cells are of a semiconductor material such as silicon. The center-to-center spacing of the solar cells from each other, on the flat surface, is a function of the solar cell size and other design parameters. The center-to-center spacing of the solar cells is determined by the formula:

$$(r_o)^{1/2} \geqslant (C)(r_f)^2$$

where $r_o$ is the radius of the solar cell, C is a constant determined by the heat dissipation capabilities of the heat sink and $r_f$ is one-half the spacing between the centers of the solar cells.

A grid contact **30** is on the incident surface of the solar cell. Typically, the grid contact is of metal having good electrical conductivity properties, such as gold. Solar radiation striking the grid contact may be reflected from the grid contact and never impinge the incident surface. Therefore, the grid contact should not cover a significant percentage of the incident surface area, usually no more than 5% of the incident surface area.

A plurality of refractory lens **28**, circular in shape, are spaced from the solar cells and are in a plane parallel to the heat sink. Each refractory lens is spaced from a solar cell so that the cell is approximately at the focal point of a lens. The focal point is the point on the optical axis of the lens at which converge light rays which prior to impinging the lens were parallel to the optical axis of the lens.

The heat sink of the photovoltaic device is shown mounted on a pivotal mount **42** having a pivotal axis **44**. The fins are oriented on the device so that they are perpendicular to the pivotal axis. This orientation of the fins does not inhibit heat transfer from the fins by means of natural convection.

In the operation of the photovoltaic device, solar radiation, designated as **40**, is incident on the refractory lens. After traversing the lens, the solar radiation is concentrated onto the solar cells in a concentration in the range of 500 to 1,600 suns. That is to say, the lens magnifies the solar radiation in the range of 500 to 1,600. The power density on earth of solar radiation without any magnification is 1 sun and it is equal to 0.1 watt per cubic centimeter. Magnification of a lens is the ratio of the area of the lens to the area on which the lens is focused. Since both the incident surface of the solar cell and the refractory lens are circular in shape, the magnification of the lens is equal to the ratio squared of the radius of the lens to the radius of the incident surface.

This concentration of solar radiation onto the solar cell can result in the increased absorption of the solar radiation into the solar cells. Solar radiation focused on the solar cells at the incident surface may pass through either the first or second regions and be absorbed therein. If the solar radiation is absorbed in either the first or second regions, an electron-hole pair is created at the point of absorption. Electron carriers which are created in a P-type region, such as the first region, and hole carriers which are created in an N-type region are termed minority carriers in these respective regions. If these minority carriers are generated within their diffusion length of the P-N junction so as to make possible their diffusion across the P-N junction, a usable current will be generated within the solar cell.

The efficiency of the solar cells is a function of the operating temperature of the cells. When operated at high temperatures, in the order of 200°C, the efficiency of a solar cell is reduced to the order of about 0%. In the photovoltaic device, the operating temperature of the solar cell can be maintained at a temperature of about 20°C above the ambient, to assure little degradation in solar cell efficiency, even with solar radiation concentration in the range of 500 to 1,600 suns.

One factor utilized by this photovoltaic device to maintain a suitable solar cell operating temperature is to use a plurality of smaller solar cells and lenses instead of one large solar cell and one large lens. For purposes of explanation: assume one lens of 1 m in diameter was to focus solar radiation of a power density of 100 W/cm$^2$ onto one solar cell of 3 cm in diameter as compared to ten lenses of 10 cm in diameter focusing solar radiation of a power density of 100 W/cm$^2$ onto ten solar cells of 3 mm in diameter. Although the same power density was focused on both the one larger 3 cm solar cell and the ten smaller 3 mm solar cells, the temperature rise in the ten smaller solar cells would be much lower, as much as a factor of 10 to 100 lower. Thus, the solar cell operating temperature can be lowered by using a plurality of smaller cells and lenses.

A type of construction developed by *E. Abrams; U.S. Patent 4,003,756; Jan. 18, 1977; assigned to Solar Dynamics Corporation* is one in which a number of silicon solar cells are disposed in facing relation such that sunlight in concentrated form, as provided by a cooperating sunlight-gathering lens, impinges thereon and also is interreflected between the cells, so that the electrical output is significantly enhanced.

A scheme developed by *J.R. Dettling; U.S. Patent 4,021,267; May 3, 1977; assigned to United Technologies Corporation* permits the use of a large proportion of the solar spectrum in the conversion of solar energy to electricity by means of photovoltaic cells. The apparatus comprises a collecting element which concentrates the incident radiation, a collimating element which forms the concentrated incident radiation into a beam of parallel photons, a spectral separation element, such as a prism, prism plate or diffraction grating which spectrally separates the solar radiation in the collimated beam, and a plurality of photovoltaic cells disposed in the separated spectrum, the energy gap of the cells being matched to the energy of the photons in that portion of the spectrum in which the cells are located.

The elements of such a device are shown in Figure 102, on the following page. The apparatus consists of a concentrator element **1** which may be a lens or any other light-concentrating device such as a cylindrical or parabolic mirror or an array of mirrors. The collimator **2** is a lens or mirror having a configuration, determined by the configuration of the concentration that will cause the convergent or divergent beam from the concentrator element to be rendered parallel. The beam of paralleled light from the collimator is then passed through a suitable spectral separator **3** such as a prism, prism plate or diffraction grating to separate the light into its various components. Prisms or prism plates are preferred since they operate on a refraction principle where losses are less than in the interface-based diffraction gratings. Finally, photovoltaic cells **4a, 4b, 4c** having the appropriate energy gaps are disposed in the spectrally separated light at a location in the spectral region where the energy of the incident radiation closely approximates the energy gap of the cell.

Since, in passing through the spectral separation, infrared radiation is less refracted than the visible and ultraviolet, it is possible by appropriate location of the photovoltaic cells to almost completely reject the low-energy infrared photons. These infrared photons are undesirable since they are not converted to electricity and merely heat the cells and decrease their efficiency. It is also possible to utilize certain of the higher energy ultraviolet radiation which may be difficult to match to a photovoltaic material by coating a lower energy photovoltaic material with thin layers of fluorescent material which will produce, upon exposure to ultraviolet, light responsive to the energy gap of the cell.

**Figure 102: Solar Cell Device for Using All Portions of Spectrum of Incident Sunlight**

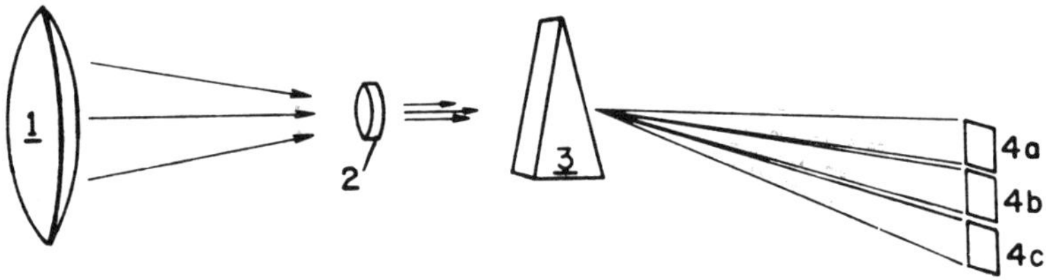


Source: U.S. Patent 4,021,267

From a consideration of the geometry of the simple system described, it becomes readily apparent why an efficient, high-energy system requires both a concentrator and a collimator in conjunction with the spectral separator. If, for example, it were attempted to employ the system without a concentrator, the total area of the photovoltaic cells must necessarily be greater than the area of the spectral separator which receives incident radiation. Thus, although spectral separation is obtained permitting the advantages associated with matching the energy gap of the photovoltaic cell to the energy of the incident radiation, the intensity of the radiation to each cell is substantially lowered and the required total area of the photovoltaic cells becomes uneconomically large. If an attempt is made to use the system with a concentrator without some form of collimation, the spectral separator will not efficiently separate all the light since the incident radiation to the spectral separator does not consist of paralleled beams.

As a result, a substantial portion of the light passing through the separator will still be "white." Thus, while it is still necessary that the area of the photovoltaic cells be greater than the area of the spectral separator upon which the incident radiation impinges, the use of concentrator and collimating elements permits much higher intensity radiation to be passed to the photovoltaic cells through the spectral separator. The ability to reject the low-energy infrared photons substantially overcomes the heat problems associated with the use of a concentrator.

In the schematic arrangement set forth in Figure 102, it is suggested, for example, that the low energy infrared, IR, ($<$0.6 eV) be rejected, the near IR (0.6 to 1.1 eV) be converted in a germanium photovoltaic cell **4a** having a threshold of 0.6 eV, the near IR and visible portion of the spectrum (1.1 to 2.4 eV) be converted in a silicon photovoltaic cell **4b** having a threshold of 1.1 eV, a gallium arsenide cell having a threshold of 1.4 eV, or a CdS-CuS cell having a threshold of 1.2 eV. The higher energy portion of the spectrum ($>$2.4 eV) could be converted by a CdS cell **4c** having a threshold of 2.4 eV, or a gallium phosphide, all having a threshold of 2.26 eV.

It is also apparent that other different materials can be used and that as more photovoltaic materials become commercially available, it may be possible to more precisely and efficiently utilize various portions of the spectrum by using a larger number of different photovoltaic cells.

A type of construction developed by *R. Kaplow and R.I. Frank; U.S. Patent 4,042,417; August 16, 1977; assigned to Massachusetts Institute of Technology* includes a plurality of unit solar cells each having one or more P-N junctions. An optical light-focusing system, which includes an array of lens elements, focuses the incoming radiation into a series of preferably narrow beams that are incident on the surfaces of the unit solar cells at locations lying immediately adjacent but spaced from the P-N junctions.

Such a design is shown schematically in Figure 103, on the following page. The array **10** is formed of a number of solar cells **12** which may comprise a plurality of slices of N-type silicon, which are separated by vertical conductive electrodes **14** made of any conducting material, such as aluminum, which forms an ohmic contact to the adjacent silicon regions. Each of the silicon slices has formed therein, such as by diffusion, a vertical $P^+$ region and an $N^+$ region. A vertical P-N junction **20** is created at the interface of the $P^+$ and N-type substrate which gives rise to the photovoltaic effect, as is known.

The P-N junctions of the plurality of the unit solar cells in the array lie in planes that are parallel to one another and, as shown, may also for certain conditions of incident light, be parallel to the direction of incident radiation. The upper or light-receiving surface of the solar cell array may be coated with an antireflection coating **24**. The incident light is focused by an optical lens structure **26** interposed between the incident light source and the light-receiving surface of the solar cell array, into a series of narrow focused beams which are incident immediately adjacent, but slightly spaced toward the N-type substrate material, from the planes of the $P^+$-N junctions. As stated, the focusing of the incident radiation in this manner onto the surface of the unit solar cells has been found to significantly increase the efficiency of conversion of incident light to electrical energy, resulting from the creation of carriers near the $P^+$-N junctions which are collected by suitable, conventional means (not shown).

To this end, the optical lens structure may be in the form of a series of miniature cylindrical lenses formed of a suitable transparent material such as acrylic or polystyrene. As shown, the lens structure includes a planar surface that rests on, and is secured to, the upper light-receiving surface of the array of solar cells, such as by means of a transparent cement.

**Figure 103: MIT-Developed Photovoltaic System Including a Lens Structure**

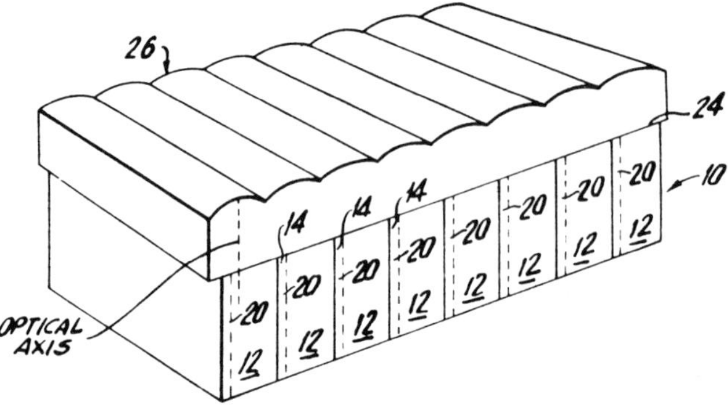


Source: U.S. Patent 4,042,417

The use of optical concentrators in high-intensity solid state solar cell devices has been further developed by *R. Kaplow and R.I. Frank; U.S. Patent 4,110,122; August 29, 1978; assigned to Massachusetts Institute of Technology* and in *U.S. Patent 4,131,984; January 2, 1979; assigned to Massachusetts Institute of Technology.*

A concentrator design developed by *A.I. Mlavsky and R. Winston; U.S. Patent 4,045,246; August 30, 1977; assigned to Mobil Tyco Solar Energy Corporation* is one in which the solar energy concentrators are each characterized by having a chamber with a solar radiation transmissive entrance wall, sidewalls adapted to concentrate solar radiation, one or more solar cells disposed in each chamber, and means for passing a dielectric, transparent cooling fluid through each chamber. The cooling fluid has an index of refraction which promotes solar energy concentration onto the solar cells in addition to that provided by the sidewalls.

A design developed by *C.R. Russell; U.S. Patent 4,052,228; October 4, 1977* incorporates an optical concentrator and cooling system in which a photovoltaic cell array is immersed in a liquid inside an elongated tube having a curved transparent wall for incident radiation, the liquid having a refractive index suitable for concentrating the incident radiation onto the photovoltaic cell array. Such a design is shown in simplified transverse section in Figure 104.

**Figure 104: Optical Concentrator and Cooling System for Photovoltaic Cells**

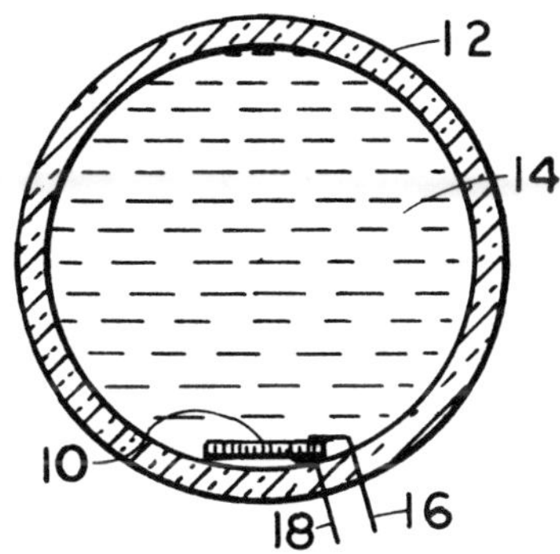


Source: U.S. Patent 4,052,228

In the above figure, the numeral **12** designates a curved transparent surface of plastic or glass. A clear liquid **14**, which is compatible with the other material used in constructing the system, has a refractive index suitable for concentrating the entering sunlight on the photovoltaic cells **10** mounted inside the elongated tube formed at least in part by the curved transparent surface. These solar cells are immersed in the liquid for cooling. The electrical energy produced by illumination of the photovoltaic cells is connected to an external load through power leads **16** and **18**. The photovoltaic cells are electrically interconnected by connectors. The cells may be interconnected in parallel or in series or in some combination of parallel and series so as to provide the desired voltage at the power leads.

The elongated tube formed at least in part by the curved transparent surface and containing the liquid is closed on the ends by end plates. The curved transparent surface may be coated by a nonreflective coating to reduce reflection of the sunlight striking this surface and thereby increase the amount of sunlight entering the tube. The curved transparent surface may be cylindrical as shown or it may be eliptical in cross section or some other shape to produce a higher concentration of sunlight on the solar cells.

A type of construction developed by *A. Meulenberg, Jr.; U.S. Patent 4,053,327; October 11, 1977; assigned to Communications Satellite Corporation* provides a cover slide for a solar cell which comprises a plurality of converging lenses arranged to focus the incident light so that it does not fall on the grid lines of the front electrode of the solar cell. Such a cell design is shown in Figure 105.

Figure 105: COMSAT Design for Light-Concentrating Solar Cell Cover

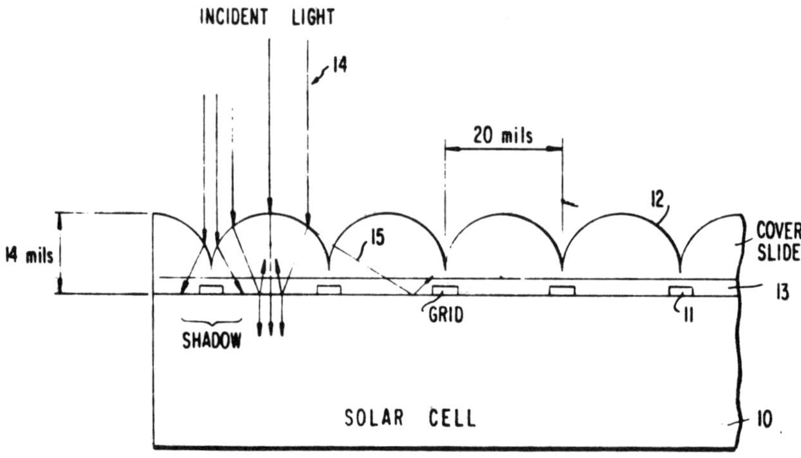


Source: U.S. Patent 4,053,327

The above figure illustrates in cross section a conventional solar cell **10** having a plurality of grid lines **11** forming the top electrode on the top surface of the solar cell. The cover slide **12** is in the form of a lens arrangement comprising a plurality of hemi-cylinders. As illustrated, the diameter of the hemi-cylinders is equal to the repeat period of the grid lines. The cover slide could be extruded in plastic or glass, for example, for terrestrial use, or formed in fused silica, quartz or sapphire for space applications. The cover slide may be supported above the solar cell, attached directly to the top surface thereof with a suitable adhesive **13**, or integrally bonded directly to the top surface thereof.

The hemi-cylinders of the cover slide act as converging lenses which concentrate the incident light **14** on the exposed surface of the solar cell and away from the grid lines. An additional advantage of the cover slide structure is illustrated by the light ray **15**. Because of the multiple reflection which occurs between the individual hemi-cylinders, surface reflection is reduced, thereby increasing the light trasmitted into the solar cell.

A modification of the hemi-cylinder construction is a "waffle" lens wherein each lens element is generally hemispherical in shape but terminates in a square base rather than a circular base. In the case of a waffle lens, Figure 105 is a cross-sectional view taken along either of two perpendicular directions. An advantage of this lens structure is the greater flexibility in the grid structure permitting cross-connecting of grids to increase reliability and decrease the distance carriers would have to travel in the surface region of the solar cell.

A device developed by *R.L. Bell; U.S. Patent 4,115,149; September 19, 1978; assigned to Varian Associates, Inc.* is one in which a lens or mirror is employed to concentrate sunlight onto a photovoltaic cell. The cell employs a broad conductor around the periphery of its active surface, and narrow higher resistance conductor strips over the inner part of the surface. The cell and mirror are sized, positioned and shaped such that the image of the sun on the cell's surface is nonuniform, with a proportionately higher concentration of light falling near the periphery of the cell adjacent the peripheral conductors. Thus, the generated current is greater at the periphery, so that less energy is lost due to series resistance of the surface conductors, for more efficiency compared to a uniformly illuminated cell.

The nonuniformity can conveniently be obtained by utilizing the natural spherical aberrations of a simple spherical mirror or lens. Also, the cell can be positioned relatively close to the mirror outside the focal plane, so that the illuminated area on the cell is larger than the focused image of the sun in the focal plane. Thus, for given diameters of cell and mirror, the arrangement is more compact, and much less sensitive to inaccuracies in mirror curvature.

The cell and concentrator designs are shown in Figure 106. The typical photovoltaic cell **10** is indicated in the upper right of the figure and comprises an outer, relatively wide and low resistance bus contact **11**; and inner, relatively high resistance fine contacts **12** which form a mesh over the active area or inner part of the cell. On the underside of the cell (not shown) a broad, uniform-area contact is usually provided and an output lead **14** is attached thereto. Another output lead **16** is attached to the bus contact.

The photovoltaic cell is preferably fabricated of compounds of the III-V elements and may be up to 2 to 3 centimeters in diameter and about 100 to 150 microns thick. Its radius is indicated by the letter **r**.

In operation, when solar energy falls directly on the top surface of the cell, an opto-electrical conversion in accordance with well-known principles occurs, as a result of which a potential difference will appear across the output leads and electrical current may be drawn therefrom. The amount of energy generated is roughly proportional to the amount of illumination falling on the cell's surface.

If a given amount of optical energy E is made to fall uniformly on the surface of the cell, a uniform energy conversion will take place throughout the cell and uniform currents will be supplied to each and every portion of fine mesh conductors **12**. Those currents supplied to the mesh conductors at the center of the cell will obviously have a longer path length to travel than those supplied to the mesh conductors near bus **11**, so that more energy of the current generated at the center of the cell will be lost in resistance heating of conductors **12** than is lost due to currents generated at the portion of the cell near bus **11**. Also, the currents will cause voltage drops or differentials across mesh conductors **12** which will in turn create variations in the voltage across different parts of the cell. Some areas will then have voltages which are not optimum for the generation of power, thereby further reducing efficiency.

**Figure 106: Varian Associates Concentrator Arrangement for Photovoltaic Cell**

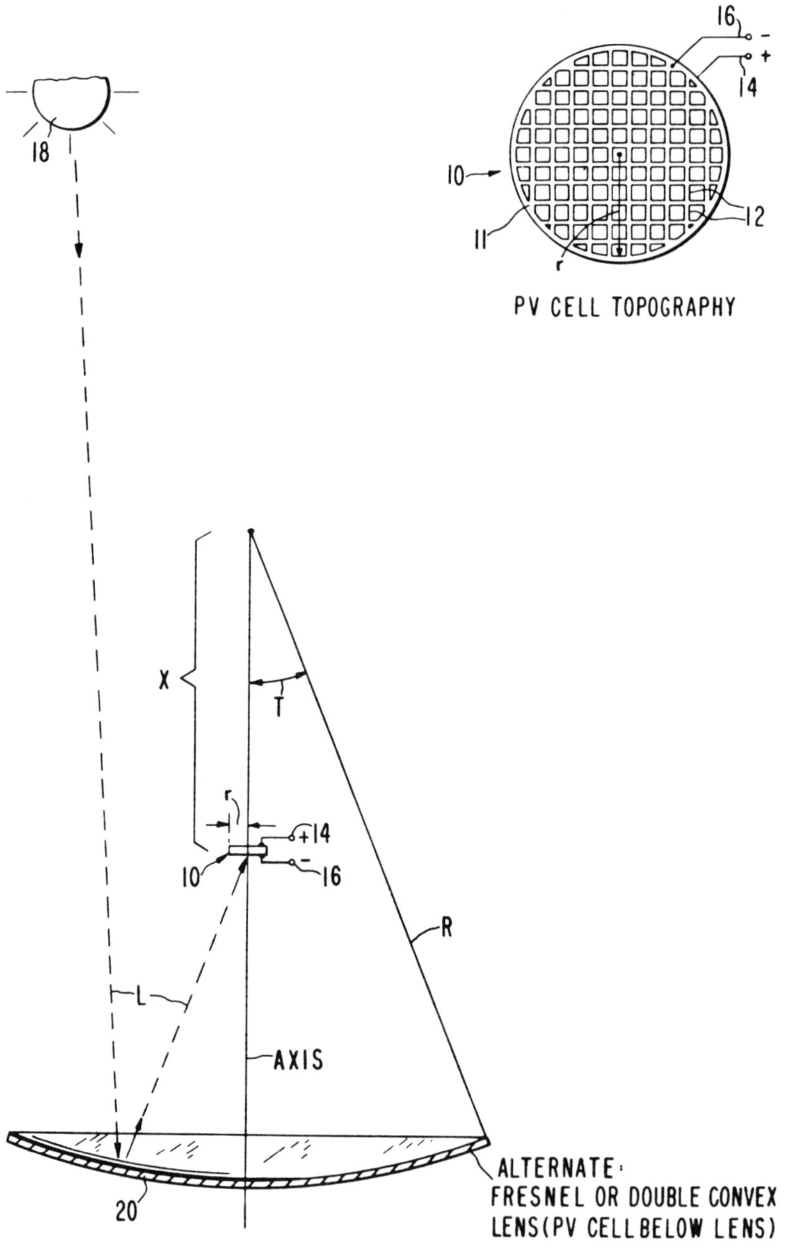


Source: U.S. Patent 4,115,149

In the present device, if energy E is made to fall upon the cell in a nonuniform pattern, with a greater proportion of energy falling near the bus conductor at the periphery of the cell than near the center of the cell, it will be obvious that less energy will be lost in heating of conductors **12** because smaller currents will be generated at the center of the cell. In addition, the effective series resistance of the cell will be reduced, and the voltage nonuniformity due to flow of the output current will also be reduced. The larger currents generated by the greater illumination intensity at the periphery of the cell near bus **11** will travel shorter distances, and thus will not encounter as high a resistance. Thereby, the amount of energy generated by the cell for the same amount of input energy E will be increased.

Figure 106 also shows a typical mirror concentrator cell arrangement. While a concave mirror concentrator **20** is shown, it will be apparent that the device is also applicable to Fresnel or double-convex lens concentrators, which would be placed between the light source and the cell, as indicated by the printed notation in the figure.

Light **L** from the sun or other optical energy source **18** is allowed to shine upon the reflecting surface of mirror **20**. The axis of the mirror should be aligned with or pointed directly at the source, but the source is shown off-axis to facilitate illustration. It is reflected from the mirror and directed to the surface of photovoltaic cell **10**, which is mounted on the axis of the mirror at a point between the mirror and its center of curvature. The mirror is most economically made a section of a sphere having a radius **R**. The angle **T** subtended by the mirror axis and any radius extending from the center of curvature point on the axis to the edge of the mirror is termed the angular half aperture of the mirror.

A device developed by *A.B. Meinel and W.B. Meinel; U.S. Patent 4,131,485; December 26, 1978; assigned to Motorola, Inc.* is one in which a non-imaging modular optical system is utilized to collect and concentrate solar energy. The system employs two reflective surfaces with a nominal ray cross-over point in front of the receiving element which distributes the available energy uniformly across the receiving element to avoid hot-spot problems. The system also is provided with a third reflective surface which serves to direct stray energy to the receiving element whether the stray energy be the result of low quality reflective surfaces or less than perfect tracking of the system with respect to the energy source.

A device developed by *A.A. Dormidontov, E.M. Zykov, T.A. Litsenko, B.A. Nikitin, V.I. Polyakov, D.S. Strebkov, V.A. Unishkov and V.V. Chernyshov; U.S. Patent 4,140,142; February 20, 1979* is one utilizing optical concentrators which focus the radiation energy in a focal spot and are disposed on the operating surface of the cells so that the absorption band of the focused radiation in the focal spot is located in the base region and is spaced from the P–N junction by a distance not exceeding the diffusion length of the minority carriers in the base region.

A type of concentrator developed by *G.R. Johnson, M.G. Miles, I.O. Salyer and E.E. Hardy; U.S. Patent 4,143,234; March 6, 1979; assigned to Monsanto Company* utilizes a solid dielectric of suitable shape rather than a metallized reflector whereby total internal reflectance can be used to increase the concentration of solar energy reaching the photoelectric cell or to increase the angle of acceptance of the incident energy into the collector.

Open-trough or dish collectors have two general weaknesses: their angular acceptance angle, for acceptance of incident radiation, is limited; and a substantial light loss, typically 20%, is caused by imperfect metal reflections within the trough. The use of total internal reflection, through the filling of the collector with a solid material having a suitable index of refraction, helps to solve the problems with the added advantage of ease in large-scale fabrications using plastics or glasses.

The incident light is reflected at the entrance end of the total-internal-reflectance-type collector, thereby substantially increasing the acceptance angle of the system. After refraction, the light impinges on the side surfaces which are surrounded by low refractive index media, air, for example, and suffers total internal refraction, thereby eliminating the light losses encountered on metallic reflection.

A type of solar cell device construction developed by *A.I. Mlavsky; U.S. Patent 4,144,095; March 13, 1979; assigned to Mobil Tyco Solar Energy Corporation* consists of at least one hermetically sealed envelope in which a trough-shaped radiation collector, having a reflective surface, is disposed so as to receive incident solar radiation through the wall of the envelope. Radiation conversion means comprising at least one solar cell is positioned inside the trough to receive radiation concentrated by the collector. The collector and the radiation conversion means may be formed as a unit for joint positioning inside the envelope. Means may be provided for orienting the envelope toward the sun and a cooling fluid may be circulated through the envelope in order to maintain the temperature below a suitable level.

An apparatus developed by *T.A. Litsenko, V.N. Potapov and D.S. Strebkov; U.S. Patent 4,146,407; March 27, 1979* is one in which a solar energy concentrator also performs the function of a cooling means for the photovoltaic solar energy converter. With that object in view, the concentrator is made as a hollow hermetically sealed vessel filled with a transparent heat-transfer agent, while the photovoltaic solar energy converter is placed inside the hollow vessel.

## SOLAR TRACKING DEVICES

A mechanism developed by *R.L. Samuels; U.S. Patent 3,515,594; June 2, 1970; assigned to TRW Inc.* is one for automatically orienting a solar cell array toward the sun. A sensing panel having an absorbing surface to be exposed generally toward the sun and a radiating surface shielded from the sun but thermally connected to the absorbing surface is variably covered by a sensor shutter which is controlled by passive, bimetallic, radiation-direction-sensitive means connected to the solar cell array.

A power-drive unit including a thermally expansive fluid-filled cylinder and piston connected therewith is mounted on the panel and drives an array-orienting mechanism in response to the temperature of the sensing panel as determined by its angle of exposure toward the sun, the degree of its shielding therefrom as by the sensing shutter, and the rate of thermal radiation from the sensing panel. The power-drive element may also drive the second shutter for variable shielding of the panel for additional feedback control of the system.

A solar energy concentrator developed by *W.M. Hubbard; U.S. Patent 3,977,773; August 31, 1976; assigned to Rohr Industries, Inc.* is an improved concentrator having its reflective surface constructed from a plurality of side-by-side mirror segments which provides an inexpensive method of construction.

The concentrators in common usage are constructed of heavy and expensive material, such as various types of metals. The inner reflective surface is often constructed from various polished metals, such as aluminum, stainless steel, silver, and the like. This type of construction requires continuous maintenance as there is a tendency for all metals to oxidize or corrode after continued exposure to the elements, thus reducing the efficiency of the collector surface.

Other inner reflective surfaces are made from formed one-piece silvered or aluminized glass mirrors. This is the least expensive with regard to initial material costs and upkeep, but requires a considerable labor expense for forming the required parabolic shape as a glass mirror is very fragile and is often broken during the forming process. Additionally, large glass structures are easily broken during handling as well as being a prey for vandalism.

A device developed by *B.H. Beam; U.S. Patent 3,988,166; October 26, 1976; assigned to Beam Engineering, Inc.* consists of an array of photovoltaic cells, a parabolic concentrator for concentrating solar energy onto the cells and a watertight chamber including a solar energy pervious window adjacent the focus of the parabolic concentrator. The solar cell array is disposed within the chamber in alignment with the window. A quantity of water is disposed in the chamber; the quantity being sufficient to absorb heat energy so as to limit the temperature rise of the solar cell array during periods of solar energy impingement thereon. The watertight chamber has sufficient external surface area that the heat energy stored therein is transferred away during non-solar-energy-producing periods of the diurnal cycle.

Such a device is shown in Figure 107, on the following page. A solar cell array identified generally at **14** is mounted in alignment with an opening **16** formed in a parabolic reflector wall **18** which constitutes a part of a Cassegrain concentrating system. The system also includes a secondary hyperbolic reflector **20** which is supported in fixed space relation to the parabolic reflector wall by radially-extending struts **22**.

**Figure 107: Tracking and Concentrating Apparatus for Enhancing the Output of Photovoltaic Solar Cells**

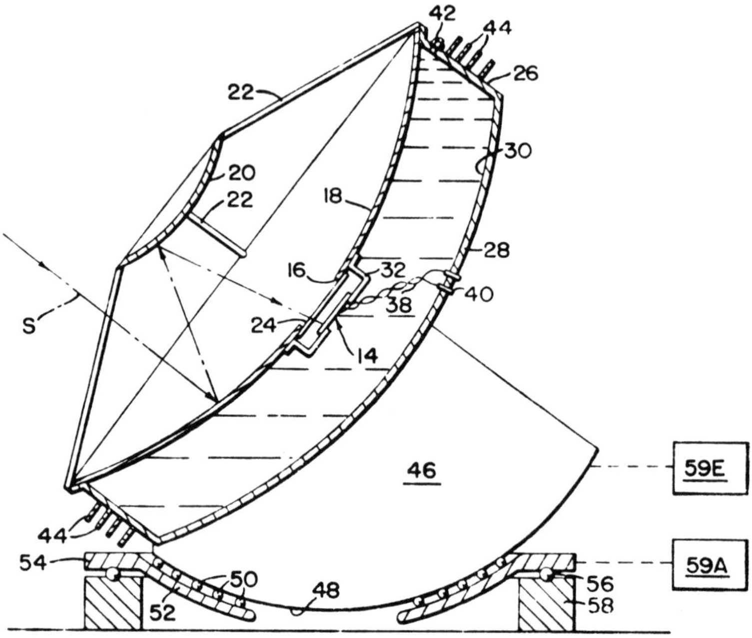

Source: U.S. Patent 3,988,166

The Cassegrain concentration (disclosed in U.S. Patent 2,985,783), for example, exemplifies any suitable structure for concentrating solar energy onto a relatively small area. Solar energy traversing a path **S** is reflected by the concave surface of the parabolic wall to the convex surface of the reflector and thence through the opening to the solar cell array.

Spanning the opening in the parabolic wall is a window **24** that can be glass, Plexiglass, or like material that is solar-energy-pervious-liquid impervious. At the periphery of the parabolic wall and rearward of the reflective surface thereof is an impervious cylindric wall **26**. The cylindric wall terminates rearward of the parabolic wall where it is spanned by an impervious rear wall **28** so as to form a watertight reservoir **30**. The reservoir is bounded by the parabolic wall, the window, cylindric wall and the rear wall. The rear wall is of parabolic shape; such shape improves the balance or symmetry of the structure but is not critical.

The solar array is supported within the chamber by means of brackets **32**; the solar cell array is supported in alignment with the window and spaced therefrom so that the image of the solar rays emanating from the secondary reflector is focused on the array.

The solar cell array can be formed, for example, of seven identical solar cells which are hexagonal shaped so as to afford closely spaced mounting thereof. The solar cells are preferably about 2 inches in maximum dimension to facilitate construction of the same from a standard 2-inch boule of crystal material. The shape of the cells and their number is not critical, but will be determined from cost and convenience. The cells will have conducting grids on their active surfaces, which is known to present art and practice, but which is especially important in this device to minimize series resistance at high currents which would reduce efficiency. The spacing between the solar cell array and the window is approximately one-eighth inch to 1 inch to allow circulation of the coolant medium but to avoid substantial absorption of the solar radiation by the coolant medium.

The solar cells are adhesively secured to an insulative ring which is supported at the inner ends of the brackets. The solar cells are connected in series, by conductors not shown, and because each solar cell produces an output voltage of about 0.4 volt the voltage output of the series combination in the example considered is about 2.8 volts. The output voltage of the series combination is connected via conductors **38** to a pair of output terminals **40** which extend through wall **28** in a watertight mount of conventional form.

There is an inlet opening to the reservoir, such as a valved opening **42** in cylindric wall **26**. The valved opening permits the reservoir to be filled with a suitable nonconductive heat-absorbing medium such as water. The volume of the reservoir is sufficient that the temperature rise of the heat-absorbing medium and of the solar cells is confined within a range of efficient operation of the cells. For more efficiently transferring the heat energy stored in the medium in the chamber, there is a plurality of radially-extending cooling fins **44** mounted on the exterior of the cylindric wall.

A support structure **46** is secured to the rear wall. The support structure has a lower arcuate surface **48** which forms one race of a bearing having rollers, schematically shown at **50**, and a lower race **52**. The lower race is rigid with an azimuth ring **54**, the lower surface of which forms a race for a bearing that includes balls **56** and a rigid base ring **58**. Conventional means schematically shown at **59E** and **59A** are provided for positioning the elevation and azimuth of the structure so that it is pointed toward the sun during daylight periods in order that solar energy is maximally concentrated through the window onto the solar cell array.

A mechanism developed by *L.W. Brantley and B.D. Lawson; U.S. Patent 4,011,854; March 15, 1977; assigned to U.S. National Aeronautics and Space Administration* provides a mount for continuously orienting a collector dish relative to the sun in a system adapted to perform both diurnal and seasonal solar tracking. The mount is characterized by a rigid, angulated axle having a linear midportion supporting a collector dish, and oppositely extended end portions normally related to the midportion of the axle and received in spaced journals.

The longitudinal axis of symmetry for the midportion of the axle is coincident with a seasonal axis, while the axes of the journals are coincident with a diurnal axis paralleling the earth's polar axis. Drive means are provided for periodically displacing the axle about the diurnal axis at a substantially constant rate, while other drive means are provided for periodically indexing the dish through 1° about the seasonal axis, once during each of the earth's successive rotations about its polar axis, whereby the position of the dish relative to the axle is varied for accommodating seasonal tracking as changes in the angle of inclination of the polar axis occurs.

# Solar Cell Mountings

While the cost of mounting cells on tracking collectors is clearly a major concern, when such devices are contemplated, there is also cause for concern about the cost of mounting flat plate arrays. If array prices fall below \$300 to \$500/kW, the cost of supporting and installing the arrays could begin to exceed the cost of the arrays themselves (3). Thus the general problem of solar cell mountings is not just a routine mechanical problem but a vital cost ingredient in an area where cost reduction is the name of the game.

## FLAT PLATE ARRAYS (PANELS)

Since each individual, commonly known, solar cell generates only a small amount of power, usually much less power than is required for most applications, the required voltage and current is realized by interconnecting a plurality of solar cells in a series and parallel matrix. This matrix is usually referred to as a solar cell array, and generates electrical energy from solar radiation for a variety of uses.

As a practical matter, the size of the wafer that constitutes the photovoltaic cell, which may advantageously be formed from silicon, is generally 2¼ or 3 inches in diameter, circular cells being quite common in commercial use. As each individual cell generates a voltage, for example, of about 0.4 to 0.6 V, that is ordinarily insufficient to generate the proper current desired for a particular job. Thus solar cells, connected in series or in parallel, are mounted on a frame in a more or less permanent way so that the resulting unitary structure, which may be referred to as a solar panel, can be transported and erected, thereby transporting and erecting a plurality of individual solar cells.

In a mode of construction developed by *D.C. Miley; U.S. Patent 3,527,619; September 8, 1970; assigned to Itek Corporation* each individual solar cell of the array is formed in such a manner that two corners, in the upper surface at one end of the cell, are not covered with the solar sensitive surface (i.e., do not have an underlying P-N junction) and, instead, have terminal areas electrically separated from the solar sensitive surface. To make a connection between the generally conductively coated lower cell surface and the separated contact areas, the lower surface is connected along the cell side to the terminal areas.

Freeing the uncovered corners from the underlying P-N junction is accomplished by etching the solar sensitive surface at the two corners to a depth which exceeds the P-N junction depth. The other two corners of the upper surface form terminal areas for the current pickup conductors overlying the upper (solar sensitive) surface of the cell.

The array is formed by arranging the cells in rows and columns, the columns being formed by opposing the terminal areas connected with the upper surface of one cell with the terminal areas connected to the lower surface of the other cell. In this manner, connecting cells in one column constitute a series connection and in one row constitute a parallel connection. Actual connections are made by interconnecting any four cells by a metal tab in contact with the four immediately adjacent corners of the four cells. In this manner, a different corner of each cell is connected with a different corner of each other cell.

A technique developed by *R.K. Riel, K.S. Tarneja, F.G. Ernick and P.M. Kisinko; U.S. Patent 3,546,542; December 8, 1970; assigned to Westinghouse Electric Corporation* is one in which a body of semiconductor material is employed as a continuous substrate upon which an epitaxial layer is grown. The epitaxial layer is divided into a plurality of isolated areas effectively resulting in individual solar cells. The solar cells are then electrically joined together by evaporated metal electrical contacts to form a high voltage solar cell panel.

A type of construction developed by *J. Lebrun; U.S. Patent 3,553,030; Jan. 5, 1971; assigned to U.S. Philips Corporation* comprises a flexible insulating support having on at least one of its faces metallic zones at least some of which are cut across the thickness of the support to form tags for the electric interconnection of the attached radiation-sensitive elements and at least partially for fixing the elements to the support.

A solar cell assembly developed by *R.E. Blevins; U.S. Patent 3,562,020; Feb. 9, 1971; assigned to TRW Inc.* comprises a substrate, and a grid of electrically insulating material adhesively joined to a surface of the substrate. The grid is formed with an array of openings therein which are shaped to accommodate one or more solar cells along the lateral extent of each opening. Thus, the grid serves as a jig for locating and assembling a multiplicity of solar cells in a desired array.

The solar cells are arranged within the openings so as to substantially fill the lateral extent of the openings, and are adhesively joined to the substrate surface. Means are provided for electrically connecting the solar cells in circuit with each other.

A process developed by *P.S. DuPont; U.S. Patent 3,565,719; February 23, 1971; assigned to U.S. National Aeronautics and Space Administration* involves the fabrication of solar cells on a substrate consisting of mounting solar cells face down in a flexible mat. The mat is then bent to the configuration the cells will have in final assembled form, and then a substrate is bonded to the backs of all the solar cells at one time.

A technique developed by *A.F. Forestieri, J.D. Broder and D.T. Bernatowicz; U.S. Patent 3,780,424; December 25, 1973; assigned to U.S. National Aeronautics and Space Administration* involves mounting an array of silicon solar cells on a flexible substrate to form a module. Large arrays of solar cells are required for space vehicles having power levels in the multikilowatt range. By way of example, it is contemplated that a space station will require about 25 kW of power. Such large solar cell arrays may utilize flexible substrates to enable them to be rolled or folded for storage during the launch phase.

Protective covers are also required for photovoltaic devices that are used in space. For example, silicon solar cells are covered with quartz or other transparent glasses to aid in the dissipation of heat from the illuminated cell and to minimize damage from bombarding particles. Such cells and covers are generally rigid which makes them undesirable for flexible arrays where a large number of cells must be stored during launch and subsequently deployed in space.

These problems are overcome in this device which utilizes a heat sealable transparent plastic film, such as a fluorinated ethylene-propylene copolymer, both as the protective cover and as the adhesive for mounting solar cells to a flexible substrate. A laminate comprising the substrate, a plastic film adhesive layer, the solar cell array, and a plastic film cover layer is bonded in a heated press.

A device developed by *C.Z. Leinkram, W.D. Oaks, J.A. Eisele and B.J. Faraday; U.S. Patent 3,833,425; September 3, 1974; assigned to the U.S. Secretary of the Navy* features (1) the use of a metallized wafer between the solar cells and the metal mounting chassis, which wafer is a good heat conductor and an electrical insulator; (2) the use of metallic electrically conductive solders between the solar cells and the wafer and between the wafer and the mounting panel; and (3) the use of a metallized tape to cover the nonactive parts of the cell array and the mounting panel, the tape being nonabsorptive to incoming solar radiation but transmissive to infrared energy (heat) arising in the solar cell array.

The silicon cells in a solar cell assembly for powering an earth-circling satellite in prior art devices were mounted on the aluminum panel according to the following sequence: (1) a 5 mil layer of fiber glass was secured to the aluminum chassis panel by means of epoxy; and (2) the cell array was then secured to the fiber glass layer by either a silicone adhesive or an epoxy.

This approach suffered from the poor thermal conductivity of both the fiber glass and the epoxy mounting adhesive. In addition, because of the thermal coefficient of linear expansion between the aluminum, fiber glass, epoxy and silicon, high mechanical stresses occurred and potential failure modes existed.

This device provides an excellent thermal path between the panel and the cells, a low absorptivity-to-emissivity ratio coating on the exposed nonelectrical parts of the cells and panel, and eliminates the potential failure mode caused by the linear thermal expansion mismatch.

A process developed by *J.G. Haynos; U.S. Patent 3,849,880; November 26, 1974; assigned to Communications Satellite Corporation (Comsat)* provides a method of forming an array which is simpler to carry out and requires less handling of the cells than the prior art method. A rigid or flexible substrate is first prepared by printing, or depressing, an outline of the positions of the cells in the array.

Apertures are punched in the substrate in regions which correspond to the area of the bottom electrode that is to be welded to the interconnecting means. Each cell is placed in the preprinted position of the substrate by means of an adhesive. The adhesive is placed to adhere the bottom of each cell to the substrate but is not placed in the region of the aforementioned apertures. The adhesive holds the cells to the substrate and retains them in their proper matrix positions. The interconnectors are inserted so as to extend from the top electrodes of one column of cells to the bottom electrodes of the adjacent column of cells. The interconnectors are then welded or soldered to the top electrodes.

Next, the substrate with the cells thereon is turned over and the interconnectors are welded or soldered to the bottom electrodes through the apertures which are prepunched in the substrate. Unlike the prior art, the matrix arrangement of cells is never handled in the absence of the supporting substrate, and therefore the interconnectors may be extremely thin when compared with the interconnectors of the prior art.

Very similar ground is covered by *J.G. Haynos; U.S. Patent 3,874,931; April 1, 1975; assigned to Communications Satellite Corporation (Comsat).*

A process developed by *A.C. Baskett and P.M. Riddle; U.S. Patent 3,957,537; May 18, 1976; assigned to Imperial Chemical Industries Limited, England* is one in which modules comprising photocells are encapsulated in laminates comprising at least one rigid member and at least one transparent member laminated together by a transparent hot melt adhesive such as plasticized polyvinyl butyral or an ethylene copolymer.

The preferred members are glass sheets and the preferred ethylene copolymer is a terpolymer of ethylene, methylmethacrylate and methacrylic acid. In a modification of the process, the hot melt adhesive need not be transparent provided that in the laminate it merely surrounds the photocell and does not cover its light-sensitive surface.

A technique developed by *C.Z. Leinkram, W.D. Oaks, J.A. Eisele and B.J. Faraday; U.S. Patent 3,964,155; June 22, 1976; assigned to the Secretary of the Navy* involves mounting silicon solar cells in a planar array that not only yields electrical insulation between cells but allows for a multifold increase in thermal dissipation of the cell array. The process involves metallizing a wafer of beryllium oxide on each side so that the outer surface is copper, etching the wafer on one side so that the only metallized parts which remain are those on which the solar cells are to be mounted or wiring is to be attached, soldering the solar cells on the unetched copper prominences, coating the aluminum panel on which the cells are mounted with a copper layer, soldering the underside of the wafer on the upper surface of the aluminum panel with soft solder such as indium, and covering all remaining passive surfaces with a Teflon FEP tape the underside of which carries a layer of silver and then a layer of Inconel metal.

A process developed by *P.B. Kennedy; U.S. Patent 3,973,996; August 10, 1976; assigned to The Boeing Company* is one for mounting and interconnecting solar cells to form a solar cell array particularly suited to space applications. The solar cells, which are already silver plated on the underside, are diffusion welded to a suitably plated polyimide film substrate. The attachment area and current conductor patterns are provided in the plating on the substrate and electrical connection of the solar cells is accomplished by soldering or diffusion welding directly to the plated pattern on the substrate.

A technique developed by *W.J. Biter; U.S. Patent 4,042,418; August 16, 1977; assigned to Westinghouse Electric Corporation* is one in which an integrated array of solar cells is produced in continuous layers of photovoltaic junction-forming semiconductor materials. Adjacent solar cells are sufficiently isolated by virtue of a relatively high resistivity in the semiconductor layers. The solar cells are connected in series by shorting the junction at selected points.

Ideally, a solar panel that houses a plurality of silicon solar energy cells will have a life expectancy of at least ten and, more advantageously, twenty years. Consequently, the materials from which the panels are formed will be selected not only so that they will protect the encapsulated cells over the anticipated life span but so that the cells within the panels will operate with proper integrity over the anticipated lifetime.

If one of the basic components of the solar panel fails, the life of the panel can be substantially decreased regardless of the fact that the silicon solar energy cells that are a part of the panel have not been damaged and are capable of continued, optimal function. In particular, a problem that has been encountered relates to the resinous material that is generally utilized to form at least the base member of the tray in which the solar cells are mounted.

It has been known in the past that an epoxy type material often called epoxy fiber glass, can form an excellent and long-lasting base member to one planar surface of which the solar cells may be adhered. The remainder of the panel is composed of a frame with a top opposite the flat base open for the entrance of light. After the solar cells have been mounted on the epoxy base of the tray, the remainder of the space bounded by the frame and base is filled with a silicone rubber of a type that is known to have an extended life expectancy under varying ambient conditions. Both epoxy resin and silicone rubber are known to have desirable life expectancies.

A substantial problem that has arisen in constructing such trays, however, has been to maintain proper adherence of the silicone rubber to the other materials from which the tray is constructed. Manufacturers of silicone rubber, therefore, prescribe that a primer should be used between the silicone rubber and the material to be coated with that rubber in order to obtain desired adherence of the rubber.

A process developed by *J. Lindmayer; U.S. Patent 4,057,439; November 8, 1977; assigned to Solarex Corporation* addresses the problem and provides a method of assembling a solar energy cell panel in which improved adherence of the solar cells to a base or support member of the panel is obtained. This result is accomplished by the expedient of utilizing a one part silicone resin in lieu of the primer prescribed by the manufacturer.

A solar array design developed by *J.W. Bond, Jr.; U.S. Patent 4,063,963; Dec. 20, 1977* is one in which solar cells are interconnected by filaments stretched crosswise between the edges of a rectangular frame to form a gridwork. This produces a lightweight inexpensive support that is not susceptible to damage from winds, snow and ice. Additionally, the top of the cells have thin resistive wires mounted thereto for preventing snow and ice accumulation.

Figure 108 is a diagram of such an array. It shows an array or panel of photovoltaic solar cells **1** which is mounted to a support **2** which positions the panel so as to face the sun at about 45° to the horizontal. The support is only schematically illustrated, as any suitable supporting leg arrangement can be used and, preferably, the arrangement selected should have means to adjust the relative angular position of the panel.

The panel consists of a rectangular or other geometric frame **3** which can, for example, be constructed of ⅜ inch diameter hollow aluminum or plastic tubing. To the frame is attached the supporting gridwork **4** having openings preferably of 1 to 4 inches which can suitably be formed from either filaments (of nylon, for example) stretched crosswise between the sides, from thin rigid rods or a preformed grid of molded plastic.

Coated photovoltaic solar cells **5** of 0.030 inch thickness and approximately 2 to 4 inches in diameter are then attached to the gridwork at proper intervals. The cells may be suitably attached by an epoxy cement, and a stronger bond can be achieved by securing the cell to a backing such that a gridwork portion is sandwiched between them.

In those areas where high winds are a problem, added strength can be achieved by various gridwork arrangements. For example, a second gridwork **8** with elements thereof disposed in substantially orthogonal relation to the elements of the gridwork **4**, as shown in the figure, may be employed for added strength.

**Figure 108: Terrestrial Photovoltaic Solar Cell Panel**

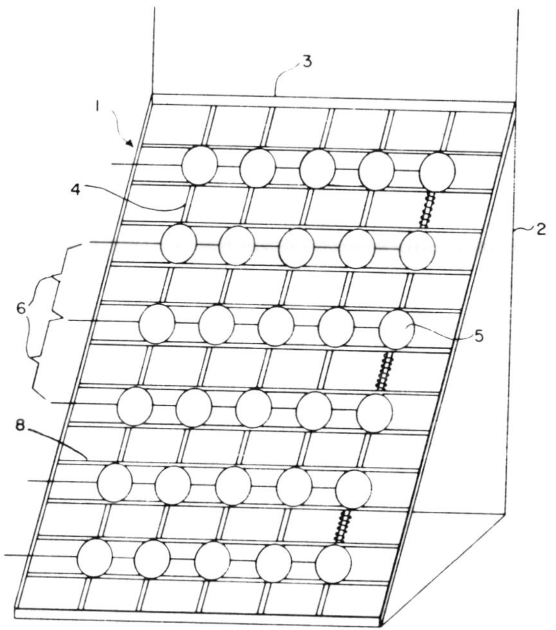


Source: U.S. Patent 4,063,963

Moreover, as strength can be enhanced by sandwiching the cells between two gridworks that are both attached to the frame **3**, the gridworks **4** and **8** each may include selectively spaced elements and the gridworks may be spaced with respect to each other with selected orientation of elements thereof to permit such sandwiching of the solar cells **5**.

For example, the elements of the two gridworks may be in substantially orthogonal relation and the cells may be edge mounted in sandwich relation to the gridworks as shown in the figure, with added strength obtained thereby. The cells are then interconnected in series by leads **6** and,in areas having cold weather conditions, thin resistive wire of, for example, tungsten can be applied to the upper surface of the cells, such that a small amount of current will cause ice and snow to melt and drop off the array.

As can be plainly seen, solar panels constructed in this manner are light in weight, aiding transportation in portable modes and reducing roof loads in fixed emplacements, while at the same time the fact that surface area is reduced to a minimum with openings throughout reduces likelihood of wind damage and in combination with resistance heating, ice and snow damage.

A type of solar panel construction developed by *I. Rubin; U.S. Patent 4,089,705; May 16, 1978; assigned to U.S. National Aeronautics and Space Administration* is one in which the cells are wafer thin and of two geometrical types, both of the same area and electrical rating, namely hexagon cells and hourglass cells. The hourglass cells are composites of half hexagons. A nearly perfect nesting relationship of the cells achieves a high density packing whereby optimum energy production per panel area is achieved.

A type of solar cell panel construction developed by *P.J. Caruso and W.T. Kurth; U.S. Patent 4,132,570; January 2, 1979; assigned to Exxon Research & Engineering Co.* is one in which the support structure is formed of a lightweight high strength plastic material having integral rib stiffeners to provide longitudinal and lateral stiffness but allow for torsional flexibility to take support warpage. The support is also provided with a perimeter skirt for mounting supports and especially completed modules to each other or to an appropriate standard.

Optionally and preferably flanges are also provided in the support structure for versatility in the ways in which a module having such a support may be mounted for receiving isolation. At least one junction box, and preferably two junction boxes, are provided on the support structure to allow terminations to be encapsulated or otherwise protected from harm by the environment and/or animals.

Posts also are provided at the intersection of the integral rib stiffeners to allow installation of a cover for electronic packages mounted in the cavities formed by the rib stiffeners. The support structure has a perimeter dike on the top surface of the structure which serves to contain encapsulating material in the top surface when an encapsulating material is employed. The perimeter dike also protects the edges of a cover sheet of rigid transparent material, when such material is used in forming the module.

## SHINGLE ARRANGEMENT

According to a report by the Office of Technology Assessment (3), the General Electric Company has proposed using a photovoltaic array as a shingle and argues that the devices should be given a credit as a roofing material. This could improve the attractiveness of solar cells for household installation.

In addition to the dual use of solar cells as power producers and roof covering in household application (as also discussed at the beginning of this volume under "Applications"), it is possible to use overlapping or shingled arrays in extraterrestial uses, in satellites, for example.

The latter possibility has been discussed by *C.Z. Leinkram and W.D. Oaks; U.S. Patent 3,769,091; October 30, 1973; assigned to the U.S. Secretary of the Navy.*

It provides a method of mounting solar cells in the array in which the cells are arranged on a thermally conductive but electrically insulative wafer and prominences so that they overlap each other in a shingled structure but their bottom surfaces are supported and remain parallel to the top surface of the mounting wafer, the array structure permitting series wiring of the cells and providing a rugged structure having a thermally conductive path from the solar cells through the mounting elements.

Very similar ground is covered by *C.Z. Leinkram and W.D. Oaks; U.S. Patent 3,811,181; May 21, 1974; assigned to the U.S. Secretary of the Navy.*

## TUBULAR SURFACE MOUNTINGS

According to *H. Weinstein and R.H. Lee; U.S. Patent 3,990,914; Nov. 9, 1976; assigned to Sensor Technology, Inc.* high efficiency, low cost solar energy conversion is facilitated by using tubular photovoltaic solar cells situated at the focus of a line-generated paraboloidal reflector. Advantageously, each solar cell comprises a pair of concentric glass tubes that are hermetically sealed at the ends. A photovoltaic junction is formed over the entire inside surface of one of the concentric tubes.

For example, this may comprise an inner electrically conductive film, contiguous layers of $Cu_2S$ and CdS forming a heterojunction, and an outer film of optically transparent but electrically conductive material. The conductive films provide electrical connection to the junction via external contacts that are symmetrically disposed at the ends of the tubular cell.

In other embodiments the photovoltaic junction is formed in a crystalline silicon layer that is grown in situ on one of the glass tubes.

A type of construction developed by *A.I. Mlavsky; U.S. Patent Reissue 29,833; Nov. 14, 1978; assigned to Mobil Tyco Solar Energy Corp.* is one in which tubular solar cells are provided which can be coupled together in series and parallel arrays to form an integrated structure. Solar energy concentrators are combined with the solar cells to maximize their power output. The solar cells may be cooled by circulating a heat exchange fluid through the interior of the solar cells and the heat captured by such fluid may be utilized, for example, to provide hot water for a heating system. The coolant circulating system of the solar cells also may be integrated with a solar thermal device so as to form a two-stage heating system, whereby the coolant is preheated as it cools the solar cells and then is heated further by the solar thermal device.

# Photovoltaic Cogeneration Hybrid Systems

The attractiveness of photovoltaic devices can be increased significantly if effective use can be made of the thermal energy carried away by water pumped over the back surfaces of collecting cells. Such systems are the photovoltaic analogs of cogeneration devices, and an analysis of the opportunities presented by such devices is much the same as those conducted for conventional systems. In both cases, the net efficiency of systems can only be understood by examining the combined demands for electricity and thermal energy in each proposed situation. There are clearly many useful applications for the thermal energy with temperatures in the 50° to 100°C range which can easily be extracted from most photovoltaic cogeneration devices. About 23 to 28% of the primary energy consumed in the United States is used at temperatures below 108°C (3).

With photovoltaic systems, the critical question is whether the electric generating efficiency which is lost because of operating the cells at higher temperatures is compensated by the value of the thermal energy produced. Cell performance degrades almost linearly with temperature at temperatures of interest primarily because of a drop in the operating voltage of the cell (at high temperatures, thermally excited electrons begin to dominate the electrical properties of the semiconductor device).

GaAs and other high efficiency cells are less affected by high temperature operation than are silicon devices. A commercial silicon cell operating with an efficiency of 12% at 27°C has an efficiency of only 8% if operated at 100°C, while a GaAs cell with an efficiency of 18.5% at 28°C can operate with 16% efficiency at 100°C and about 12% efficiency at 200°C.

It is possible to operate flat-plate collectors at elevated temperatures, but cogeneration will probably be easier to justify for concentrating systems. Care must be taken to cool concentrator cells even if waste heat is not employed.

If photovoltaic cogeneration proves to be attractive, concentrator photovoltaic systems may continue to be economically attractive even if the price goals for flat-plate arrays are achieved (3).

A device developed by *J. Shigemasa; U.S. Patent 3,956,017; May 11, 1976; assigned to Sharp KK, Japan* is one incorporating a metal layer made of a high heat conductivity material such as silver or aluminum at the rear surface of a solar battery with the use of vacuum evaporation technology. A thermoelectric converter is attached to the metal layer in a manner to convert solar energy into electric energy by means of not only the optoelectric effect but also the thermoelectric effect. The solar battery can be maintained at its preferred operative temperature by the metal layer and the thermoelectric converter.

Such a device is shown schematically in Figure 109. The device shown comprises a silicon solar cell **1** of the conventional P-N junction type, a heat conduction metal layer **2** and a thermoelectric converter **3**. The silicon solar cell can be of a conventional construction and includes a P-type semiconductor layer **11** and an N-type semiconductor layer **12**. The silicon solar cell can be of a columnar configuration of which the diameter is about 5 cm and the height is about 0.4 mm.

The upper surface **101** of the solar cell is exposed to radiation. A comb shaped electrode **130** is provided on the upper surface of the solar cell with the use of vacuum evaporation technology in order to enhance the optoelectric conversion efficiency. The pair of lead wires **13** are connected with the comb shaped electrode and the heat conduction metal layer, which serves as an ohmic contact electrode for the solar cell. A load resistance **131** is connected to the pair of lead wires.

The heat conduction metal layer is made of a high heat conductivity material such as silver or aluminum and is provided at the rear surface of the solar cell with the use of vacuum evaporation technology. Thickness of the heat conduction metal layer is preferably 10 to 20 $\mu$m.

The thermoelectric converter comprises a P-type semiconductor **31** including bismuth, selenium and tellurium at proper concentration, and an N-type semiconductor **32** including antimony, bismuth and tellurium at proper concentration as is well known in the art. The thermoelectric converter is attached to the heat conduction metal layer with the use of soldering technology. The length of the thermoelectric converter **3** is 2 mm to 2 cm and preferably about 1 cm. Electrical energy generated by the thermoelectric converter is led out through a pair of lead wires **33** respectively which are soldered to the P-type semiconductor and the N-type semiconductor. A load resistance **331** is provided between the pair of lead wires **33**.

When the upper surface of the solar cell is exposed to radiation, the solar cell provides output voltage through the pair of lead wires **13** by means of the optoelectric effect.

**Figure 109: Combined Thermoelectric and Solar Cell Device**

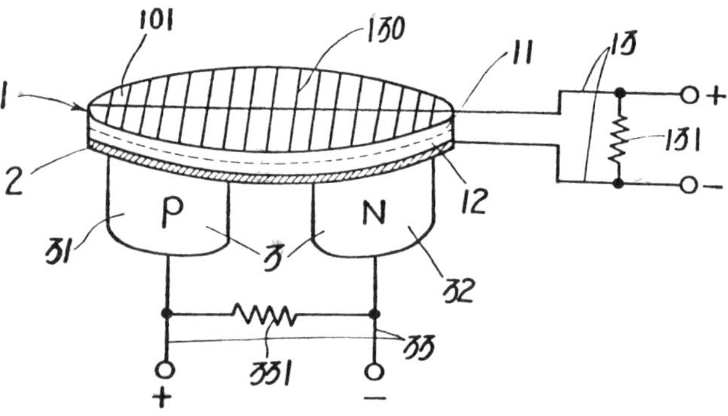


Source: U.S. Patent 3,956,017

At the same time the solar cell **1** is heated and the heat energy created within the solar cell is transferred to the thermoelectric converter **3** through the heat conduction metal layer **2**. The solar cell must be maintained below 170°C. The solar cell is maintained at about 90°C in this process because of provision of the heat conduction metal layer and the thermoelectric converter. The heat energy created within the solar cell can be converted into electric energy by the thermoelectric converter by means of the Seebeck effect and is led out through the pair of lead wires **33**.

A system developed by *R.L. Bell; U.S. Patent 4,002,031; January 11, 1977; assigned to Varian Associates, Inc.* employs gallium arsenide photovoltaic cells to convert light to direct current. Optical concentrators reduce the needed area of cells. Gallium arsenide retains high conversion efficiency up to several hundred degrees, so the waste heat may be used to produce mechanical power in a Rakine cycle engine.

Such a system is shown schematically in Figure 110. The conversion system has a number of collectors **10** for concentrating the sun's electromagnetic radiation **11**, including infrared, visible and near ultraviolet. Collectors **10** are shown as lenses, but could equally well be mirrors. They should ideally cover as much of the area of a collecting field as possible. The collectors may be driven rotationally to follow the sun for maximum focal efficiency or they may be stationary, spherically or cylindrically focused collectors.

The concentrated radiation **12** falls on an array of photovoltaic receptors **13**. Each receptor comprises a junction cell **14** having negative and positive electrical contacts **15, 16** on opposing sides. The contacts of an array of cells are electrically connected in series by connecting wires **17** to negative and positive output terminals **18, 19**.

**Figure 110: Solar Energy Converter with Waste Heat Engine**

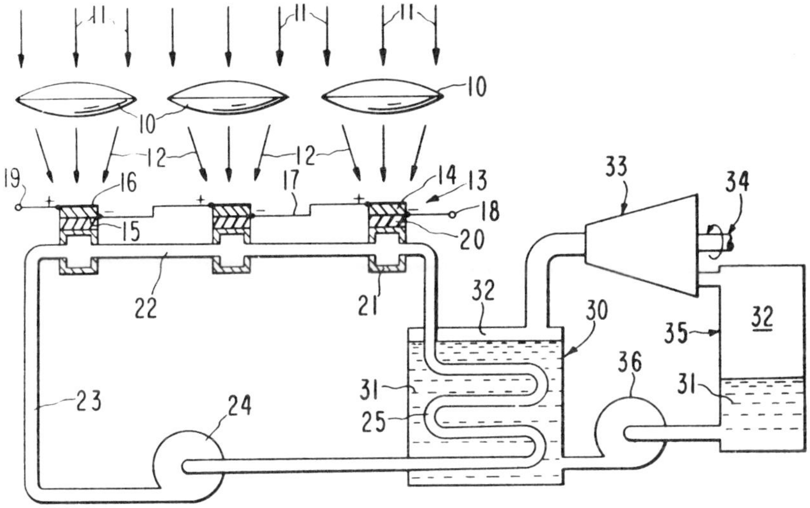

Source: U.S. Patent 4,002,031

In this way the electromotive forces of the cells, on the order of 1 V, are added to form a commercially useful dc voltage.

Each cell **13** is thermally bonded to a thermally conducting insulator **20**, as of beryllia ceramic. Insulators are in turn bonded to metallic heat exchangers **21**. A cooling fluid **23** circulating through pipes **22** flows through the heat exchangers, removing the portion of absorbed solar energy which is not converted to electricity, typically 80 to 85%. If the cooling fluid is a good insulator, such as gaseous helium or a silicone liquid, the insulators will not be needed, but corresponding insulation in the connecting pipes would be required.

The cooling fluid may be water or a liquid metal. It also may be a vaporizable working fluid for the heat engine, the vaporization taking place directly in the heat exchangers. For maximum thermal efficiency the exchangers may be piped in parallel instead of the series piping **22** shown.

The hot coolant is circulated by a pump **24** through the input coil **25** of a heat exchanger-boiler **30**. The output side of the boiler contains a vaporizable liquid **31** which serves as the working fluid of the heat engine. Its vapor **32** passes through a steam turbine **33** to drive an output shaft **34** from which useful mechanical energy is extracted. The expanded and cooled vapor enters a condenser **35** whence condensed liquid **31** is fed back to the boiler by a feed pump **36**.

An apparatus developed by *K.F. Griffiths; U.S. Patent 4,095,997; June 20, 1978* is a combined solar cell and hot air collector apparatus. In a solar collector having at least one solar cell, an air retaining plate is located parallel to and immediately behind the unlit side of the cell. The plate is spaced from the back of the cell to permit passage of substantially all the air that is to be treated by the cell. The retaining plate is in communication with that portion of the collector through which the air is withdrawn.

A device developed by *J.E. Kelly; U.S. Patent 4,137,097; January 30, 1979* consists of a solar energy collector and heat exchanger assembly. The apparatus further consists of a housing having a transparent top and a transparent wall within the housing parallel to the top to divide the housing into upper and lower chambers, the upper chamber having an air inlet and an air outlet, the lower chamber having a liquid inlet interconnected by a conduit to a liquid outlet.

The photocells are disposed in the lower chamber and adapted to provide heat to the conduit, and a reflector structure is disposed in the lower chamber for directing sunlight to the photocells and for reflecting sunlight to the upper chamber.

A device developed by *R.L. Field; U.S. Patent 4,137,098; January 30, 1979; assigned to the U.S. Secretary of the Navy* is an energy absorbing venetian blind type device for generating electricity, providing heat, and serving as a sunshade. A plurality of slats covered with an array of photovoltaic cells are enclosed between two panes of glass of a window housing. A heat removal system using forced air cools the photovoltaic cells and collects heat for heating purposes elsewhere. The electricity generated by the photovoltaic cells is collected for immediate use or stored in storage batteries for later use.

A hybrid solar cell system developed by *J. Lindmayer; U.S. Patent 4,149,903; April 17, 1979; assigned to Solarex Corporation* is a device adapted to generate an electrical current from sunlight and also to collect in an electrically insulating fluid thermal energy generated by such cells.

The solar energy collecting device comprises a duct adapted for guiding the flow of an electrically insulating fluid such as air, and one or more photovoltaic cells mounted on an electrically insulated portion of the exterior surface of the duct. At least one of the photovoltaic cells has a heat sink which extends into the interior of the duct and is adapted to be contacted by the electrically insulating fluid.

# Photoemissive Cell Devices

As is well known, the phenomenon of photoemission, wherein electrons are ejected from a surface located in a vacuum and exposed to light energy, is exhibited by many materials, although the efficiency of this conversion varies with the material and with the wavelength of the light. Since light of shorter wavelengths contains a higher level of energy, more materials will exhibit photoemission in applications where ultraviolet light is present, as in outer space where the ultraviolet rays are not screened by the earth's atmosphere. However, some metals, such as cesium, as well as metal alloys and numerous semiconductive materials photoemit electrons upon exposure to light radiation in the visible wavelength range, with cesium cells being responsive, for example, to infrared radiation as well.

To efficiently convert solar energy to electrical energy by photoemission then, it is necessary to select a material which will serve as a cathode to emit electrons in the presence of light energy and to provide an anode for receiving the emitted electrons. The cathode must be highly illuminated by light energy of the appropriate wavelengths to insure emission of a large quantity of electrons, and the anode must be so located as to intercept a high proportion of the electrons.

However, the geometrical arrangement of the anode and cathode must be such that the anode will not cast a shadow on the cathode so that maximum use of the cathode surface is obtained. Further, the anode itself must be so positioned that light does not fall on it, lest photoemission from the anode surface reduce the effective forward current flow alternately, the anode must be of a nonemitting material.

Numerous solar energy converters have been developed to take advantage of the ability of materials to photoemit electrons, but all have suffered disadvantages. Many are complex and time consuming to manufacture, many require exotic materials which are expensive and are difficult to handle, and many are limited in their applications and cannot be used, for example, in both outer space and terrestrial environments. In addition, many of the prior art devices encounter space charge limitations which prevent a significant flow of current, thus severely restricting their usefulness as power sources.

Some attempts have been made to utilize the photoemissive characteristics of metals and metal alloys to produce electrical currents of usable magnitude, but principally such photoemitters have been used in the detection or measurement of light, rather than in the generation of power.

A device developed by *C.W. Geer; U.S. Patent 3,510,714; May 5, 1970; assigned to Research Corporation* is a photoemissive device to convert solar energy to electrical current.

A parabolic cathode is employed so that any light not absorbed is concentrated at its focus. An anode is disposed at the center of the cathode with a hole being provided in the anode at the parabolic cathode focus so that light reflected from the cathode does not strike the anode. The anode is constructed of low work function material so it readily absorbs electrons, and there is included a reflective shield on the anode to prevent light from striking it by directing it instead at the cathode.

An improved photoemissive solar energy converter has been developed by *G.J. Williams; U.S. Patent 4,094,703; June 13, 1978; assigned to Cornell Research Foundation.* It consists of a number of electrically conductive, photoemitting electrodes geometrically arrayed so that light will strike only one surface of each to produce electron emission. The electrodes are in a spaced, generally parallel relationship so that electrons emitted by the illuminated surface of one electrode will travel to, and be collected by, a nonilluminated surface of a next adjacent electrode.

A vacuum surrounds the electrodes so that only space conduction occurs. Each electrode has one surface which acts as a photoemitting cathode and one surface which acts as an electron collector, or anode, and thus each pair of adjacent electrode surfaces is analagous to a voltage cell, the plurality of electrodes forming a solar energy battery.

# Solar Cell Energy Storage

One of the significant problems encountered with solar cells is the storage of energy produced by these cells. It is desirable to have storage capabilities associated with solar cells since these cells are able to produce more energy than is needed at any particular time. Thus, a solar cell in a home heating application should be able to store the energy collected during the day for use overnight. Conventional silicon or gallium arsenide cells must be connected to storage batteries since they do not have intrinsic storage capabilities.

Another possibility for solar cell energy storage lies in the generation of a compound by electrochemical means during peak output periods; that compound is then separately reacted to produce energy during off-peak periods. That possibility is discussed in the following section of this volume on "Photoelectrochemical Energy Conversion."

A system developed by *L.A. Ule; U.S. Patent 3,696,286; October 3, 1972; assigned to North American Rockwell Corporation* is one where in a power solar cell array consisting of many solar cells connected to deliver useful electrical power, there is imbedded a smaller reference solar array consisting of solar cells connected in series with a Zener diode and load resistor so devised that the voltage that appears across the load resistor is equal to or a constant fraction of the voltage at which the power array, operating at the same temperature and solar exposure as the reference array, delivers maximum electrical power.

The voltage difference between the large solar array or the given fraction thereof and the reference solar array is used directly as means to constrain the large array to operate at the voltage of maximum power, typically any excess power being used to charge a storage battery.

A system developed by *A.F. Hogrefe and R.M. Sullivan; U.S. Patent 3,740,636; June 19, 1973; assigned to the U.S. Secretary of the Navy* includes an electronic system for providing fully automatic control of a spacecraft solar cell/battery electrical power system, including the provision of continuous telemetry monitoring of the battery charge state. Circuitry accurately measures, on an ampere-minute basis, the charging and discharging of the battery and functions to maintain the battery in a fully charged condition, with a provision for automatic reduction of the charge current to a temperature-dependent trickle value when the proper amount of charge has been returned to the battery after a previous discharge.

A bipolar charge quantizer circuit is utilized to monitor battery charge and discharge operations and includes a finite time integrator circuit capable of supplying an output signal pulse each time the battery is charged or discharged by a predetermined amount. The quantizer is designed with a predetermined offset factor to account for battery inefficiency. Temperature dependent maximum battery voltage limiting and maximum charge current limiting are also provided, along with means which operate to automatically shunt excess electrical power from the solar cell array into the spacecraft instrument package in order to provide temperature control therein.

A system developed by *A. Lamprecht; U.S. Patent 3,860,863; January 14, 1975; assigned to Rowenta-Werke GmbH, Germany* consists of solar cells connected across a storage device providing a direct voltage input to a dc transformer, which converts its direct voltage input into a higher voltage alternating or intermittent output and comprising an oscillator or switching circuit having a photosensitive device connected in its timing circuit so that the frequency of oscillation or switching varies with the illumination level of the photoelectric cells.

A circuit developed by *S. Takeda and T. Otake; U.S. Patent 3,979,656; Sept. 7, 1976; assigned to Suwa Seikosha KK, Japan* is a battery charging circuit for effecting efficient charging of a battery by a solar energy source. A voltage detecting circuit is coupled in parallel with a battery for detecting each state of charge of same, the detecting circuit including a constant voltage element. Transistor by-pass circuitry includes first and second current path electrodes connected in parallel with the battery. The transistor by-pass circuitry further includes a third control electrode coupled to the detecting circuit, and in response to the state of charging of the battery detected thereby, respectively effects either an increase or a decrease in the current carried by the current path electrodes.

A device developed by *E.J. Mahoney and P.S. Natanson; U.S. Patent 3,982,963; September 28, 1976; assigned to Solar Power Corporation* is a device for converting sunlight into electrical energy which can be used to maintain a charge in a battery. The device includes at least one translucent panel having a plurality of individual electrically connected solar cells mounted therein. The translucent panel is mounted in a frame which can support the panel on a mounting surface. This frame includes a peripheral support wall and a panel support surface extending therebetween. The peripheral support wall has an upper edge and the support surface is located at a level below the upper edge of the peripheral wall thereby to support the panel within the confines of the wall.

A type of construction developed by *S.K. Deb; U.S. Patent 4,115,631; Sept. 19, 1978; assigned to Optel Corporation and Grumman Aerospace Corporation* consists of a photogalvanic cell having an electrode and a counterelectrode positioned in spaced registry. A charge storage layer containing a transition metal oxide covers the electrode while a semisolid electrolyte covers the counterelectrode. An overcoat layer containing tin oxide is disposed in intervening contact between the charge storage layer and the semisolid electrolyte. The cell converts light energy to electrical energy and also stores electrical charge for use after the cell is no longer exposed to light.

The construction of such a cell is shown in Figure 111. The cell includes a light passing substrate **10** covered by a light passing conducting electrode **12**. The substrate and electrode are available as a prefabricated material, NESA glass. A charge storage layer **14** covers the electrode. The charge storage layer may be an evaporated tungsten oxide ($WO_3$) film formed by conventional vacuum evaporation techniques. The charge storage layer may also be formed by molybdenum oxide, vanadium oxide, niobium oxide or tantalum oxide. It should be noted, however, that these materials must be chemically compatible with the compensator layer.

The overcoat layer **16** must be permeable to the electrolyte cation, which in the preferred devices is hydrogen. The layer must also be more resistant to attack from the semisolid electrolyte **18** than the charge storage layer.

It has been found that the preferred overcoat layer is a layer of tin oxide ($SnO_2$) from 100 to 1,000 Å thick. The tin oxide may also be doped with antimony, for example. Other suitable layers include silicon oxide, silicon oxide doped with gold, titanium oxide and chromium nitride, as well as mixtures of the above. These layers are preferably formed using well known sputtering techniques.

The overcoat layer is covered with the semisolid electrolyte which is theorized to function as a source of charge compensating ions and is typically an acidic medium in the form of a liquid or gel. Suitable electrolytes may be found, for example, in U.S. Patent 3,708,220. A preferred electrolyte is a sulfuric acid solution wherein the sulfuric acid may be mixed with materials such as water, glycerin, polyvinyl alcohol, or combinations thereof so as to obtain the desired consistency and conductivity.

The counterelectrode **20** covers the semisolid electrolyte and may be either transparent or opaque and may be selected from a conductive material which is nonreactive with the layer in contact with it. Typical counterelectrode materials include noble metals, acid-resistant metals and carbon. Leads **22** and **24** are respectively connected to the electrode **12** and counterelectrode **20**. The leads will exhibit a voltage measurable by voltmeter **28** when the device is exposed to light as shown.

Figure 111: Energy Storing Photogalvanic Cell Having Dielectric Overcoating

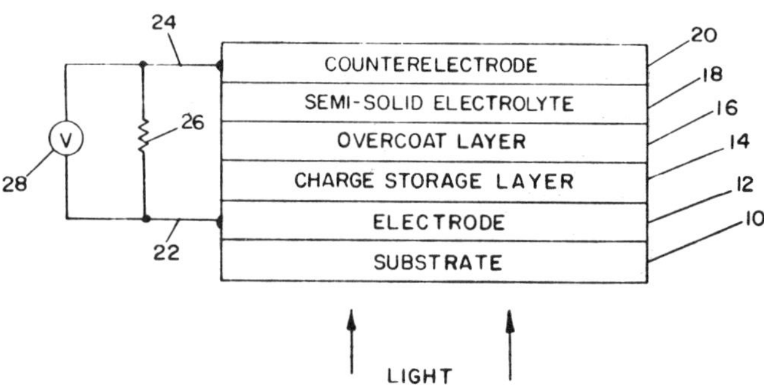


Source: U.S. Patent 4,115,631

Thus, as in the case of conventional photogalvanic cells, a load **26** placed across the leads **24** and **22** will draw current during the conversion of light energy to electrical energy. However, the present device will further store electrical charge for use after light is removed thereby acting as a storage battery capable of driving the load after the light has been removed.

Normally, it would be expected that the placement of an insulating dielectric layer, such as the overcoat layer **16** separating the semisolid electrolyte **18** and the charge storage layer **14**, would inhibit the necessary current flow of the device and thus be detrimental to the normal operation of the device. Contrary to this, however, it has been found that the insertion of the overcoat layer substantially increases the life of the device by inhibiting attack of the charge storage layer.

As an additional design consideration, the $TiO_2$ used in the overcoat layer has been used in the rutile form. However, the anatase form may similarly be used.

A cell designed by *M.L. Bayard; U.S. Patent 4,119,768; October 10, 1978; assigned to Texas Instruments Incorporated* is a photovoltaic cell which includes a photoactive body sandwiched between first and second electrode layers of ionically conductive material, wherein the photoactive body is effective to generate electric energy in response to its exposure to a light source, while enabling the storage of energy in the cell which comprises a photovoltaic battery.

A device developed by *H. Witzke; U.S. Patent 4,152,490; May 1, 1979; assigned to Optel Corporation* is a device capable of converting activating radiant energy into electrical energy and storing the electrical energy. The device comprises:

(1) first and second electrodes, at least one of which is transparent to activating radiant energy;

(2) a compensator layer of electrolyte material in contact with one of the electrodes, the latter being compatible with the electrolyte material;

(3) a transparent layer of charge storage material comprising a refractory metal oxide positioned between and in contact with the other electrode, which is transparent, and the compensator layer;

(4) means for irradiating the device with light of a wavelength of about 4000 A; and

(5) means attached to the electrodes for connecting the device to a load to be driven by a voltage developed in the device. The device is characterized by creation of electrons upon irradiation by the activating radiant energy. The created electrons are stored in the charge storage layer; the electrons flow through the electrolyte material and a load when a load is connected across the connecting means.

# Photoelectrochemical Energy Conversion

A number of techniques have been proposed for using photochemistry to generate hydrogen and other chemicals directly in solar collectors with chemical reactors driven by sunlight (3). The chemicals produced could then be stored or burned much like natural gas. Several preliminary tests have demonstrated the feasibility of the approach, although the efficiency of processes is quite low.

It is also possible to convert light energy directly into electricity by exposing electrodes immersed in chemicals to sunlight. If the materials are properly chosen, a current can be produced without any net chemical change in the materials used. Conversion efficiencies greater than 5% have been reported with polycrystalline CdSe-based photoelectrochemical cells (3).

Photochemical energy conversion has many of the advantages of photovoltaic energy conversion and, in addition, the advantage that the chemicals may be stored for later use. However, the technology is not as well developed as photovoltaics, and much research remains to be done. The nascent state of photochemical energy conversion is compounded by the fact that public awareness and government funding are both low, and thus research has been for the most part confined to a few universities.

In 1972, Fujishima and Honda announced the discovery of a process by which light energy incident on an electrode suspended in water could be used directly to convert the water into hydrogen and oxygen without any noticeable degradation of the electrode. The original Japanese work has generated a considerable amount of interest and a number of other materials have been examined (3).

A system being studied by research teams of the Allied Chemical Corporation and Massachusetts Institute of Technology is illustrated in Figure 112.

**Figure 112: Photochemical Energy Conversion**

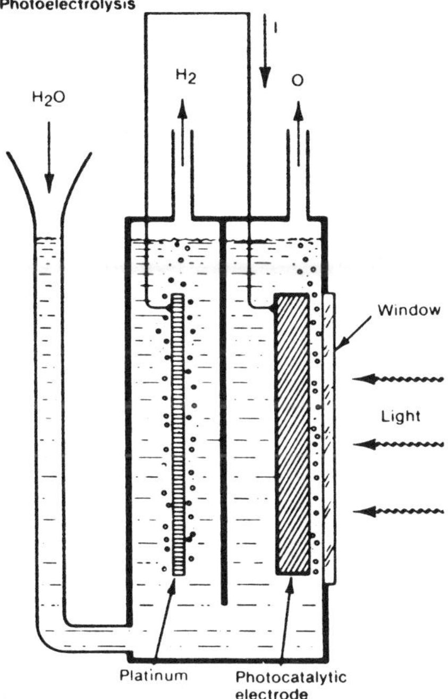


Source: Reference (3)

In this system, hydrogen is produced at the platinum cathode and oxygen is generated at an anode made from semiconductor material such as $TiO_3$ or GaAs. Some work has also been done on systems in which both electrodes are semiconductors. Other techniques for generating hydrogen have also been demonstrated.

Researchers at the California Institute of Technology have demonstrated that hydrogen is released when light strikes a complex rhodium compound. Research at the University of North Carolina at Chapel Hill has led to the development of a ruthenium compound which can split water into hydrogen and oxygen when applied in a monolayer to a glass surface and exposed to sunlight. All of these processes exhibit very low efficiencies, and long-term stability has not been demonstrated. Almost no work has been done to determine the potential cost of such systems.

The photosynthetic process used by plants to convert sunlight into storable fuel sources is clearly a photochemical system with great potential. Work is underway to better understand the basic chemistry of the process and, of course, there is increasing interest in the use of plant materials as a fuel source (3).

A device developed by *I.E. Theodorou and R. Payne; U.S. Patent 3,879,228; April 22, 1975; assigned to the U.S. Secretary of the Air Force* is a photoelectrochemical device for converting incident radiation energy to electrical energy through the utilization of a photochemically reversible galvanic cell. Basically, the device is constructed of a sealed transparent container into which is placed a chemically inert anode coated with a photoresponsive material and a chemically inert cathode.

A compatible electrolytic solution is also disposed within the container and provides a conductivity path across the electrodes to complete the electrochemical circuit. Decomposition of the photoresponsive material is induced by incident radiation passing through the transparent container. The products of the photoreaction recombine electrochemically by passing an electric current through a resistance load connected externally to the anode and the cathode.

Such a device is shown in Figure 113. The converter comprises a sealed transparent glass container **10** containing a 0.01 molar potassium chloride aqueous electrolyte **12**. An anode, composed of a platinum base **14** with a silver halide coating **16**, having a thickness of about one micron is immersed within the electrolyte together with a platinum cathode **18**. The coating **16** is deposited on the base in a nonporous, uniform manner. The electrodes **16** and **18** are positioned adjacent and at a distance of about 0.5 cm from each other. They are connected to an external resistance **20** by means of terminals **22** and **24**.

The platinum anode **14** was coated prior to assembly of the cell. It was first cleaned chemically by immersion in aqua regia and then washed in distilled water. After cleaning, the anode was heated to red heat in a gas flame. The cleaned anode was then immersed for about 20 seconds in a bath of molten silver chloride maintained at about 700°C. The electrode was withdrawn from the bath and allowed to cool. It was then inserted into the container **10** together with the cathode **18** and the electrolytic solution **12** and the container was sealed in a conventional manner not shown. It has been found that the addition of free halogen molecules to the electrolyte increases the cell's overall efficiency.

The specific cell shown in the figure is based on a simple galvanic cell in combination with a photodecomposable substrate. The cell is initially in the fully discharged state. Incident radiation penetrates the glass container **10** and impinges on and is absorbed by the silver chloride including a photochemical reaction. The silver remains in situ on the anode **14** while the chlorine molecules diffuse into the electrolyte **12** and eventually cross to the cathode **18** where they become available for subsequent electrochemical reduction. The following electrochemical half cell reactions take place at the anode and cathode respectively:

(1) $$Ag + Cl^{-} = AgCl + e$$

(2) $$\tfrac{1}{2}Cl_2 = Cl^{-} - e$$

The overall net chemical reaction is:

$$(3) \qquad Ag + \tfrac{1}{2}Cl_2 = AgCl$$

accompanied by the passage of 1 faraday of electric charge through the external load.

**Figure 113: Photoregenerative Electrochemical Energy Converter**

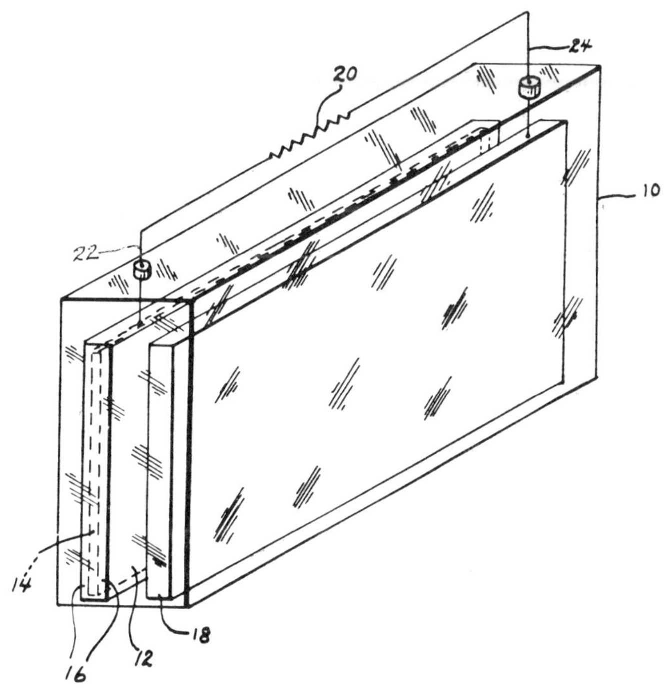


Source: U.S. Patent 3,879,228

The electrochemical reaction at the electrodes is initiated by connecting terminals **22** and **24** to a resistance load **20** and drawing electric current from the cell. The cell reactants are consumed when current is drawn from the cell according to the half cell reactions above. At the cathode, chlorine is reduced to chloride ions which diffuse into the electrolyte. At the anode, chloride ions are discharged on the silver to form silver chloride so that the cell returns precisely to its original condition. The electrochemical reaction occurring during discharge is reversed by photochemically decomposing the silver chloride in accordance with the following reaction:

$$(4) \qquad AgCl + h\nu = Ag + \tfrac{1}{2}Cl_2$$

During the operation of the device, the anode electrode becomes very efficient by becoming a composite electrode composed of specks of photolytic silver embedded in the silver chloride substrate. The entire device remains in a state of equilibrium during shelf storage since the electrochemical reaction is not initiated unless a current is withdrawn; nor does a photochemical reaction take place without the effect of incident radiation. Consequently, the device is corrosion-free when not functioning but can be quickly utilized to convert radiant energy to electrical energy through the intermediary of incident radiation.

Energy conversion is accomplished automatically and in a continuous manner without the need for periodic recharging from an external voltage source. The device is especially valuable and useful as a solar cell for aerospace applications since it is simple, lightweight and operates continuously in an unattended manner.

A system developed by *J.S. Kilby, J.W. Lathrop and W.A. Porter; U.S. Patent 4,021,323; assigned to Texas Instruments Incorporated* is one in which solar energy conversion is provided by a structure formed of a plurality of photovoltaic sources. An electrolyte wets the sources. Upon exposure to light the photovoltaic sources cause a current to flow in the electrolyte producing an electrochemical reaction. The products of this reaction may then be collected and stored. In a preferred example the electrolyte is an aqueous solution of hydrogen iodide, and the hydrogen produced by the electrochemical reaction may be stored, burned as a fuel or used in a fuel cell to produce electrical energy.

Such a system is shown in Figure 114. A series of tubes **50** are mounted with their axes inclined. Strips **52** repose on the bottom of each tube. In a preferred example, the electrolyte is an aqueous solution of hydrogen iodide, and the reaction within each tube may be written as:

$$(1) \qquad 2HI + H_2O + \text{electrical energy} \longrightarrow H_2 + I_2 + H_2O$$

The $I_2$ thus produced may be considered to exist as triiodide ($I_3^-$) which forms by reaction with the iodide ($I^-$) ions from the HI. In any event the products of this reaction, hydrogen and triiodide ions, are removed by manifolds **56** and **58**. The hydrogen collected in manifold **56** is transported to a storage unit **60** where it may either be compressed and stored as a gas or in the form of a hydride. The iodine is effectively stored in the electrolyte and may be transported by tubes **57** to manifold **58**.

If desired, electrical energy can be obtained from the system by recombining the hydrogen and iodine in a fuel cell **61**. In the fuel cell, the reaction will then be:

$$(2) \qquad H_2 + I_2 + H_2O \longrightarrow 2HI + H_2O + \text{electrical energy}$$

The hydrogen is transported to the fuel cell by means of tube **59**, while the iodine is transported in an aqueous solution by tube **62**. Because the solution in the array of tubes **50** will also be heated by the rays **54** of the sun, energy in the form of heat may be taken from the system by a conventional heat exchanger **63**.

Figure 114: Photoelectrochemical Energy Device Producing Hydrogen from Hydrogen Iodide

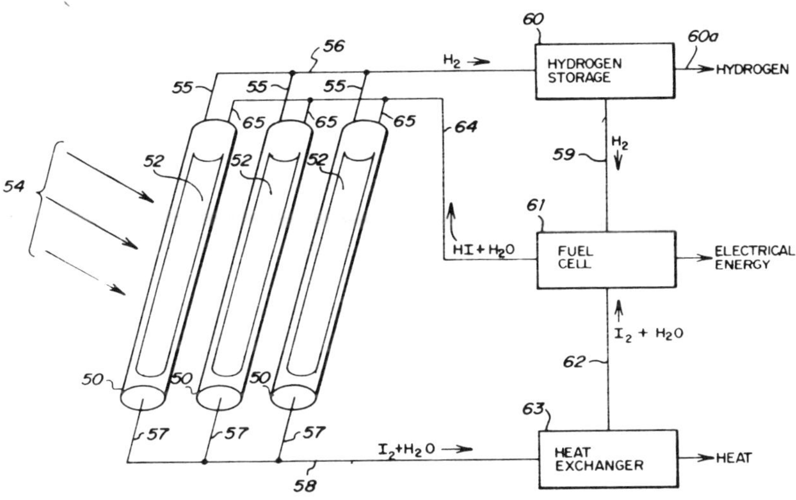


Source: U.S. Patent 4,021,323

The recombined products from the fuel cell **61** may then be returned to tubes **50** by means of manifold **64** and tubes **65**.

The system is a closed system capable of continuous operation. With materials such as hydriodic acid, it is efficient because the charge reaction can be carried out without detectable electrode overpotential and the discharge reaction can be carried out without detectable electrode polarization. As a result the reactions

$$(3) \qquad H_2 + I_2 + H_2O \longrightarrow 2HI + H_2O + \text{electrical energy}$$

can be made to approach thermodynamic reversibility very closely, thus providing an efficient energy storage and supply system.

For other applications, it may be desirable to use a system of this type for the production of hydrogen. If desired, hydrogen may be extracted from manifold **56** or from storage unit **60** as indicated by tube **60a**. If hydrogen is removed from the system, it will be necessary to modify the system. This may be done by replacing fuel cell **61** by a unit in which hydrogen sulfide is bubbled through the triiodide ions, forming new hydriodic acid which is returned to the solar converters through manifold **64**. If desired, the hydrogen sulfide may be formed from hydrogen from manifold **56**, or may be supplied from an external source.

A similar system is described by *J.S. Kilby, J.W. Lathrop and W.A. Porter; U.S. Patent 4,136,436; January 30, 1979; assigned to Texas Instruments Inc.*

A device developed by *F.K. Fong and N. Winograd; U.S. Patent 4,022,950; May 10, 1977; assigned to Purdue Research Foundation* is a device to convert solar energy into electricity, making use of the reversible photogalvanic principle. This principle is based on the spontaneous light and dark reactions between two electrochemical half cells constructed from a reversible electrochemical reaction. The device uses photosensitizers that operate in a broad band infrared spectral region.

The specific photosensitizer used is chlorophyll a coated on a platinum electrode suspended in an ionic salt solution on one side of the cell and a platinum electrode is also suspended on the other side of the half cell in a hydroquinone solution, and the electrodes are electrically connected. These solutions are interconnected by a salt bridge which permits the flow of cation charge reversibly between the first and second half cells upon on-off light irradiation of the chlorophyll coated electrode.

Compositions developed by *N.N. Lichtin and P.D. Wildes; U.S. Patent 4,052,536; October 4, 1977; assigned to The Trustees of Boston University* are electrolytes which are useful in solar energy conversion and which have a wider range of wavelength response and enhanced activity compared to the photoredox system contained therein. These electrolytes contain one or more photosensitizing dyes which luminesce within the range of wavelengths absorbed by the photoredox system.

A cell design developed by *S.K. Deb, S. Chen, H. Witzke, M.A. Russak and J. Reichman; U.S. Patent 4,117,210; September 26, 1978; assigned to Optel Corp. and Grumman Aerospace Corp.* includes a conducting $SnO_2$ electrode upon which is deposited a semitransparent film of Ti. A metal oxide thin film, such as $TiO_2$ is in turn deposited upon the semitransparent Ti thin film. An aqueous (acid or base) electrolyte contacts the metal oxide thin film to form a photoactive site for converting light to electrical energy. The semitransparent film reduces the internal resistance of the cell by assisting charge transfer between the metal oxide film and the electrode. Also, use of a semitransparent film permits bidirectional irradiation of the cell to increase photoconversion efficiency.

Such a cell design is shown in Figure 115. The cell includes a light transparent glass substrate **1** having a thin film conductive material serving as electrode **2** deposited thereon. Typically, the electrode may be fabricated from materials such as conducting $SnO_2$. The electrode-glass substrate combination is available in prefabricated form and is known in the industry as Nesa glass.

A semitransparent metal film **3** is appropriately deposited on the electrode by means such as thermal evaporation, sputtering or chemical vapor deposition. In the preferred example of this device, the semitransparent film is Ti. Titanium is disclosed by way of illustration and not of limitation.

**Figure 115: Photogalvanic Cell Having Transparent Photoactive $TiO_2$ Thin Film**

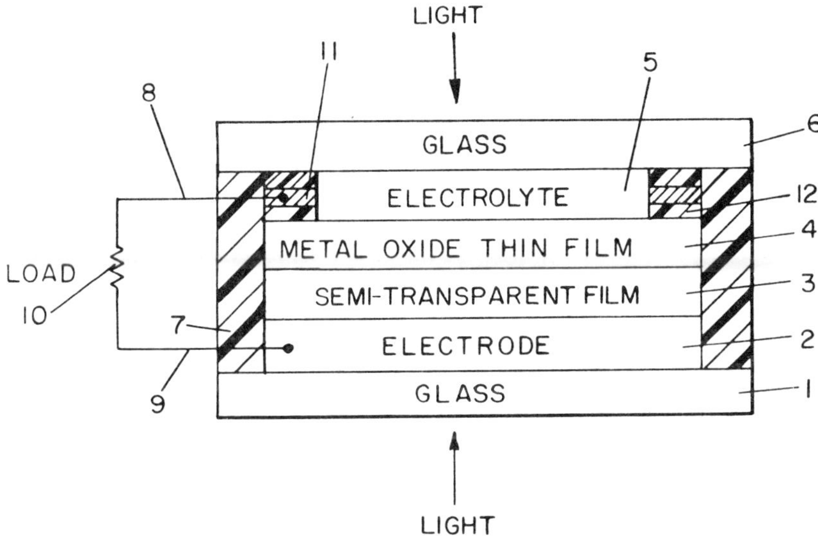


Source: U.S. Patent 4,117,210

Means which are at least semitransparent and positioned in an abutting relationship between the electrode **2** and a metal oxide thin film **4** are provided, for facilitating charge transfer between the metal oxide and the electrode. This material serves as an intermediate layer between a subsequently deposited metal oxide thin film **4** of material such as $TiO_2$ and the electrode. This intermediary layer facilitates electron transfer between the thin film and the electrode.

It is again noted that the Ti film is semitransparent which permits bidirectional irradiation of the cell. Thus, both Ti and $TiO_2$ are used by way of illustration and not of limitation.

As shown in the figure, above the metal oxide thin film is an aqueous electrolyte which may, for example, be a solution of $H_2SO_4$ and water. The upper end of the cell is sealed with a transparent glass disc **6**. However, a transparent plastic disc would suffice.

In order to permit the withdrawal of electrical charge from the cell, a counterelectrode is necessary. Since most known efficient counterelectrodes, for a photogalvanic cell of this type, are opaque, it is necessary to mount the counterelectrode in such a way that the interface between the glass disc and the electrolyte **5** may be irradiated with light through the disc. Also, it is necessary for the counterelectrode to form an interface with the electrolyte so that charge transfer may occur from the electrolyte to the counterelectrode **11**.

The figure illustrates one means of positioning the counterelectrode so that both of the aforementioned conditions are met. However, a number of alternate physical arrangements for the counterelectrode **11** will be obvious to one skilled in the art. As shown, the counterelectrode is an annular ring, fabricated from a material such as carbon or platinized carbon. The counterelectrode has an interface with the electrolyte and does not substantially block the passage of irradiating light through the glass disc **6** and through the electrolyte **5**. The counterelectrode is insulated from the metal oxide thin film **4** and is properly supported by a grooved annular insulating ring **12** which may be formed from a suitable inert insulating material, such as epoxy.

An outer wall **7** seals and supports the outward edges of the inner cell components. The wall may be fabricated from an appropriate inert insulating material such as epoxy.

Wires **8** and **9** are respectively connected to the counterelectrode **11** and electrode **2** thereby enabling a load **10** to be connected between the wires and to draw electrical power from the cell during irradiation by a light source.

Typically, a thickness for the Ti semitransparent film **3** and the $TiO_2$ metal oxide thin film are approximately 500 A and approximately 2000 A, respectively.

In operation of the device, when it is irradiated with ultraviolet light corresponding to the absorbing frequency of $TiO_2$ either by exposing the $TiO_2$ thin film through the Ti semitransparent film or through the electrolyte, an emf is generated across electrodes **2** and **11**.

A cell design developed by *H. Witzke, S. Chen, S.K. Deb and M.A. Russak; U.S. Patent 4,118,546; October 3, 1978; assigned to Optel Corporation and Grumman Aerospace Corporation* includes a glass substrate with a transparent electrode which receives irradiating light energy. A second electrode is positioned in spaced relationship from the first electrode and has a thin film of charge storing tungsten oxide deposited thereon. Spaced from both the transparent electrode and the tungsten oxide thin film is a counterelectrode. An electrolyte having $TiO_2$ powder mixed therein forms a photoactive site at the surface of the transparent electrode. By physically separating the tungsten oxide thin film from the transparent electrode, more light irradiates the $TiO_2$ thereby increasing the photoconversion of the cell.

A cell design developed by *H. Witzke, S. Chen, S.K. Deb, S.R. Jost, J. Reichman and M.A. Russak; U.S. Patent 4,118,547; October 3, 1978; assigned to Optel Corporation and Grumman Aerospace Corporation* is a sealed device which includes an electrode having a semiconductor thin film coating. A liquid electrolyte contacts the thin film to form a photoactive interface which converts light energy to electrical energy. A counterelectrode is positioned in spaced relation to the electrode and also contacts the electrolyte. Leads are connected to the electrode and counterelectrode so that a load may be driven by the device when the device is exposed to light.

A type of construction developed by *K. Miyatani and I. Sato; U.S. Patent 4,124,464; November 7, 1978; assigned to RCA Corporation* features an N-type $TiO_2$ semiconductor anode for a water photolysis cell having one or more grooves which exhibits increased oxygen evolution and greater photocurrent with minimal photon reflection losses.

Liquid semiconductor photocells using chalcogenide semiconductors have been advanced to the point where they compete favorably with silicon devices for solar power conversion. However, in common with silicon devices, the semiconductor needs to be a single crystal.

In a process developed by *A. Heller and B. Miller; U.S. Patent 4,127,449; November 28, 1978; assigned to Bell Telephone Laboratories, Incorporated* the chalcogenide semiconductor is made by anodizing cadmium or bismuth in a sulfide, selenide or telluride electrolyte. The anodized element, when operated photovoltaically in an electrolyte similar to the anodizing solution, produces useful power conversion and is relatively inexpensive.

A system developed by *H. McKinzie, M.S. Wrighton and J. Lester; U.S. Patent 4,128,704; December 5, 1978; assigned to GTE Laboratories Incorporated* is a photochemical energy storage system which includes at least one electrolytic solution containing a reduction-oxidation couple, one species of which can be stored in a second phase. The device includes a charging system and an energy delivery system.

The charging system includes an N- or P-type photosensitive electrode at which, under illumination, an oxidation or reduction reaction occurs. In the charging cycle, one species is oxidized while the other is reduced, and one of these products is stored in a phase other than the phase in which the reaction occurs. The energy delivery system transfers electrons to or from the high energy product on demand for delivery through an electrical load to an electrode at which the original reactants can be reconstituted.

# Economics of Solar Cells

The economics of solar cells have been reviewed in a study by the U.S. Office of Technology Assessment (3). They point out that computing the cost of solar energy systems is treacherous because of differing estimates of the cost and performance of solar equipment which may become available in the next decade.

They go on to point out (3), that it is also extremely difficult to establish for comparative purposes the cost of conventional energy systems which may be operating during the next few decades.

The following are the major conclusions of the American Physical Society Study Group on Solar Photovoltaic Energy Conversion as they affect future costs and markets.

1. It is unlikely that photovoltaics will contribute more than about 1% of the U.S. electrical energy produced near the end of the century. Central power production is the most extensively studied and clearly perceived long-term, large-scale application of PV for this country. Barring further rises in the cost or availability of fuels, prices for 12-16% efficient flat plate modules or concentrator arrays of about 10-40¢ per peak watt ($W_p$) in 1975 dollars will be required to compete with the projected cost of coal-generated electricity (about 45-70 mills/kWh levelized bus bar cost in the year 2000), according to the APS Study Group (1).

2. It is anticipated that only a small fraction of the electricity generated after 1990 will be based on gaseous and liquid fuels. Therefore photovoltaics will not significantly reduce the use of these fuels. The major effect of photovoltaic generation will be a displacement of some combination of coal and nuclear fuels.

3. Because of the costs associated with encapsulation, foundations, support structure, and installation of PV array fields, there is a large economic penalty for the use of low-efficiency cells. To compete in the U.S. central power generation market, even zero cost PV cells must have a limiting minimum efficiency. The use of modules with efficiency as low as 10% will probably require substantial reduction in these other costs, even if the modules themselves are inexpensive.

4. The minimization of materials usage in the overall PV system must be emphasized. New designs for PV systems should be tested against the availability of materials at high rates of construction of generating capacity, because political or economic developments might produce a need for accelerated rates of deployment. Some current designs, in addition to being extremely expensive, would make major demands on U.S. ability to supply certain materials in sufficient quantity.

5. Energy storage would be necessary if photovoltaics came to supply more than about 10-20% of the electrical energy in a typical region.

## MATERIALS AND PRODUCTION COSTS

The most satisfactory technique for anticipating the future cost of photovoltaic devices would be to anticipate future manufacturing techniques and develop a precise cost estimate for each processing step; this approach is taken in the next section. Unfortunately, the results of such analyses are inconclusive since many future manufacturing processes capable of dramatic cost reductions are based on techniques which are now only laboratory procedures or which anticipate progress in research. It is possible, however, to make some estimates of potential cost reduction by examining the history of cost reductions achieved in similar types of manufacturing (3).

The vast majority of the photovoltaic cells now being sold are single-crystal silicon cells in flat arrays. The bulk of Federal funding to reduce the cost of cells is being directed to silicon technology. The Federal Low-Cost Silicon Solar Array (LSSA) project, managed by the Jet Propulsion Laboratory (JPL) has made a careful analysis of the component costs of each step in the manufacturing process and is systematically examining techniques for reducing costs in four major areas:

- Production of the pure silicon feedstock,
- Preparation of a thin sheet of silicon,
- Fabrication of cells, and
- Arrangement of the cells in a weatherproof array.

JPL objectives for cost reductions in each area are shown in Table 4. In reaching these goals, maintaining high cell efficiency will be critically important since many costs in cell manufacture and in the installation of photovoltaic arrays are proportional to overall cell area and not to power.

Table 4: The Distribution of Costs in the Manufacture of Silicon Photovoltaic Devices Using Contemporary Technology (all costs in $/peak W)*

| | ....... Ingot Technology ....... | | | | Noningot Technology | |
|---|---|---|---|---|---|---|
| | Representative Data for 1976 | 1978 | 1980 | 1982 | 1984 | 1986 |
| Polysilicon | 2.01 | 1.16 | 0.76 | 0.28 | 0.14 | 0.06 |
| Crystal growth and cutting | 4.18 | 2.50 | 1.43 | 0.67 | 0.23 | 0.10 |
| Cell fabrication | 5.97 | 1.87 | 1.01 | 0.58 | 0.34 | 0.19 |
| Encapsulation materials | 0.78 | 0.22 | 0.12 | 0.09 | 0.07 | 0.03 |
| Module assembly and encapsulating | 3.99 | 1.25 | 0.68 | 0.38 | 0.22 | 0.12 |
| Price | 16.93 | | | | | |
| Goal | 20.00 | 7.00 | 4.00 | 2.00 | 1.00 | 0.50 |

*Does not include inflation–prices in 1975 constant dollars.

Note: The detailed price goal allocations are for silicon ingot technology up through 1982 and for silicon sheet technology in 1984 and 1986. These allocations for use within the LSSA Project are subject to revision as better knowledge is gained and should not be construed as predicted prices.

It is expected that after 1982, ingot and sheet technology modules will be cost-competitive, with the possibility that the $0.50/W goal can be achieved by either technology. Technology developments and future production cost parameters will be major factors in determining the most cost-effective designs.

Source: Jet Propulsion Laboratory. Low-Cost Silicon Solar Array Project. Received by OTA, June 1977 as quoted in Reference (3).

### Silicon Purification

The purified polycrystalline silicon used as the raw material of commercial cell manufacture now costs about $65/kg. The Jet Propulsion Laboratory estimated that, if the goal of $0.50/watt is to be reached, the silicon material cost must be reduced to about $10/kg and the amount of silicon wasted in the manufacturing process considerably reduced. Perhaps even more importantly, current techniques for manufacturing silicon are extremely inefficient in their use of energy; approximately 7,000 kWh of energy is required to manufacture a cell with a peak output of 1 kW (assuming that cells are 100 microns thick and 82% of the silicon entering the manufacturing process is wasted). This means that the device must operate in an average climate for about 4 years before it produces as much energy as was consumed in manufacturing the component silicon (3).

Several promising techniques for improving the purification have been experimentally verified, and it should be possible to develop cell arrays capable of producing all the energy used in their manufacture in 3 to 4 months. A significant amount of chemical engineering and process development is needed, however, to demonstrate that these laboratory experiments can be scaled-up by many orders of magnitude to form the basis of a commercial facility.

It is possible to produce photovoltaic cells with silicon less pure than the "semiconductor grade" material now used in cell manufacture, but a careful analysis must be made to determine whether the lower silicon costs would compensate for the additional system costs which would be incurred by the reduced cell efficiency which results. Manufacturing very high efficiency cells for use in concentrators may even require silicon which is of higher purity than the material now used to produce most semiconductors.

Silicon costs probably represent the single greatest technical barrier to meeting JPL's cost goals for nonconcentrating arrays in the early 1980s. This is because construction of plants capable of manufacturing silicon in quantities large enough to achieve the required cost reductions would need to be established very quickly—probably within the next year—and the investments required will be large, compared to previous spending in photovoltaic manufacturing.

New silicon plants are likely to require more capital investment per unit of cell production than any other stage in the cell production process—between $20 and $40 million for a single plant. There is no incentive to invest in such equipment of this magnitude solely for the purpose of selling silicon to the semiconductor industry because material costs for these devices are already a small part of the device cost. Silicon prices, therefore, are unlikely to fall by 1982 unless the government takes some action (3).

Improved sawing techniques may be able to cut silicon material requirements by two-thirds by producing thinner silicon cells and reducing the material lost as sawdust. Improving the techniques used to grow crystals could also reduce losses. Development of a ribbon or thin-film, crystal-growing process could greatly reduce wastage. Development of an amorphous silicon cell with adequate performance would dramatically reduce silicon requirements in cells since these cells would probably require less than 1% of the silicon used in commercial cells. Silicon requirements can also be greatly reduced if concentrator devices are used.

**Formation of Silicon Sheets**

Growing pure silicon crystals and sawing them into thin wafers now represents about 25% of the price of arrays. A number of active programs exist for improving the batch processes in which crystal ingots are currently being produced and sawed into wafers; techniques have also been designed for drawing single-crystal sheets or ribbons directly from molten silicon, but a considerable amount of engineering work must be done before a commercial process is available. Development of an ingot technique adequate to meet the $1 to $2 per watt cost goal appears to be assured, and improved ingot techniques may even be adequate to meet the 1986 cost goal.

The problems remaining in this technology appear to be largely ones of improving mechanical designs; this is another area where the program could be accelerated by the Government. Although crystal growing equipment is relatively expensive, unsubsidized commercial interest in this kind of equipment in the next few years is likely to be greater than commercial interest in advanced silicon refinement processes. Research progress which makes it possible to use polycrystalline or amorphous materials would substantially reduce the cost of this step in production.

Table 4 indicates that if current techniques are converted to mass production, silicon wafer blanks can be produced for about $112/m$^2$ ($0.80/watt if the cell is 14% efficient) if polycrystalline material were purchased for $35/kg. This could be reduced to approximately $78/m$^2$ if polycrystalline material were purchased for $10/kg. A recent analysis conducted for DOE by Texas Instruments estimated that it would be possible to produce "solar-grade" polycrystalline silicon for $10/kg (or $214/kW at 14% efficiency), and for approximately $5/kg if arrays are to be manufactured for $100/kW.

### Cell Fabrication

The process of converting silicon wafers into an operating photovoltaic array with 1 kW peak output involves a number of individual steps which now require many hours of hand labor, adding about $6 to the price of a 1-watt cell. Processes include the creation of the photovoltaic junction, the addition of electrical contacts, and the application of antireflective coatings. Studies of mass-production techniques conducted for JPL by Motorola, Texas Instruments, and RCA all indicate that this cost could be reduced to $0.30 to $0.90/watt using known mass-production apparatus, if plants capable of producing about 5 to 50 MW annually were constructed. The costs associated with each step in the fabricating process are shown in Table 4.

Projecting further price reductions, however, requires a fair amount of optimism. It can be seen from Table 5, however, that cell prices must fall to $1 to $2/watt before there will be markets large enough to support several competing fabricating plants of the size envisioned in the JPL studies. It is likely that manufacturers will show greater desire to invest in fabrication equipment than in the more capital-intensive devices required to manufacture silicon wafers. This is because significant cost reductions can be accomplished with plants much smaller than 50 MW/year, and it is much less likely that fabricating plants would become obsolete—even if breakthroughs dramatically reduce the cost of manufacturing silicon wafers.

The laborious and expensive processes for making photovoltaic cells are also extremely inefficient in their use of energy. Silicon material, for example, is melted when it is purified, cooled, and shipped to the crystal-growing facility where it must be melted again to grow a crystal, again cooled, heated at least once more during cell fabrication, and then cooled to ambient temperatures for the third time. No attempt is made to recover any of the heat wasted when the cells are cooled. The efficiency of this process could clearly be improved significantly to reduce both the cost of the cell and the time required for the cell to "pay-back" the energy used in its manufacture once it is installed.

**Table 5. Market Forecasts for Photovoltaic Devices at Different Prices (in megawatts of annual sales)**

| Marketing Study | . . Array Prices in Dollars per Peak Watt . . 10 | 3 | 1 | 0.5 | 0.1-0.3 |
|---|---|---|---|---|---|
| BDM/FEA | | | | | |
| DOD market | 10 | 75 | 100 | – | – |
| Worldwide commercial market | 1.5 | 20 | 70 | 100 | – |
| Intertechnology Corporation | 0.5 | 13 | 126 | 270 | – |
| Motorola | 1.5 | 20-30 | – | – | – |
| Texas Instruments | 0.4 | 2.6 | 30 | 100 | 20,000 |
| RCA | 0.8 | 13 | 200 | 2,000 | 100,000 |
| Westinghouse | – | – | – | – | – |
| DOE planning, objectives | 1.0 | 8 | 75 | 500 | 5,000 |

Data from: ERDA briefing on the "Photovoltaics Program," Sept. 9, 1977.
BDM Corp., "Characteristics of the Present Worldwide Photovoltaic Power Systems Market," May 1977, p. 1-1.
Intertechnology Corp. "Photovoltaic Power Systems Market Identification and Analysis," Aug. 23, 1977.
BDM Corp., *DOD Photovoltaic Energy Conversion Systems Market Inventory and Analysis, Summary Volume,* p. 17.
P.D. Maycock and G.F. Wakefield (Texas Instruments), "Business Analysis of Solar Photovoltaic Energy Conversion," *The Conference Record of the Eleventh IEEE Photovoltaic Specialists Conference-1975.* Scottsdale, Ariz., May 1975, pp 252-255.
The Motorola Corp.

Source: Reference (3)

### Module Assembly and Encapsulation

Finally, the process of connecting cells together into arrays and encapsulating them to protect the cells from the weather must be reduced by a factor of 10 from current costs to about $0.12/watt. A variety of techniques have been proposed and the cost reduction seems feasible, but the exact technique which will be used is not yet clear, although some sort of thin film process seems promising at this time.

The nonsilicon thin-film technologies which show the greatest potential for reaching the cost goals early in the 1980s are all based on the $CdS/Cu_2S$ heterojunction cell. Much more work has been done on the problem of reducing the cost of manufacturing $CdS/Cu_2S$ than on any other thin film (3).

Cadmium sulfide cells are produced commercially in the United States by Solar Energy Systems (SES) of Newark, Del. These cells are produced in batch processes and hermetically sealed in glass. Their efficiency is 3.2% with an array efficiency of 2.5%. SES is attempting to keep their array prices competitive with the market price of silicon arrays.

No technical barriers to scaling up the current manufacturing procedures are foreseen, but a significant amount of "process development" will be required to bring prices below $1,000/kW. Researchers at Westinghouse estimate that a factory based on this process capable of manufacturing cells for $1 to $2/watt could be producing commercial products within 2 years of a decision to initiate the project.

Photon Power, Inc. of El Paso, Texas is developing a process to produce cadmium sulfide solar cells directly on float glass as it comes out of a glass factory as described earlier in this volume. Such a process has an obvious low cost potential.

A number of other cells, usually based on heterojunctions between an element in Group III and Group V of the Periodic Table, are being investigated as candidate thin film cells, but few are beyond the stage of basic laboratory research. The performance of some of these experimental devices is shown in Table 1.

## MARKET FORECASTS

The free-world market for solar cells at 1976 prices was about 380 kW of which about 280 kW were sold by U.S. manufacturers. U.S. Government purchases during this period were about 108 kW, of which about 50 kW were used in satellites. Major commercial markets have appeared in communications equipment (68 kW), corrosion protection for bridges, pipelines and like applications (28 kW), and aids to navigation (20 kW). Sales during 1977 were expected to be about twice 1976 levels.

Several market surveys for photovoltaic equipment have been completed during the past several years, and some of these are summarized in Table 5. The considerable differences in the forecasts reflect differing judgments about the future cost of conventional energy and other forms of solar energy, about the rate at which an industrial infrastructure capable of supporting large-scale production can be established, and about the potential costs of support equipment (storage, controls, etc.) and installation. Some manufacturers, for example, have been skeptical about the high forecasts for small remote applications since many of these applications require a considerable amount of expensive marketing and engineering.

The surveys seem to agree that a significant fraction of sales during the next few years will occur in developing countries. Photovoltaic equipment is ideally suited to places where no utility grid is available and where labor for installing the equipment is relatively inexpensive. Consumers in the capital cities of many developing nations now pay as much as $0.20 to $0.25/kWh for electricity, and prices in more remote areas are often higher (if power is available at all). The modular nature of photovoltaic equipment has the additional advantage of allowing functioning power sources to be installed quickly and in sizes appropriate for each application.

Moreover, an investment in the photovoltaic power source does not commit a nation to finding a reliable source of fuel or to maintaining a highly trained group of operators—two serious problems for developing countries.

Other possible areas where sales of photovoltaic equipment may increase rapidly during the next few years include applications by the U.S. Department of Defense (large, cost-effective purchases appear possible during the next few years) and the armed forces of other nations; the agricultural sector, for irrigation and other pumping applications; and the transportation industry, for highway markers and lighting.

If prices fall below about $0.50/watt, an explosive growth in sales could occur since at this price photovoltaic equipment might provide electricity which is competitive with residential and commercial electricity rates in many parts of the United States (3). By the time prices fall to $0.10 to $0.30/kW, the photovoltaic electricity may be competitive with electricity sold at bulk rates to large industrial consumers. Estimating sales at these levels is extremely speculative since generating large amounts of power from photovoltaic devices would require a fundamental change in the ways in which the nation now supplies and consumes electric energy.

Moreover, when array prices reach these low levels, the overall cost and attractiveness of photovoltaic systems are likely to be dominated by factors other than the cost of the cells themselves.

The use of photovoltaic energy by industry in the future is a difficult thing to estimate as pointed out by the Office of Technology Assessment (3).

One difficulty encountered in reviewing the future value of solar-generated heat for industry is that as energy prices increase, industries undoubtedly will find many places where low-temperature heat can be recovered from existing manufacturing processes at relatively low cost, possibly narrowing the market for solar equipment. Conventional cogeneration also will become increasingly attractive as fuel costs rise. Solar cogeneration systems were able to compete with conventional cogeneration systems used in industry in sunny regions only if it was assumed that oil prices increase to more than $16/bbl by the year 2000 (the more expensive systems required prices near $30/bbl to compete). In less-favored climates, it was necessary to assume that oil prices rose to more than $20 to $25/bbl before solar compared favorably.

It can be seen, therefore, that while a market for solar heat and electricity for industry may develop by the mid- to late-1980s, the major near-term use of solar energy in these applications is likely to occur in situations where conventional fuels are not readily available or inconvenient to use, or where increased use of these fuels is forbidden by national standards for air and water quality.

# Bibliography


(1) Ehrenreich, H. et al, *Solar Photovoltaic Energy Conversion–Principal Conclusions of the APS Study Group,* New York, The American Physical Society (Jan. 1979).

(2) Hovel, H., "Solar Cells–Where are We?," *Chemtech* 9, No. 3, 191-200 (March 1979).

(3) Office of Technology Assessment, *Application of Solar Technology to Today's Energy Needs,* Vol. 1, Wash, DC (June 1978).

(4) Office of Technology Assessment, *Application of Solar Technology to Today's Energy Needs,* Vol. 2 (designated Prepublication Draft), Wash, DC (June 1977).

(5) Division of Solar Technology, U.S. Dept. of Energy, *Photovoltaic Program: Program Summary,* Report DOE/ET-0019/1, Wash, DC (Jan. 1978).

(6) Ranney, M.W., *Solar Cells,* Park Ridge, NJ, Noyes Data Corp. (1969).

(7) Paul, J.K., *Solar Heating and Cooling,* Park Ridge, NJ, Noyes Data Corp. (1977).

(8) Jet Propulsion Laboratory, California Institute of Technology, "Low Cost Solar Array Project Quaterly Report No. 8 for Jan.-March 1978"; Pasadena, CA (1979).

(9) Jet Propulsion Laboratory, California Institute of Technology, "Low Cost Solar Array Project Quarterly Report No. 9 for April-June 1978"; Pasadena, CA (1979).

(10) Commoner, B., "The Solar Transition," *The New Yorker,* April 23 and April 30, 1979.

(11) Anon., *Chem. and Eng. News*, p. 15 (May 7, 1979).

# Company Index

The company names listed below are given exactly as they appear in the patents, despite name changes, mergers and acquisitions which have, at times, resulted in the revision of a company name.

# Inventor Index

# U.S. Patent Number Index

## NOTICE

# PASSIVE SOLAR ENERGY DESIGN AND MATERIALS 1979

**Edited by J.K. Paul**

*Energy Technology Review No. 41*

Passive solar energy is a dynamic system whereby a building collects and stores energy to heat and cool itself. Unlike active solar systems employing complicated technology, passive systems collect, store and distribute thermal energy by natural radiation, conduction and convection. Sophisticated design and wise selection of building materials eliminate the need for fans, pumps or heat exchangers.

This volume describes over 100 buildings using specially designed components for the 3 passive systems, direct gain, indirect gain and isolated gain. Movable insulation for nocturnal shielding of glass areas and polymer ceiling tiles containing material to store and release heat are innovations discussed. Hybrid arrangements, utilizing simple mechanical aids to operate an otherwise passive system, are included.

A select group of case studies gives the reader an in-depth view of passive solar design. This book will interest producers of architectural and building materials, chemicals, and components, as well as architects and designers.

A partial and condensed table of contents with chapter headings and **examples of some** subtitles follows. Over 200 illustrations.



**ISBN 0-8155-0746-1** **386 pages**

# FEDERAL ENERGY INFORMATION SOURCES AND DATA BASES 1979

## by Carolyn C. Bloch

Determining the proper agency as a source of energy information is often more perplexing than perusing that information once it is obtained. Federal agencies amass myriad amounts of data and systematically store it away for reference. This information can serve as an easy, economical tool for scientists, managers, and engineers concerned with all phases of energy studies and projects, provided they have the know-how to obtain it. This time-saving directory of sources offers the assistance they need.

The directory comprises all agencies that deal in some capacity with energy—civilian, military, and legislative. It supplies the address and a capsule description of each source, delineating its field of emphasis, services offered and, many times, availability of publications.

The following cites main source categories and **examples of some** specific sources.

*I. CABINET DEPARTMENTS*

1. **DEPARTMENT OF ENERGY**
   Libraries
   Government-Owned/Contractor-Operated Laboratory and Facility Libraries
   DOE Centers and Systems
   Bonneville Power Administration
   Lawrence Livermore Laboratory
   Energy Information Administration
2. **DEPARTMENT OF AGRICULTURE**
3. **DEPARTMENT OF COMMERCE**
   National Technical Information Service
   Foreign Translations
   Patent Office
   Off. of Technol. Assessment & Forecast
   District Offices
   National Bureau of Standards
   National Standard Data Reference System
   Nat. Oceanic Atmospheric Admin.
4. **DEPARTMENT OF DEFENSE**
   Library and Analysis Centers
   Navy
   Army
   Army Mobility R&D Command, Ft. Belvoir
   Air Force
   Weapons Lab., Kirtland AFB Tech. Lib.
   Tri-Service Industry Information Center
   How DDC Serves the General Public
   Service to Legal Profession
5. **DEPARTMENT OF HOUSING AND URBAN DEVELOPMENT**
   Nat. Solar Heating & Cooling Inf. Center
6. **DEPARTMENT OF INTERIOR**
   U.S. Geological Survey
   Energy Resources Data Systems
7. **DEPARTMENT OF TRANSPORTATION**
   Information Systems, TRISNET

*II. ADMINISTRATIVE AGENCIES*

1. **ENVIRONMENTAL PROTECTION AGENCY**
   Information Sources and Systems
   Air Pollution Technical Center
   Solid Waste Inf. Retrieval System
2. **GENERAL SERVICES ADMIN.**
   Data Bank on U.S. Resources
3. **NASA**
   Library Network (NALNET)
   Industrial Application Centers
   National Space Science Data Center
4. **NATIONAL SCIENCE FOUNDATION**
   Directorate for Scientific, Technological and International Affairs
5. **NUCLEAR REGULATORY COM. LIB.**

*III. QUASI-GOVERNMENT AGENCIES*

1. **SMITHSONIAN SCIENCE INFORMATION EXCHANGE**

*IV. CONGRESSIONAL OFFICES*

1. **CONGRESSIONAL BUDGET OFFICE**
2. **CONGRESSIONAL RESEARCH SERVICE**
3. **GENERAL ACCOUNTING OFFICE**
4. **GOVERNMENT PRINTING OFFICE**
5. **LIBRARY OF CONGRESS**
   Science & Technology Div., Ref. Section
6. **OFFICE OF TECHNOLOGY ASSESSMENT**

*V. INDEX*

**ISBN 0-8155-0764-X** **115 pages**

# WIND POWER 1979
## RECENT DEVELOPMENTS

**Edited by D.J. De Renzo**

*Energy Technology Review No. 46*

Wind power could become a sizeable supplier of clean, inexhaustible energy. Heretofore, high cost has hindered this growth, but wind energy conversion systems are becoming more cost-competitive with other systems. Solutions to the attendant technological problems are also evolving.

This new book presents developments occurring since 1975, when Noyes Data Corporation published its first book on wind power. While the first chapter reviews the status of wind power conversion in 1975, subsequent chapters give an up-to-date assessment of geographical distribution of wind power in the United States, of technological advances, and of large- and small-scale use.

An appendix provides names and addresses of many manufacturers, researchers, and distributors dealing with wind energy conversion systems.

The partial table of contents below lists chapter headings and **examples of some** subtitles.



**ISBN 0-8155-0759-3**

**347 pages**

# ENERGY FROM SOLID WASTE 1979
## RECENT DEVELOPMENTS

**Edited by Francis A. Domino**

*Energy Technology Review No. 42*
*Pollution Technology Review No. 56*

The emphasis in the field of solid waste has shifted from disposal to utilization. Vast stores of energy are waiting to be tapped through competent technology.

When we consider that every day each resident of the United States generates about four pounds of waste from products of great diversity, it is apparent that solid waste may someday serve as a sizeable source of energy. But first technical and environmental aspects, as well as social, legal and economic factors, will have to be examined in depth. This book reviews the many phases of waste disposal and energy recovery, discussing recently tested technology. It describes plants presently producing power from refuse while also salvaging material. Technological difficulties are also discussed. It also presents a proposed waste utilization system for the city of New York, as an example that could be applied to other cities.

The final chapter offers greatly detailed recommendations for municipal officials on waste processing and energy recovery plants.

**Examples of some** important subtitles are found along with chapter headings in the partial, condensed table of contents given below:



**ISBN 0-8155-0750-X** **321 pages**

# HEAT PUMP TECHNOLOGY FOR SAVING ENERGY 1979

**Edited by M.J. Collie**

*Energy Technology Review No. 39*

With energy costs escalating and fossil fuel supplies diminishing, the heat pump is becoming attractive for residential space heating and cooling because of its high efficiency. This book compares its overall efficiency to that of electrical resistance and fossil fuel heating, covering air-source and water-source pumps, and single-package and split-system units.

It reports on computer-simulated studies of residential heating in 9 cities in which a heat pump replaced a gas or oil furnace, and on tests of heat pumps in buildings in Washington, D.C., Hanover, New Hampshire, and Albuquerque, New Mexico. Data on the Annual Cycle Energy House, in which a heat pump, thermal storage and solar assistance provide space heating, cooling and water heating, attest to energy saved, as do the solar-assisted heat pumps in other areas.

Because early heat pumps had design deficiencies, improvements were mandated. The latter chapters examine modifications such as capacity control to alter refrigerant flow.

Following is a condensed table of contents with **examples of some** important subtitles.



**ISBN 0-8155-0744-5**

**315 pages**

# COGENERATION OF STEAM AND ELECTRIC POWER 1978

**Edited by Robert Noyes**

*Energy Technology Review No. 29*

The cogeneration concept, as presented in this book, refers to the generation of both process steam and electricity on site by heating water with a primary fuel, such as coal, oil or gas.

When fuel was cheap, industries were not interested in capital outlays necessary for cogeneration equipment. Now, however, the picture has changed. Refineries, chemical plants, paper mills and other big energy users can get maximum value from their fuel by a "topping cycle," viz. producing steam and running it through a turbine generator before using the steam for processing operations. Also, cogeneration can be accomplished through a "bottoming cycle," utilizing process steam prior to electrical power generation. Both methods are discussed in detail.

In the past few years, the U.S. Government has been placing much emphasis on the need for conserving energy. One of the ways of saving considerable amounts of energy is the development of cogeneration plants by those industries which use significant amounts of electricity and thermal energy for process heat and/or space heating. Since electric power generation produces heat as a by-product, any thermal energy which can be used produces a fuel savings of from 10 to 30%, depending on the applications and methods involved. The studies indicate that for the greatest potential development of cogeneration, government support may be required.

Various technological studies, such as material prepared for the National Science Foundation and the Federal Energy Administration, were the basis for this review. A partial, condensed table of contents follows here.



**ISBN 0-8155-0706-2** **258 pages**